Astronautics

**A Historical Perspective of Mankind's Efforts
to Conquer the Cosmos**

Book 1 – Dawn of the Space Age

Ted Spitzmiller

Astronautics

**A Historical Perspective of Mankind's Efforts
to Conquer the Cosmos**

Book 1 – Dawn of the Space Age

By
Ted Spitzmiller

An Apogee Books Publication

Dedication

This book is dedicated to the memory of my mother
Helen Elizabeth Eckert
who encouraged me to pursue my interest in aviation and space.

Acknowledgements

Many people have moved through my life and helped me along the way. Mr. O.F. Libby, of the Curtiss-Wright Corporation, took the time to answer queries from an eight-year-old boy. Mrs. Florence Rader and Mrs. Elizabeth Frazer were high school teachers who recognized that some kids do not perform to their potential and need a little straight talk. My wife, Donna, has been a constant source of strength through the years. Les From gave me the first opportunity to exercise my technical writing skills. Sid Gutierrez used his valuable time to review excerpts and provide important historical insights. Ramu Ramakesavan gave a critical assessment from an international perspective and Charles Gray provided a technical evaluation. Finally, my appreciation to Penny Rogers, who supplied the grammatical guidance to make up for my years of neglect of the English language.

We acknowledge the financial support of the Government of Canada through the Book Publishing Industry Development Program for our publishing activities.

Published by Apogee Books, Box 62034, Burlington, Ontario, Canada, L7R 4K2, http://www.apogeebooks.com Tel: 905 637 5737
Printed and bound in the USA

Astronautics (Book One) by Ted Spitzmiller
ISBN 9781-894959-63-6

Contents

Astronautics is the definitive history of mankind's exploration of space. It examines the epic events that shaped the era and provides appropriate insight into the wide-ranging impact that this endeavor has had on technology, politics, and society. *Astronautics* provides enough detail to satisfy the serious enthusiast, but without the minutia that so often blocks the casual reader. To accomplish this, *Astronautics* is presented as two books, each covering a particular phase of man's progress.

Book 1 — Dawn of the Space Age chronicles mankind's desire to know more about the cosmos and his dreams of reaching into its depths. It describes the initial discoveries, inventions, and engineering innovations that became the foundation of rocket technology. It follows the two preeminent countries in their quest for the 'ultimate weapon' that would provide the path to space. It describes the decisions that resulted in the first artificial satellite programs in the United States and the former Soviet Union. It follows the events that shaped the initial thrust into space as represented by the first Soviet Sputniks and the shocked response by the Americans. It details the belated and often failure prone launches that humbled a great nation. The book describes the first attempts to reach the Moon and the planets and explains the techniques and physics involved. It illustrates the engineering requirements of the first manned spacecraft and the selection and experiences of the first spacefarers.

Book 2 — To the Moon and Towards the Future examines the events leading to a commitment by the American President John Kennedy to land a man on the Moon within the decade of the 1960s and the affect of that venture on future space exploration. It details the Gemini, Voskhod, Soyuz and Apollo programs and the exploration of the Moon. It reviews the development of the most complex machine devised by man—the Space Shuttle and covers the evolution of the space station. It highlights the effort to find extraterrestrial life and the exploration of the outer planets. It examines advanced propulsion technologies and speculates on what might lie ahead in space exploration.

Each chapter of each book analyzes a topic so that readers can achieve a relatively complete understanding of a special interest area without the need to ferret information from multiple chapters. However, each chapter and each book is linked to the whole by a careful interconnection of a set of themes.

Writing a concise history of space exploration is a daunting task. There are scores of books that detail various aspects of man's path to the stars, and there are two primary attributes that stand out when evaluating them: scope and technical accuracy. Scope is typically sacrificed, often because the subject matter offers a bewildering array of areas that intimidates all but the truly knowledgeable. In those few books where scope has successfully captured the essence, the level of technical detail is often overwhelming for all but the serious reader. *Astronautics* attempts to simplify and clarify technology, politics, and events to make them easier to comprehend.

The reader may note that the spelling of some names, particularly Russian, may be different from that seen elsewhere. The more common English spelling has been used here. Likewise, dates may vary by a day depending on the time and the location of the event.

Prologue

The 20th Century has been identified as the first in which technology made dramatic changes in the cultural, political, and economic standards of man. Electricity, the automobile, the airplane, and the computer are but a few of the "future shock" elements that made this one-hundred year period of progress almost unbelievable. However, it was man's escape from this small blue planet into the reaches of outer space, which set the standard of excitement that has eclipsed all other advances.

To convert dreams into reality, mankind had to establish the properties of that reality. To reach into the heavens above the earth meant understanding not only the physical environment but also the laws of science that govern those boundaries. The first verse of the Bible provides the essence of that environment when it states, *"In the beginning God created the heavens and the earth"*. From this simple sentence the four essentials that establish these bounds are presented: *beginning* denotes the domain of time, *created* implies the application of force, *heavens* asserts the dimension of space, and *earth* declares the substance of matter. These are called the Genesis Factors. The consequences and interaction of these four fundamental attributes were discovered and engineered over more than two millennia by a handful of philosophers, theorists, and experimenters. *Dawn of the Space Age,* the first book in the *Astronautics* series*,* takes the reader step-by-step through the discoveries that established the known universe. It identifies the visionaries, theoreticians, and experimentalists that set the stage for mankind's advance into space.

Understanding the sky above and the lights that dominated both the day and night presented the first challenge. Names such as Ptolemy, Aristotle, Eratosthenes, Copernicus, Galileo, and Brahe contributed to that understanding. Determining the laws that govern the application of these attributes brought forth the findings of men like Kepler and Newton, while Tsiolkovsky, and Oberth made significant contributions to establish the essentials of modern rocket science. Feeding the imagination were the science fiction writers such as Jules Verne and H. G. Wells, and engineering the hardware were Robert Goddard, Esnault-Pelterie and Wernher von Braun. Countless others played a role in bringing rocket science into the forefront of technology that today contributes greatly to our daily lives and enriches our intellectual curiosity.

Astronautics is not simply a history book. It is a story of triumph and tragedy, of hopes and dreams of mankind that have become a reality as well as those that yet remain unfulfilled.

Mankind's eyes searched the heavens for countless millennia, gazing in awe at the magnificent display of lights that decorate the night sky. Likewise, the sun, that brilliant ball of fire that dominates the daylight and could blind a person if they were to look directly at it for but a few seconds, was a mystery. The moon too, presented an enigma as it proceeded through its monthly phases and could periodically be viewed in both nighttime and daytime skies. The imagination sparkled with wonder as man attempted to determine what these illuminated objects above him were and what meaning they might have for his own existence. So in awe was man of these heavenly bodies that many cultures ascribed god-like powers to them and some even were, and still are, worshipped as gods.

One thing was certain for early man: the earth, upon which his foot trod, was solid and sure and must be the center of God's creation. Taunting man's earthbound existence was the bird that flew, seemingly without effort, into the air above. Oh, if only man could fly, if he could reach-out for the heavens above, what wondrous sights might he behold? With these visions crowding the imagination, man began his search to understand his environment and to explain the expansive creation that God had set before him.

Cosmology, from the Greek *cosmos*, is the study of the universe by observation and implies an orderly system of relationships. Ancient cosmologists in Mesopotamia and Babylonia recorded the unchanging relationships of stars more than 1000 years before Christ. The Chinese were early admirers of the night sky, identifying and naming more then 280 constellations (groups of stars) as opposed to the 55 defined by the Greeks. These early cosmologists also noted what they referred to as "wandering stars"—those few that changed their relationship with the others. Initially five of the wandering stars were identified and named; Mercury, Venus, Mars, Saturn and Jupiter using the Greek word for wandering star—planet. The occasional shooting star or the ominous solar or lunar eclipse took on powerful meanings that often created fear among those who did not understand what was happening.

Astronomy carries the study of cosmology forward to formulate theories and laws to which the heavenly bodies are subject. The Greek philosopher Aristotle (350 BC) proposed an arrangement that held that the Sun, the Moon, and the planets all revolved around Earth on a set of perfect, transparent crystalline celestial spheres. The outermost sphere was composed of stars in fixed positions. A second sphere, inside the sphere of stars, held the wandering planets. The Sun and Moon occupied the two innermost spheres. The four basic elements (earth, air, fire, and water) made up everything below the innermost sphere of the Moon. However, the Greek astronomer Aristarchus of Sámos (250 BC) was the first to propose that the Earth moved around the Sun, but Aristotle's model of the universe prevailed for almost 1,800 years.

With the assumption that the Earth was the center of the universe, early astronomers such as Ptolemy (150 AD) attempted to apply mathematical equations to establish positions and predict future movements of the heavenly bodies. Because he was able to produce some mathematical models that supported the theory, this structure of the universe was often referred to as the Ptolemaic system. Thus, until the 16th century, most people (including early astronomers) considered Earth to be at the center of the universe.

Even the shape of the Earth itself was argued. Biblical Old Testament references define the Earth as "suspended in space" (Job 26:7) in a comparison to the spherical sun and moon and refer to "the circle of the earth" (Isaiah 40:21). Pythagoras, in the sixth century BC, also suggested a spherical Earth. Eratosthenes of Alexandria (about 200 BC) calculated the circumference of the Earth to within 50 miles of its actual size. The Greeks also defined positional relationships on the Earth by defining meridians and parallels. They identified the poles, equator, and tropics. However, the spherical Earth concept did not prevail with the common man who was far too busy trying to scratch out a living and could care less if it was flat or round. With the rise of the Roman culture, the concept of the Earth as a flat disk with oceans around it emerged as the dominant theme. Nevertheless, most scholars understood that the Earth was indeed a sphere by the time of Columbus. Moreover, those who lived by the sea saw empirical evidence of this curvature (that amounts to about eight inches per mile) when a ship sailed away to disappear over the horizon with the lower portions of the hull gradually vanishing from view first.

One particular problem for cosmologists since ancient times was the apparent backward (*retrograde*) motion of some of the wandering stars (Mars, Jupiter, and Saturn). Periodically the nightly motion of these planets, as viewed from the Earth, appears to stop and then proceed in the opposite direction. To account for this retrograde motion, medieval cosmology stated that each planet revolved in a circle called the *epicycle*, and the center of each epicycle revolved around the Earth on a path called the *deferent*.

Copernicus: Establishing the Center

Nicolaus Copernicus (1473-1543), a Polish astronomer and one of the most learned men of his time, had the opportunity to study philosophy, medicine, astronomy, and mathematics at several universities throughout Europe. About 1515, Copernicus, influenced by Domenico Maria, a critic of Ptolemy, wrote a short astronomical paper in which he proposed that the Sun occupied the center of the universe, and that the Earth, spinning on its axis once daily, and the planets revolved around the sun. This is called the heliocentric, or sun-centered system. His major work, *De Revolutionibus Orbium Coelestium* (On the Revolutions of the Celestial Spheres), was completed in 1530 but not printed for another 13 years. Perhaps because of its controversial subject matter, it was published by a Lutheran printer in Nürnberg, Germany.

Copernicus also noted the tilt of the Earth's axis. He retained some elements of the Ptolemaic system, including the planet-bearing spheres, with the outermost sphere bearing the fixed stars. However, Copernicus's heliocentric theory accounted for the daily and yearly motion of the Sun and stars, and it explained the retrograde motion of Mars, Jupiter, and Saturn and the fact that Mercury and Venus never vary more than a predetermined distance from the sun.

Copernican theory also provided for the distance of the planets from the Sun according to their period of revolution. The greater the radius of a planet's orbit, the longer the planet takes to make one trip (orbit) around the sun. However, most 16th-century scholars who understood Copernicus's claims had a difficult time accepting them, perhaps mostly from a philosophical point of view. As a result, while parts of his theory were adopted, most were ignored or rejected.

The Danish astronomer Tycho Brahe in 1588 formulated a compromise theory in which the Earth remained in the center, and all the planets revolved around the sun as it revolved around the Earth. One aspect of the Copernican theory that held Brahe at odds was that the period of orbit for a perfect circle for the planets contradicted the mathematical solution. The German astronomer Johannes Kepler resolved this problem when he discovered that the orbits were not circular but elliptical.

Galileo's Heresy

Galilei Galileo (1564-1642) was an Italian physicist and astronomer, who, along with Brahe and Kepler, initiated a scientific revolution that culminated in the work of English physicist Sir Isaac Newton. Galileo studied medicine, philosophy, and mathematics. In 1609 he became aware that a method of bringing far objects closer by the use of curved glass had been invented in Holland. He was able to quickly fashion such devices, and the first application of a "spyglass", as it was known, was its value for naval (maritime) operations. His benefactors handsomely rewarded him for this invention.

By December of that same year, Galileo built a telescope with a magnification power of 20 times. Turning it to the heavens, he immediately discovered mountains and craters on the moon. His eyes now allowed him to walk among the craggy features of this far away place drawn near. He saw that the Milky Way, a bright area in the dark night sky, was actually composed of tens of thousands (if not millions) of stars. He discovered sunspots and the four largest satellites of Jupiter. It was Kepler, however, who first used the French word *satellite* (meaning guard or attendant) to describe the small objects around Jupiter that turned out to be its moons. He published these findings in March 1610 in *The Starry Messenger*. By December 1610 he had observed the phases of Venus, which contradicted Ptolemaic astronomy and confirmed his trust in the Copernican system that he openly advocated.

Galileo's discoveries were derided by some philosophers of the day. One particularly influential professor of philosophy convinced the Medici (the ruling family of Florence as well as Galileo's employer) that belief in a moving Earth was heretical. In 1614 a Florentine priest was prodded to denounce Galileo's writings from the pulpit prompting Galileo to defend his work in an open letter. He declared that there was no disagreement between Biblical passages and his scientific observations. He stated that Biblical understanding should be viewed in light of increasing human knowledge, because scientific understanding changes, while the Biblical foundation is immutable, no scientific position should ever be made an article of Roman Catholic faith.

By 1616 Copernican books were censored, and the Jesuit Cardinal Robert Bellarmine demanded that Galileo not defend nor believe the concept that the Earth moves. Cardinal Bellarmine had previously cautioned him to treat this subject only hypothetically and not as literal truths, nor should he attempt to reconcile them with the Bible. Galileo acquiesced for a time and remained silent on the subject for several years. He spent this time formulating a method of determining longitude at sea by predictions of the positions of Jupiter's satellites, and other scientific endeavors.

Galileo astutely recognized that tidal motions appeared to be influenced by the position of the moon and sun. He published a book in 1632 called *Dialogue on the Tides* in which he assessed the Ptolemaic and Copernican theories relative to the physics of ocean tides—coming down on the side of Copernicus. This was the final straw, and Galileo, summoned to Rome, stood trial for suspicion of heresy. He was sentenced in 1633 to life imprisonment, which was quickly commuted to a permanent house arrest. Rome ordered all copies of the *Dialogue on the Tides* to be burned, and the result of his heresy to be read publicly in every university, lest anyone consider embracing the Copernican theory. In an effort to smooth the damage done to the reputation of the Catholic Church by the Inquisition, Pope John Paul II convened an investigation into Galileo's trial in 1979. Thirteen years later a papal commission acknowledged the Vatican's 17th century error and Galileo was exonerated.

Despite the Catholic Church, Galileo was not finished as an astronomer, and his last book, *Discourses Concerning Two New Sciences,* published in 1638, refines earlier studies of planetary motion. The book proved pivotal in directing Isaac Newton's thoughts on the laws of gravitation and motion that were to connect Kepler's planetary laws with Galileo's mathematical physics.

While the Copernican theory was suppressed as a result of the ecclesiastical trial of Galileo in 1633, some remained secret followers of Copernicus. Others adopted the geocentric-heliocentric system of Brahe. Serious students of astronomy in the emerging Protestant sections of England, France, the Netherlands, and Denmark were Copernicans, but anti-Copernican views prevailed in other portions of the world for at least another century.

Although the heavens may seem just a curious set of lights that could enchant or engender fear, to those who plotted their positions, these lights in the sky could be very useful to man. A set of tables, constructed by the astronomers of the day (such as Galileo), allowed mariners to determine their latitude (north/south) relative to the position of the sun at noon during the day. At night, navigation by the stars was enhanced with the ability to view the moons of Jupiter (using the newly developed telescope), to provide an approximation of longitude (east/west). However, it wasn't until the late 18th century that an accurate time piece (perfected by James Harrison) allowed for precise navigation of longitude at sea.

Newton's Apple

By the late 17th century the system of celestial mechanics proposed by the English natural philosopher, Sir Isaac Newton, fully supported Copernicus. Born in Woolsthorp, England, on Christmas Day 1642 (the same year Galileo died), Newton first published his laws in *Philosophiae Naturalis Principia Mathematica* (1687) and used them to prove the results concerning the motion of objects. For most people, Isaac Newton is best remembered for his first three laws of motion.

The first of Newton's laws states "A body remains at rest, or moves in a straight line (at a constant velocity), unless acted upon by a net outside force." The second law reveals that "The acceleration of an object of constant mass is proportional to the force acting upon it: $\mathbf{F} = \mathbf{ma}$."

While these first two laws are equally important to rocketry and space travel, it is the third law that provides the pivotal ingredient for the rocket: "Whenever one body exerts force upon a second body, the second body exerts an equal and opposite force upon the first body." It is often expressed as *"For every action there is an equal and opposite reaction."* This reactive force is what generates the thrust of a rocket motor.

Newton, though not a humble man, was humbled by the work of his predecessors and produced the now famous quote, *"If I have seen further* [than certain other men] *it is by standing on ye shoulders of giants.*" He was referring to the efforts of those scholars who had gone before and laid the foundations on which he was building. Newton passed away in London, March 20, 1727.

In the short space of 150 years, through the intellect, courage, and dedication of Copernicus, Brahe, Kepler, Galileo and Newton, mankind had defined the mechanics of the cosmos. By the 19th century astronomers came to realize that stars were in fact glowing spheres of fire like our Sun but perhaps many times larger. Moreover, our Sun, whose radius is over 100 times that of the Earth, is, in fact, only a medium size star. More than one million Earths could fit within the volume of the Sun.

In addition, stars were separated by distances so vast that a large 'yardstick' had to be defined. The speed of light (186,000 miles per second) became a measure of remoteness when defined as the distance light will travel in 365 days (the light-year) or 5,880 billion miles. At that speed the light from our Sun takes about 7 minutes to reach the Earth. The nearest star, Proxima Centauri (part of the triple star Alpha Centauri), is 4.3 light-years from our Earth, meaning that Proxima Centauri is 25 billion miles from Earth.

As the 19th century dawned, the dreamers and visionaries now had a stable environment that was understood and could be proved mathematically. Now only the imagination limited what was possible, and for some that imagination would lead mankind into space.

Space travel is the marriage of many centuries-old disciplines—astronomy, physics, chemistry, and mathematics—engineered into specialized devices. Reaction motors working on the principles of motion defined by physicist Isaac Newton were actually demonstrated almost two thousand years earlier by Heron of Alexandria (Egypt), a Greek mathematician and engineer of his time. The actual date of the invention as well as the specific configuration of the device, now called the *aeolipile*, is not known, but it may accurately be traced to the time of Christ. The device, a small hollow copper sphere suspended over a fire, allowed water within to be boiled and the resulting steam expelled (action) to spin the device (reaction). It was simply a steam driven rotor that, while providing a demonstration of reactive motion, was apparently carried no further in application. Use of action-reaction engines would have to wait for another millennium and a new age of visionaries.

The First Rockets

A form of gunpowder, reportedly used in China late in the third century AD, propelled the earliest rockets. Bamboo tubes filled with a mixture of saltpeter, sulphur, and charcoal and sealed at both ends were tossed into ceremonial fires during religious festivals in the belief that the noise of the explosion would frighten evil spirits. It is conjectured that some of these bamboo tubes, instead of bursting, shot into the air by the rapidly burning gunpowder being exhausted (action) from a faulty seal on one end (reaction). The next obvious step by some clever person was to deliberately produce the same effect, and *fire-arrows* were invented.

As is often the case with early tests, experimental apparatus are not always built with much forethought and planning. According to one ancient legend, a Chinese official named Wan-Hoo, about 1000 AD, attempted the first manned rocket flight, perhaps with the Moon as his ultimate goal. Using a large wicker sedan chair propelled by 47 *fire-arrows* (solid-fuel rockets) lit simultaneously by forty-seven assistants with torches, Wan-Hoo disappeared in a tremendous roar and billowing clouds of smoke, presumably on a successful flight from which he has yet to return.

Rocket fire-arrows were used to repel Mongol invaders at the battle of Kai-fung-fu in 1232 A.D. These rockets were quite large and powerful. One report, perhaps somewhat inflated, stated, *"When the rocket was lit, it made a noise that resembled thunder that could be heard for five leagues (about 15 miles). When it fell to Earth, the point of impact was devastated for 2,000 feet in all directions."* These rockets may have included the first attempts to refine the combustion chamber, which some sources describe as an "iron pot" that contained and directed the thrust of the gunpowder propellant.

The rocket found its way to Europe by the 14th century but remained as an adjunct to cannon. The powder or solid-fuel rocket was an important part of the Asian military as evidenced by the siege of Seringapatam, India, in 1799. Here the troops of Hyder Ali devastated the British by raining hundreds of iron explosive containers down upon them.

Because of this incident, British colonel Sir William Congreve began his work on improving the accuracy and power of the solid-fuel war rocket. These rockets did not use fins for stabilization in flight but rather very long sticks, called guidesticks, that tended to provide some stability. The guidestick was at first mounted to the side of the rocket, but by 1815 the guidestick was screwed to the middle of a base plate around five equi-distant exhaust holes. Rockets weighing up to 40 pounds could attain distances of 3,000 yards. Although some rockets weighed as much as 300 pounds, these were not easily transported or launched. Rockets of this type were used against Napoleon's army in 1805 and again in1806 when small boats slipped into the harbor at Boulogne and caused heavy damage to the fortress and ships there. A year later, the British attacked the Danish fleet and burned Copenhagen with an estimated 25,000 rockets. Virtually all European and Middle East countries used the Congreve rocket in the early 19th century.

During the War of 1812 a British rocket bombardment of Ft. McHenry near Baltimore, Maryland, inspired Francis Scott Key to write America's National Anthem, "The Star Spangled Banner." He observed the American flag illuminated by "the rocket's red glare." The experience of being on the receiving end of rockets inspired the American military to adopt the rocket, and they were used in the Mexican War of 1846-1850.

Englishman William Hale, looking to improve the accuracy of the Congreve rocket, devised the stickless, or rotary, rocket which he patented in 1844. This device obtained its stability in flight by imparting a spin to the rocket by slightly deflecting the exhaust gases with curved vanes in its three exhaust ports. The 24-pounder Hale rocket was 23 inches long and made from rolled sheet iron, 2.5 inches in diameter. This was one of the most popular sizes since it was light and small enough for easy transport. The average range was about 1,200 yards, although distances up to 4,000 yards were possible.

Hale rockets had an improved method of manufacture using the hydraulic press to compress the powder, making the burning more uniform and less likely to explode prematurely. However, rifled cannon that used helical groves on the inside of the gun barrel imparting a spin to the projectile appeared at this time. This technology significantly improved the accuracy, power, and safety of the cannon, and the Hale rocket was obsolete by 1890 but not completely withdrawn from most military inventories until 1919.

What Lies Beyond

One of the first problems faced by the dreamers and visionaries was the method of propulsion that would enable man to travel to distant places such as the moon or the planets. To understand the limitations on possible power sources, it was necessary to understand the composition of the atmosphere and what lay beyond it. The Italian Evangelisti Torricelli invented the *Torricellean tube* in 1643 (the first barometer), based on a theory expounded by French mathematician-philosopher Blaise Pascal. A few years later, Pascal and René Descartes, using a barometer, demonstrated that atmospheric pressure decreases with increasing altitude—the higher one went into the sky, the less air was present. It soon became accepted that the space between the planets was essentially a vacuum and that the atmosphere was a relatively thin blanket of a mixture of gases (air) that surrounded each planet. Thus, whatever means of power was used to travel in the space between the planets, it had to work without the use of ambient air. The term *outer space* was coined to identify this region beyond the Earth's atmosphere that is now established as 62 statute miles (100 km) above the Earth's surface.

Because the observable planets of Venus and Mars are similar in size and relatively close to the earth, the possibility that they might contain intelligent life began to take root in the imagination of a few, and numerous events during the nineteenth century encouraged this popular assumption. When Italian astronomer Giovanni Schiaparelli announced in 1877 that he believed grooves or channels could be seen on the Martian surface through the use of his powerful telescope, the Italian word for channel, *canali*, was literally translated to mean canals of water. This led to the obvious assumption that not only was there life on Mars but it was intelligent and had developed a technology capable of channeling water from the polar ice caps to the more arid equatorial regions to sustain their civilization. A novel, *Across the Zodiac*, written in that same year fancifully described the intelligent life that existed on our sister planet.

There were several hoaxes promoted over the years that culminated in one of the most bizarre episodes in 1938. A young radio entrepreneur, Orson Welles, decided to dramatize the H.G. Wells science fiction novel *War of the Worlds* on the night before Halloween. His efforts were so convincing that there were some instances of near panic among his radio audience who actually believed that Earth was being invaded by Martians.

American newspapers were likewise invaded by spacemen as early as 1929 when a comic strip called *Buck Rogers* first appeared. With such episodes as *The Mongols* and *Martians Invade Jupiter*, Buck Rogers created a lot of excitement among young and old alike. The strip appeared in 287 newspapers at the height of its popularity in 1934 and was translated into 18 foreign languages. Following closely, *Flash Gordon* was also a Saturday matinee serial in the local movie houses. While the technology revealed rockets and ray-guns, the space environment was often filled with clouds and a breathable atmosphere.

Jules Verne

Space travel by the inhabitants of the Earth was the subject of Jules Verne, who sent three men to the moon in his 1865 novel *From Earth to the Moon.* As with his previous novels, *Journey to the Center of the Earth* and *Twenty Thousand Leagues Under the Sea,* Verne's primary objective was a visionary story line with plausible technology as an interesting secondary appeal. He avoided the problem of on-board propulsion by using a large 900-foot-long cannon (a well understood technology) and 400,000 pounds of gun-cotton (a relatively new form of explosive). The nine-foot diameter, 20,000-pound projectile was to be made of lightweight aluminum, an element so rare at that time that chemists had produced only a few small bars weighing just ounces. Verne recognized the devastating effects of the crushing forces of acceleration (gravity forces or G-forces) resulting from the cannon on the aluminum structure and provided its occupants with "water buffers" to enclose the passengers and enable them to survive the shock of firing. While this sounded plausible to the average person and would ease the acceleration on the body by distributing the G-force evenly, the actual load factor would have killed the three travelers.

Perhaps the greatest contribution that Verne made to the dream of space travel was to acquaint the general populace with the idea that a trip to the moon (or to the planets for that matter) was primarily a question of attaining a certain velocity. This velocity had to be great enough to allow an object to escape the strong hold of the earth's gravity. Verne did make an effort to determine the velocity needed to accomplish the flight (using an *energy balance* equation). This speed is now termed *escape velocity* and is approximately 25,000 miles-per-hour (7 miles-per-second). He also recognized the absence of gravity (weightlessness) during the flight. (After the acceleration forces caused by the propellant are expended, the projectile is essentially in a free fall.) His book was a popular success, and it would awaken the dreams of many in the years to come.

Konstantine Tsiolkovsky

While there have been many theoretical prophets of space travel over the years, few have been able to transform their vision into the written word and receive a place of honor in the archives of history. One such early prophet, who was a prolific writer and who achieved that status, was the Russian Konstantine Tsiolkovsky, born into poverty September 5, 1857. Perhaps inspired by his Polish father, who was an unsuccessful educator by profession and who earned his living as a forester, Tsiolkovsky recalled being excited at an early age by a small balloon, a rare device in those days.

He lost most of his hearing by the age of 13 due to scarlet fever. It was difficult for him to attend school, where he was teased by the other children, and so he was largely home schooled and self-taught. His hearing impairment caused him to be somewhat introverted and, as many children do when their physical activities are curtailed, he turned to his imagination and dedicated himself to achieving some prominence in life.

Tsiolkovsky's interest in mathematics and physics, spurred by his reading of Jules Verne, and a high degree of intelligence allowed him to master the basics of these topics to the point that he satisfied the requirements for teaching in public schools—despite his hearing impairment. In 1876 he began employment at the school in Borovsk and later at Kaluga, just south of Moscow, and continued his quest for knowledge, successfully passing the teaching exam in 1880. Tsiolkovsky was interested in astronomy and the associated physics. He was particularly intrigued by observing the behavior of high temperature gases and measuring the velocity of light.

At the age of 23, he submitted several scientific papers to the St. Petersburg Society for Physics and Chemistry. Because Tsiolkovsky had limited access to books and papers generated outside of his cloistered world, some of his initial work appeared to duplicate much of what was already known. He was informed of this in a letter from Dimitri Mendeleyeff, the renowned physicist who structured the periodic table of the elements and the one who had reviewed his submissions. Mendeleyeff realized the problem of Tsiolkovsky's isolation and became interested in his work and provided encouragement and direction. It was through his influence that the Academy of Science in St. Petersburg provided 470 rubles (about equal to U.S. dollars at the time and a considerable sum) to enable Tsiolkovsky to continue his work.

In 1895, Tsiolkovsky mentioned space travel in an article for the first time. Recognizing that the space between the planets was mostly empty, he foresaw that a space vessel would have to be completely sealed with oxygen reserves and air purification for the human travelers. By 1898 he completed a preliminary study of the basic problems of space flight, and in 1903 he published "Exploring Space with Reactive Devices" in the Russian periodical *Science Survey Journal*. However, few scientists paid any attention to it, in part because it was written in Russian, and most scientists in other countries were unable to read his dissertation. To supplement his income, he wrote several science fiction novels, which also allowed his fertile imagination to escape the bonds of his limited resources. Unlike some of his contemporaries, Tsiolkovsky had a good relationship with the press and newspaper readers and received some modest monetary support for his work from private contributors. From 1911 to 1913 he published a series of articles in a Soviet technical magazine *Aviation Reports*.

Tsiolkovsky was the first to record the need for a reaction engine to operate in the airless environment of space. The only known mode of propulsion that would work in a vacuum was the Newtonian principal of action and reaction, the rocket. His design carried not only fuel but also its own oxidizer. While solid-fuel rockets (which combine both the fuel and the oxidizer in chemical compounds) had long been in existence, Tsiolkovsky understood that the energy available from the known chemicals of these rockets was not as efficient as energy released by liquid fuels and oxidizers.

Specific Impulse is one of several metrics that provide an indication of the efficiency of a rocket. It is a value that reflects the thrust produced by a reaction motor relative to the fuel consumed per second. Tsiolkovsky recognized that greater efficiency could be achieved with propellants that provided higher exhaust velocities, such as oxygen and hydrogen, and the resulting higher Specific Impulse. It is his recognition of this factor and other related discoveries that earned him the title of Father of Astronautics.

While Tsiolkovsky was primarily a theorist, at one point he became interested in airships and began designing a large metal structure. He built a small wind tunnel in 1891 to determine the effect of drag produced by skin friction on a large metallic airship. While this pre-dated the Wright Brothers' pioneering use of the wind tunnel by 9 years, he did not make any measurements of lifting airfoils. His meager finances and situation did not permit much in the way of experimentation.

The World War (1914-1918) interrupted Tsiolkovsky's work, but fortunately, the Russian Revolution (1917) did not appear to have any impact. In fact, the new Soviet government recognized early the value of Tsiolkovsky's work (especially because he came from a working class background) and ensured that it was not disturbed during those tumultuous years following the revolution. He frequently received sums of money in the form of honorariums for reprints of many of his articles.

His first publication under the Communist regime was a novel entitled *Outside of the Earth*, a fictional account of a journey into space. However, Tsiolkovsky's works were not widely published until after 1923 when Hermann Oberth also began publishing in Germany. Tsiolkovsky was honored by the Soviet Union on his 75th birthday in 1932 for his contributions to astronautics. He died three years later in 1935.

Robert Hutchins Goddard

The early 20th century also saw another rising star in the field of astronautics in the person of Robert Hutchins Goddard, a professor at Clark University in Massachusetts. Goddard independently discovered virtually all the basic aspects involved in space flight formulated by Tsiolkovsky and was one of the first to publish his observations and mathematical proofs in English and to move from theory to experimental applications.

Born October 5, 1882, in Worcester, Massachusetts, Goddard suffered from a variety of health problems most of his life (he was diagnosed with tuberculosis in 1913) and was fortunate to have a supporting and reasonably affluent family. Goddard folklore includes an event that occurred in 1899, while the frail 17-year-old was sitting high in a cherry tree. It was here that Goddard was struck with the desire to pursue space travel. He writes: *"On the afternoon of October 19, 1899, I climbed a tall cherry tree at the back yard of the barn...As I looked towards the fields to the east, I imagined how wonderful it would be to make a device which had even the possibility of ascending to Mars... I was a different boy then when I descended the ladder. Life now had a purpose for me."* Perhaps every young person should encounter a "cherry tree" experience early in life. This day became known as Anniversary Day and was subsequently celebrated each year by the Goddard family.

A 1908 graduate of Worcester Polytechnic Institute, Goddard became a physics instructor there and received a Ph.D. in physics from Clark University in 1911. In 1912, while a research instructor in physics at Princeton University, Goddard began to prove mathematically that, using rocket power, a device could achieve escape velocity and travel to the moon and the planets. In 1914 he received a U.S. patent for the idea of multi-stage rockets. While Tsiolkovsky theorized that the reaction engine operated on Newton's third law, Goddard, in 1915, essentially proved that the action-reaction of the exhaust gases was pushing against the inside of the rocket motor to provide the propulsive force—not against the atmosphere. A reaction motor would work in the vacuum of space. While this may seem obvious to us today, it did not during his time, even to learned people.

While his contemporaries and predecessors were primarily theorists, Goddard was a practical experimenter. This meant that he had to get equipment, materials, and facilities, as well as added support in the form of skilled assistance. He pursued and acquired funding, typically in the form of grants, from a variety of sources that included his university as well as government and private capital.

He returned to Clark University in 1916 to teach physics and received his first grant of $5,000 from the Smithsonian Institute for his work, with further grants awarded through 1934. His theoretical work to that point culminated in *Smithsonian Miscellaneous Publication No. 2540* (January 1920) "A Method of Reaching Extreme Altitudes." In this paper Goddard detailed methods of raising weather-recording instruments higher than that possible with "sounding" balloons (*sounding* is the act of probing various aspects of the atmosphere using scientific instruments). While a valuable piece of work, it was to be his most controversial, as Goddard was severely criticized by the press (no less than the prestigious New York Times) and by many in the scientific community for his views. The main point of discord was the statement made during an interview that he believed rockets could be built to reach the moon—a preposterous assumption by the scientific standards of the day. This experience led Goddard (and others such as Charles Lindbergh) to abhor the popular press for their ignorance and lack of understanding and sensitivity. He was so shaken by this experience that it influenced his interaction with the outside world from then on. (The Times would issue a somewhat humorous apology some 50 years later during the launch of the first manned expedition to the moon by Apollo 11).

Goddard became sidetracked from 1917 to 1924 when he did work for the Army and Navy on multiple-charge solid-fuel rockets. He came to the realization that commercially available solid-fuel rockets used for signal flares converted only about 2% of their potential chemical energy into thrust. By working with new compounds and the design of the throat of the rocket nozzle, he was able to raise this to 40%. This was still far from the energy efficiency necessary to send a rocket to the moon. When the First World War ended and the funding from the military ceased, he turned his attention to developing liquid-fuel rocket motors that used liquid oxygen and gasoline.

To work with cryogenics (materials at very low temperatures) required some courage as oxygen becomes a liquid at -297 degrees below zero (F). The effect of extreme cold on materials used to hold and transfer the liquid was largely unknown. Metals became brittle and lubricants cease to function at such cold temperatures. The affect on the human skin was also very destructive, and great care was taken to ensure that injuries and fire were avoided.

On December 6, 1925, his liquid fuel rocket produced sufficient power to lift its own weight for the first time. This was followed on March 16, 1926, with the first flight of a liquid fuel rocket in Auburn Massachusetts. As the flight-time was a mere 2.5 seconds and the distance traveled was 184 feet, the achievement, though a milestone, was not spectacular. This success was followed by the first launch of a scientific payload (barometer and camera) in a rocket flight in 1929.

Goddard's rockets became larger, more noisy and hazardous, and a flight on July 17, 1929, alarmed the quiet countryside of rural Massachusetts. The publicity caused Goddard's work to come to the attention of Charles Lindbergh, the noted aviator who had flown solo from New York to Paris two years earlier in 1927. Lindbergh visited Goddard to determine the focus and validity of his research and was impressed enough by his work to convince Harry Guggenheim, heir to a substantial fortune, to provide a series of grants from the Guggenheim Foundation beginning in 1930. Harry's father, Daniel (1856-1930), was a prominent American industrialist and philanthropist.

With the availability of generous funding, Goddard decided to move to a locale that provided more room for his experiments and chose a small ranch on the outskirts of Roswell, New Mexico, to continue his work. He flew his first three rockets there in 1931 and tested rockets stabilized with gyroscopes in 1932. Between 1932 and 1934, the economic depression caused an interruption in support, and Goddard had to return to teaching at Clark University, while pursuing small-scale experiments funded by the Smithsonian.

In 1934 Guggenheim funding was reinstated and continued through 1941. Goddard's testing was more programmatic at this stage, and he identified his experiments serially beginning with the letter A. On March 8, 1935, rocket A-4 reached an estimated 700 miles-per-hour, while A-5 on March 28th, under gyro-control and using redesigned exhaust deflectors, achieved an altitude of 4,800 feet. Although some accounts credit Goddard's March 8th flight with exceeding the speed of sound, it is unlikely that he achieved this distinction. Two flights in excess of one mile in altitude were made in May and July.

Goddard resolved the problem of the destructive effects of the extremely high temperatures on the rocket motor generated by the burning of the volatile combination of gasoline and liquid oxygen. He achieved this by circulating the fuel through a set of tubes that surrounded the thrust chamber before injecting it for combustion. This concept is referred to as *regenerative cooling*. The relatively cool fuel absorbs the heat resulting from the combustion. He worked to perfect this method throughout his experimental period, although his engines occasionally experienced periodic burn-through of the combustion chamber walls.

The test program became increasingly sophisticated with the use of thermocouples for measuring temperatures on the combustion chamber and hydraulic gauges to measure thrust levels that were now approaching 1,000 pounds with the K series rockets. It was during this time that he developed centrifugal pumps suitable for rocket fuels (the first rockets delivered the propellants under pressure). He successfully launched a rocket with a motor pivoted on gimbals under the control of a gyroscopic mechanism in 1937. The idea of the gimbaled motor grew out of the problems he had with the exhaust deflectors. While the deflectors accomplished the task of redirecting the rocket, they placed an excessive amount of drag on the exhaust gases, significantly decreasing the thrust of the rocket. He also experimented with multiple engines (clustering). The concepts of gimbaled engines and clustering would find their way to the moon some 30 years later. His last flight, recorded on October 10, 1941, was a disappointment when rocket P-36 jammed in the launch tower.

While Goddard did make an effort to interest the military in his work, the response was not enthusiastic. Knowledge that Germany was secretly moving ahead with the development of rockets one hundred times more powerful than those with which he was working probably would have caused heads to turn in the War Department. Budget constraints slowed progress during what should have been his most productive years. Perhaps equally restrictive was the fact that he worked, for the most part, in an intellectual vacuum. Although he period-ically entertained such luminaries as Charles Lindbergh and Harry Guggenheim, and undoubtedly probed their thoughts, his staff was mostly made up of a few skilled craftsmen who made the rocket parts from his direction. Had he had someone of equal innovative cal-iber to collaborate with (in the style of Wilbur and Orville Wright), he might not have traveled down several paths that wasted valuable time and money.

The advent of America's involve-ment in World War II on December 7, 1941, caused Goddard's expertise to be redirected by the War Department. He spent much of his time developing rock-et assisted take-off units for the Navy. From 1943 to 1945 he was an engineer-ing consultant to the Curtiss-Wright Corporation of Caldwell, New Jersey, and served as director of the American Rocket Society during 1944 and 1945. His poor health finally caught up with him, and he died of throat cancer on August 10, 1945 (he loved good cigars).

During his lifetime Goddard postu-lated and proved many of the basic tech-niques used in modern rockets. While publishing and selectively promoting his work, he was also somewhat secretive about key aspects, being paranoid that unscrupulous experimenters might steal his work (a mistrust similar to that exhib-

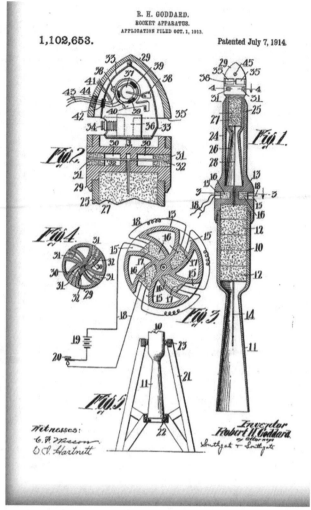

One of Goddard's rocket patents from 1914

ited by the Wright brothers). His fear of losing credit for his pioneering efforts resulted in much of his work not being built upon by others but rather being replicated independently. Thus, even though he pioneered early rocketry, critical aspects of his work were not made available to subsequent experi-menters, who had to rediscover solutions to problems he had already formulated. Dr. Frank Malina, a member of the Guggenheim Aeronautical Laboratory at the California Institute of Technology, once wrote of Goddard:

I believe Goddard became bitter in his later years because he had had no real [financial] success with rockets, while Aerojet-General Corporation and other organizations were making an industry out of them. There is no direct line from Goddard to present-day rocketry. He is on a branch that died. He was an inventive man and had a good scientific foundation; but he was not a creator of science, and he took himself too seriously. If he had taken others into his confidence, I think he would have devel-oped workable high-altitude rockets and his achievements would have been greater than they were. But not listening to, or communicating with, other qualified people hindered his accomplishments.

Goddard's wife, Esther Fiske (20 years his junior), whom he married in 1924, became his most out-spoken proponent following his death. She not only concluded several pending patents but also suc-cessfully lobbied the United States government to purchase many of them. In 1950 the first patent infringement case was heard when the German rocket expert Wernher von Braun (then working in the United States) informed his superiors that U.S. government rocket development infringed on Goddard's patent rights.

Esther Goddard was made a member of the Daniel Guggenheim Foundation and was the driving force behind the 86th Congress authorizing the issuance of a gold medal in Professor Goddard's honor on September 16, 1959. Likewise, in memory of his contributions, a major space science laboratory, NASA's Goddard Space Flight Center, Greenbelt, Maryland, was established on May 1, 1959. Esther C. Goddard died in 1982.

Robert A. C. Esnault-Pelterie

Swiss-born Robert A. C. Esnault-Pelterie (1881-1957) was a graduate of the Sorbonne (the former University of Paris) in France and a sculptor, engineer, and inventor. He was an early aviator who made his first flight in 1907 and held French Pilot License Number 4 (1908). He saw military service in the First World War and was one of the best-known early French aviation pioneers. His career as a pilot abruptly ended in a crash, but he continued to be a visionary and theorist.

In 1912, Esnault-Pelterie presented a paper to the Physics Society of France detailing various aspects of space travel. He implied that atomic energy might be a means of propulsion to reach the moon. He considered the military implications of rocketry and, in 1929, proposed a plan for the development of ballistic missiles against which he believed there could be no defense. His proposal led to government financed work on liquid fuel rocket engines. An accident during a test of an engine in October 1931 caused Esnault-Pelterie to lose four fingers. His 1930 publication, *Astronautics*, was a comprehensive review of the problems and prospects of space travel. A subsequent edition in 1934 defined interplanetary travel and included the applications of nuclear power and aerobraking—the use of atmospheric drag to slow a spacecraft for gravitational capture by a planet. The French capitulation early in World War II ended his research.

Hermann Oberth

Hermann Julius Oberth was born on June 25, 1894 in Hermannstadt, Siebenbüergen, an old German speaking colony of German Saxons in a part of Transylvania between Hungary and Romania. Two books by Jules Verne, *From the Earth to the Moon* and *Journey Around the Moon*, (given to him by his mother) aroused his interest in space travel at the early age of 11, and this passion remained strong throughout his life.

By the age of 13, Oberth discovered that many of Verne's calculations were not plucked from thin air but had some validity. His youthful unencumbered mind also grasped that interplanetary travel was not pure folly as the scientific community assumed. He computed the gravitational forces that Jules Verne's space travelers would endure during acceleration in the gun barrel. His answer—47,000 times the earth's gravity (Gs), assured that they would be squashed. He recognized the need to extend the acceleration period from the single impulse of a cannon to a longer thrusting period of several minutes and that this would require a rocket.

To confirm the action-reaction physics of Newton, Oberth took a rowboat and a pile of rocks into a nearby lake and proceeded to propel himself by ejecting (the action) mass (the rocks) in the opposite direction that he wanted the boat to move (the reaction). Beyond such a crude test as the rowboat, Oberth had no resources with which to experiment, but he continued to develop his theories, all the while teaching himself the advanced mathematics needed to prove these theories—on paper.

In 1913 Oberth went to Munich to study medicine; however, his education was interrupted by World War I. He served in the Austro-Hungarian Army in the medical corps where he decided that medicine (his father's vocation) was not for him. The laws of physics intrigued him more.

Oberth married Mathilde Hummel, known as Tilly, in 1918. As with Robert Goddard's wife Esther, Tilly would become an indispensable partner to her husband, sharing all his triumphs and tribulations. He studied physics, and in 1922 submitted a doctorial thesis about rocket-propelled space travel to the University of Heidelberg. It was rejected by the professional review committee because the idea of space travel was not realistic in academic circles. He writes: *"I refrained from writing another one, thinking to myself: Never mind, I will prove that I am able to become a greater scientist than some of you, even without the title of doctor."* He never did receive a doctorate although he had the honorary title conferred on him.

While Goddard paved the way with both theoretical and experimental activities, Oberth likewise examined space flight from a theoretical point, but only modestly with experimental activities. For example, Oberth understood that the higher the ratio between propellant mass and structural mass the faster a rocket would be able to travel. As a rocket expends fuel, its structural mass remains the same—the rocket has to carry along the dead weight of the partially empty fuel tanks and the structure that contains them. Additionally, the large and fuel-thirsty engines required to get the large vehicle off the ground are not necessarily the best choice once the initial acceleration into the upper atmosphere has been achieved. Oberth therefore reasoned that building the rocket in stages that could be separated (discarded) would improve the *mass ratio* of the rocket. He wrote: *"...the requirements for stages developed out of these formulas. If there is a small rocket on top of a big one, and if the big one is jettisoned and the small one is ignited, then their speeds are added."* This concept would become the key to building large rockets and attaining high velocities.

Oberth also sought information on the chemical energy of solid propellants and deduced that they were insufficient for flight to the moon and planets. Liquid propellants would have to be used—the most powerful combination being the cryogenic liquids of hydrogen and oxygen. This was the same conclusion that Tsiolkovsky and Goddard had reached.

Oberth, though still a student, assembled his findings for publication, but, as no publisher would accept his work, he decided to print at his own expense (and Tilly's household savings) a small 92-page booklet *Die Rakete zu den Planetenräumen* (The Rocket into Interplanetary Space). The pamphlet that appeared in 1923 contained four assertions: 1) Present technology allowed for the building of apparatus that can rise above the limits of the earth's atmosphere. 2) Further developments would permit achieving velocities such that an apparatus could leave the earth's gravitational pull. 3) Ultimately, these machines would carry man. 4) It might be possible to manufacture such apparatus for profit, and that all of these assertions would be achieved within the next few decades. Here was not only analysis and theory but also methods and practical purposes. A longer version of his work (429 pages) produced in 1929 was hailed by his contemporaries as significant in its scientific importance.

Unlike Goddard's experience with the press, Oberth's book was a success and got good publicity, despite being a very technical subject. Unlike the more practical Americans, war weary Europeans seemed more ready to embrace the idea of space travel, and serious scientific discussions resulted. Oberth was confident enough to write, *"Rockets... can be built so powerfully that they could be capable of carrying a man aloft."* Although there was some criticism of his ideas, science fiction literature that provided a utopian theme seemed to suit the war-torn continent.

The success of his books encouraged Oberth to produce a less technical account of the possibilities of space flight. Another German space flight enthusiast, Max Valier, a popular writer on scientific subjects, assisted in this endeavor by condensing Oberth's previous work and publishing it for him. This book inspired the formation of several rocket clubs in Germany. Perhaps the most significant of these was the *Verein fur Raumschiffarht* (Society for Space Travel) abbreviated as *VfR,* whose members included Oberth, Valier, and a teenager named Wernher von Braun. The nucleus for this largest and best known of the groups formed on July 5, 1927, in a backroom of the Golden Scepter—a restaurant in Breslau—with Oberth as its president. Its purpose was to popularize the idea of flight to the moon and planets and perform serious experiments with rockets.

Although the book that Valier produced failed to capture the essence of the scientific meanings that Oberth had intended, it nevertheless sold several thousand copies. It was, in fact, the apparent failure of the book's scientific impact that caused another young member of the VfR, twenty-one-year-old Willy Ley, to undertake yet another publication to clarify many of the points that Valier had failed to make. Ley would go on to become a public spokesman for rocketry and space travel over the next half century.

In 1929 Oberth, published *Wege Zur Raumschiffahrt* (The Road to Space Travel) in which he envisioned the development of ion propulsion—electric rockets. This book won an award from the French rocket pioneer, Robert Esnault-Pelterie. Oberth used the prize money to buy supplies for the VfR rocket society that had been started to move his work into the experimental stages.

Although Oberth and his Russian and American contemporaries (Tsiolkovsky and Goddard) explored many of the same aspects of reaction engines and the environment of outer space, they seemed to discover these truths independently without knowledge of each other's work. It is about this time that Oberth became aware of Goddard's publication, *A Method of Reaching Extreme Altitudes,* and Tsiolkovsky's work, and they briefly corresponded.

Because of his publications, Oberth achieved his greatest prominence as being the progenitor of several European rocket societies, such as the VfR, that ultimately spawned a team of visionaries and experimentalists who made the greatest advances in rocketry in the first half of the twentieth century. Where Goddard had been criticized for his public assertion that travel to the moon was possible, Oberth's writings ultimately caught the imagination of the public, especially when his ideas were captured on the silver screen.

Silent moviemaker Fritz Lang read Oberth's book and decided to produce an adventure story about space travel. The result was *Die Frau Im Mond* (The Girl in the Moon) released in 1929. Lange's wife, actress Thea von Harbou, wrote the script based on materials and books by Oberth and Willy Ley.

Lang wanted authenticity, so he used Oberth as technical advisor. Both Oberth and Willy Ley helped Lang build a model of a spacecraft that looked very realistic even by today's standards. To add dramatic flare to the rocket flight, Lang invented the countdown to increase the tension for the audience—an artifact that is still an integral part of today's rocket launch sequence and no less tension filled.

Oberth's unassuming manner allowed him to be somewhat manipulated by Lang's publicity agent. As a publicity stunt for Lang's film, Oberth agreed to build an actual rocket, about ten feet long, to be launched at the premier showing. With less than four months to the scheduled event, Oberth enlisted the aid of Rudolph Nebel, a former pilot in World War I who had a degree in engineering but had never practiced professionally, and Alexander Shershevesky, a Russian student of aviation who had also overstated his engineering credentials. He had never even worked on an airplane but had only written articles for aviation magazines. Between the three of these men, there was little if any engineering expertise, and efforts to produce a liquid fuel rocket quickly fell behind schedule. During experiments an explosion nearly cost Oberth the sight of one eye. He did develop a combustion chamber and nozzle for liquid propellants and liquid oxygen, the *cone jet nozzle*, and had it patented in Berlin in 1930.

Several days before the premier of the movie, Oberth realized that he would not have the rocket completed in time. At that point he returned home to Romania to ease the tension and avoid embarrassment. When it became obvious to Lang's Ufa Film Company that the rocket launch was not going to happen, they issued a statement that the launching would not occur for reasons attributed to weather. The epic film, however, enthralled moviegoers of the silent era and became a classic.

In some respects, Oberth never fully recovered from the anxiety he experienced in trying to develop a rocket for the movie. He returned to his mathematics and physics teaching position (which he held until 1938) at a school in Mediasch, Transylvania, but stayed in touch with members of the VfR. In 1938 Oberth was brought for a visit to the German rocket development center at Peenemünde by his former pupil, Wernher von Braun. Von Braun had established a facility that would have seemed like science fiction to any of the professors who had panned Oberth's doctorial dissertation almost twenty years earlier.

Oberth traveled to the United States in July 1969 to witness the launch of the Saturn V rocket that carried the Apollo 11 crew on the first lunar landing mission—the only one of the early twentieth century prophets who survived to see that momentous event. While Oberth would not play a significant role in rocket development for the remainder of his long life, his philosophical thoughts were a primary expression in his later years. At the age of 90, his book, *Primer for Those Who Would Govern,* attempted to illustrate the peril of inadequate education on the democratic process.

Hermann Oberth passed away on December 28, 1989, at the age of 95. His tombstone contains the Biblical quote that reflects his change in emphasis from space flight to his concerns for today's world: *"Blessed are those who hunger and thirst for justice."*

Defining the Essentials

The work of the great prophets of astronautics had established, by 1930, the essential physics of rocketry. The requirement for a reaction engine to operate in the vacuum of space was proved. They had defined the critical performance aspects of the rocket that would have to be achieved if space flight were to become a reality. These included the ability to use and control the energy of selected chemical propellants. This energy may be expressed as a relationship between the mass of the propellants consumed (m), the exhaust velocity (C) created, and the resulting thrust (T) generated by the engine. This is essentially Newton's Second Law which, while more commonly expressed as F = ma, may be restated as the equation T = mC. It illustrates that more thrust (T) is generated when higher exhaust velocities (C) are achieved or when more propellant mass (m) is consumed per unit time.

Thrust is the basic push provided by the engine and is approximately comparable to the horsepower rating for an internal combustion engine of an automobile. Originally measured in pounds-force, it is now expressed in a metric equivalent called Newtons (one pound of thrust equals 4.45 Newtons). [This book will use the earlier expression of pounds as it is more clearly related to the common unit of measure in the United States.]

Exhaust velocities can range from 2,000 feet per second (ft/sec) for simple solid-fuels such as gunpowder, to 8,000 ft/sec for a mixture of liquid oxygen and gasoline. The highest exhaust velocities of 12,000 ft/sec (or more) are available from liquid oxygen and liquid hydrogen. However, these exhaust velocities can only be realized if the design of the engine is optimized to take advantage of the high combustion chamber pressures and temperatures generated by the propellants. These design considerations include the size of the combustion chamber, the constriction ratio of the throat, and the flare and length of the nozzle.

Propellants with higher theoretical exhaust velocities have a higher potential Specific Impulse (Isp), which is the measure of the efficiency of the rocket's propellants. It represents the thrust generated per pound of propellant consumed per second. Measuring Isp is similar to the efficiency of a car in miles per gallon. Modern liquid propellants have specific impulses of about 300 lb/sec. and high-energy liquid propellant combinations (liquid oxygen and liquid hydrogen) typically have an Isp of up to 450 lb/sec. The theoretical maximum for chemical fuels is about 500 lb/sec. The experimentalists of the 1930's were struggling to achieve an Isp of 250 lb/sec. with liquid propellants, and solid-fuels were a distant second.

Another factor in determining the capability of a rocket is the relationship between the weight of the structure relative to the propellants. The *mass ratio* of a rocket is the weight at lift-off divided by the weight remaining after all propellants are consumed. A high mass ratio means that more propellant is pushing less rocket—resulting in higher velocity.

In a single stage rocket, the *mass fraction* is the weight of the fuel divided by the weight of the rocket. For individual stages a higher mass fraction is better, meaning that there is less non-propellent mass. A single stage rocket cannot develop enough speed to go into Earth orbit if its mass fraction is less that 90%. The best mass fraction achieved by modern rockets has been about 85%. This is why all rockets intended to orbit a payload are multi-stage—several separate rockets (typically two or three) mounted in tandem, one on top of the other. When the first stage exhausts its propellants, it is discarded and the second stage takes over, adding to the velocity imparted by the first stage. However, the mass relationships become more complex when multi-stage rockets are involved.

The main reason for multi-stage rockets is that, as the propellants are consumed, the structures which contained them are useless and only add weight to the vehicle, slowing down its future acceleration. By dropping the stages which are no longer useful, the rocket lightens itself. The thrust of the upper stages is able to provide more acceleration than the earlier stages are capable of. When a stage drops off, the rest of the rocket is still traveling at the speed that the whole assembly reached at burnout. This means that it needs less total fuel to reach a given velocity and/or altitude.

A further advantage of multiple stages is that each stage can be optimized for the conditions in which it will operate. Thus the first stage is typically operating within the lower portions of the atmosphere, while the upper stages can use engines better suited to the near vacuum conditions of space. Lower stages require more structure than upper stages as they need to support large quantities of fuel and huge engines needed to get the fully fueled rocket off the ground. They typically represent 60 percent or more of the gross weight. With these essential physics of rocketry understood, many new experimenters came forward to put the theory and equations into operating hardware.

Society for Space Travel - *VfR*

Paralleling Goddard's work but trailing it by several years was a group of experimenters in Germany who were galvanized by the writings of the visionary Hermann Oberth. Max Valier (1895-1930) proposed creating a club to raise funds to finance exploring Oberth's ideas in a series of experiments. Valier was born in Austria and studied physics at the University of Innsbruck while working as a machinist. His education was interrupted by the First World War, where he served in the Austro-Hungarian Army as an aerial observer.

After the war, Valier became a science writer and after reading Oberth's book, The Rocket into Interplanetary Space, wrote a companion book to explain Oberth's ideas in terms that could be understood by the average person. He published *Der Vorstoß in den Weltenraum* (The Advance into Space) the following year with Oberth's help. It was a success, selling six editions before 1930. He published several articles in German on the subject of space travel, including *"Berlin to New York in One Hour"* and *"A Daring Trip to Mars."*

During the summer of 1927, a group of Germans met in the back room of a restaurant to found a society that was to have a profound effect on rocketry—*Verein fur Raumschiffahrt* (Society for Space Travel), abbreviated VfR. Among the founding members were Johannes Winkler (1897-1947) and Max Valier, who were subsequently joined in 1929 by Hermann Oberth and Dr. Walter Hohmann, author of *Die Erreichbarkeit der Himmelskorper* (The Attainability of Celestial Bodies). Hohmann's mathematics on techniques for orbital changes would become the key to achieving rendezvous of two satellites 35 years into the future. The VfR grew rapidly and within a year had over 500 members. With annual dues of eight Marks (about two U.S. dollars), it was important for active members to enlist as many friends and neighbors in the society, regardless whether they were participants or not. In 1929 the VfR was able to obtain much of the equipment that Oberth had built for the *Girl in the Moon* movie rocket from Lang's Ufa film company.

Rudolph Nebel, whom Oberth had hired to help make the ill-fated rocket for the Fritz Lang movie, was a valuable member of the VfR but not for his scanty engineering attributes. His ability to scrounge facilities and material were critical to keeping the fledgling organization alive. He found a base of operations for the group at an abandoned WWI ammunition storage facility of the German Army, in the Northern Berlin suburb of Reinickendorf. On September 27, 1930, the group moved all their meager equipment to the facility, and Nebel formally opened the *Raketenflugplatz Berlin* (Rocket Flight Field –Berlin). His foresight was considerable in that he drew up a ten-year plan for the experimental activities.

In an effort to popularize rocketry, Max Valier initially instigated several activities that were not well received by much of the VfR. He persuaded Fritz von Opal, the automobile manufacturer, to develop a rocket-powered car. While some success was achieved in this area, it involved the use of solid-propellant rockets purchased commercially and did not advance the technology of liquid rocket propulsion.

Johannes Winkler, who was badly wounded in World War I, studied as a machinist at the Danzig technical college and was employed by the Junkers Airplane Company. As a founding member of the VfR, he was its first president. Working independently and financially supported by a benefactor, Hugo A. Huckel, he built a small rocket slightly more than 24 inches high that weighed about 10 lbs. and was powered by methane and liquid oxygen. Winkler's HW-I (Huckel-Winkler) achieved the first flight of a liquid fuel rocket in Europe on February 21, 1931, (some accounts record the date as March 14th) and it reportedly flew several hundred feet vertically. This was followed by the design of a more powerful rocket, the HW-II eighteen months later. Unfortunately, the rocket exploded within seconds of ignition because of a faulty fuel valve.

Nebel, for all his lack of qualifications, prodded the VfR to move forward with building a small liquid fuel rocket. Oberth had returned to Berlin following his humiliating self-exile, and, at a meeting of the principals of the VfR, it was decided to demonstrate a flight of a liquid-propellant rocket—but most believed that Oberth's design was not suitable. Klaus Riedel, an engineer and member of the VfR, produced a simple motor, designated Minimum Rocket by Nebel and shortened to "Mirak," to demonstrate the feasibility. However, work began in parallel to static test Oberth's *Kegeldüse* rocket motor.

In July 1930 the VfR demonstrated the *Kegeldüse* liquid-oxygen-and-gasoline-fueled rocket motor for the Director of the Chemisch-Technische Reichsanstalt. The test ran for 90 seconds and generated 16 pounds of thrust but failed to impress the visiting dignitaries sufficiently for them to support the research.

Equipment of the VfR experimenters was elementary at best for the first test of Oberth's steel *Kegeldüse*. A grocer's scale was used to measure the thrust of the engine, which was immersed in a pale of water from which protruded the small exhaust nozzle of the rocket. Water was used to cool the combustion chamber because these rockets were not fired with the intention of free flight but were held static—bolted to the test stand. Thus, the availability and weight of water was not a problem at this point in the experimentation process.

A Dewar flask (thermos bottle) contained the liquid oxygen and a steel cylinder of compressed nitrogen provided pressure to a small tank of gasoline to force it through copper piping to the motor. Riedel had the perilous task of igniting the engine. A metal shield had been placed several yards from the ungainly assemblage of equipment, and he was required to toss a burning gasoline soaked rag over the apparatus and take cover before the motor started with an ear splitting roar. To help overcome the destructive effects of the heat generated in the engine, the interior of the combustion chamber was lined with ceramic. It was about this time that a young teenage student named Wernher von Braun was introduced to the VfR by Willy Ley.

These early efforts required the experimenters to come to grips with the tremendous heat and pressure of the combustion process and the handling of volatile fuels and oxidizers. Liquid oxygen is a liquid only when kept at a temperature of minus 297 degrees below zero (Fahrenheit). Just as Goddard was struggling to tame these hazardous substances and environments, so was the VfR. The experimenters needed to observe a code of safety that dictated cement bunkers and steel enclosures to protect the fragile skin and bone from the explosive effects of experiments gone wrong. There were many accidents and explosions but, fortunately, few serious injuries during this period. One notable exception occurred in 1930. Max Valier was killed when an alcohol-fueled rocket exploded on a test bench.

Mirak experiments began in August 1930 when Nebel and Riedel conducted a series of tests designed to improve the motor. While progress was made, the tests ended when the motor exploded. It was almost a year before the next phase of work leading to the Mirak II was completed in May 1931. A static test turned into an unexpected flight test as the motor broke loose and rose to about 60 feet. A week later the repaired motor, installed in a rocket called the Repulsor, reached 200 feet before heading off horizontally over the Raketenflugplatz, landing almost a half mile from the launch site.

The configuration of the rocket was similar to Goddard's first effort with the engine positioned in the nose. The propellant tanks consisted of two long, thin cylinders that comprised the body allowing the exhaust to be channeled between—hence the descriptor "two-stick Repulsor". The following month, the first Repulsor, equipped with a recovery parachute, reached over 500 feet, but the parachute deployed early spoiling a longer flight.

A new design placed the propellant tanks close together forming a guide stick similar to the Congreve war rocket. It reached an altitude of over 3,000 feet and the parachute recovery system functioned perfectly allowing the rocket to be used again. Subsequent models were called the "one stick Repulsor" and reached almost a mile in altitude and two miles down range. The design was so successful (more than 80 launches were performed during 1931) that the later versions were launched with less than full propellant loads to assure they would not fly outside the perimeter of the Raketenflugplatz.

During this time Riedel determined that 60 percent alcohol (40 percent water) was a better fuel than gasoline because it was cheaper to purchase and less hazardous—although it possessed less energy. It could also be circulated around the thrust chamber to provide regenerative cooling.

In the spring of 1932, Nebel's enthusiastic salesmanship brought the *Raketenflugplatz* to the attention of certain members of the Ordnance Department of the Reichswehr, the German Army. Three inconspicuous visitors, dressed in civilian clothes, called on the VfR as representatives of the ordinance department. All three were to play important roles in the future development of liquid rockets. They were Colonel Professor Carl Becker, Chief of Ballistics and Ammunition, Major von Horstig, his ammunition expert, and Captain Dr. Walter Dornberger, in charge of powder rockets (solid-fuel) for the Army. The concept of the artillery rocket was being re-examined in light of the Versailles Treaty that prohibited artillery in the German Army.

The VfR members were quite pleased when Nebel signed a contract with the Army for the sum of 1,000 marks, contingent upon a successful firing of the Mirak II at the Army proving grounds at Kummersdorf. In July 1932 Nebel, Riedel, von Braun and several others of their team loaded their equipment into two cars and set out for Kummersdorf, 60 miles south of Berlin. The launcher was erected and the rocket fueled by 2:00 PM. On signal, Mirak II soared upward for a height of almost 200 feet. However, its trajectory then became almost horizontal so that the rocket crashed several hundred yards away before the parachute could open.

Von Braun recalls being fully aware that the toy-like Mirak II was a simplistic approach to a real liquid-fueled rocket capable of performing either a military or a scientific role. The problems of guidance and control, fuel pumps and valves, as well as the manufacturing costs, made the efforts of the VfR completely inadequate to provide an experimental program needed by the military. Only the funds and facilities of the German Army would make a large liquid fuel rocket and space travel a reality.

The winter that followed the successes of 1932 blew cold with the rise to power of Adolph Hitler. The deepening world economic depression and the new demands of the Fuehrer diminished the membership of the VfR and, consequently, the availability of funds. Von Braun, who had quickly grasped the essentials of rocketry

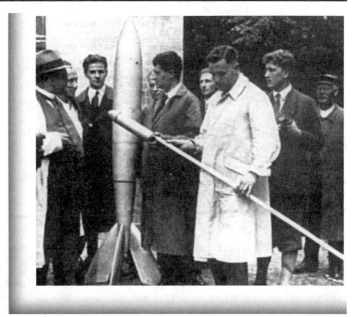

Klaus Riedel (white coat) holds a model of the *MIRAK*. Wernher von Braun stands behind him holding the conical *Kegeldüse* rocket motor. Hermann Oberth (centre) is standing next to the large unfinished rocket built for UFA film studios

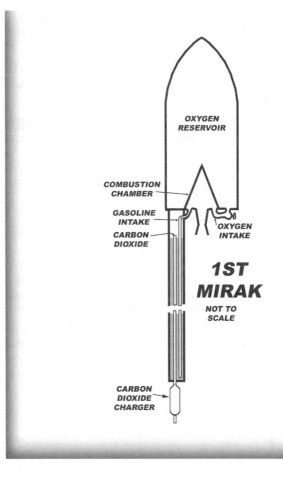

OXYGEN RESERVOIR

COMBUSTION CHAMBER

GASOLINE INTAKE

CARBON DIOXIDE

OXYGEN INTAKE

1ST MIRAK

NOT TO SCALE

CARBON DIOXIDE CHARGER

and had become the central figure in this largest group of German experimenters, continued his formal education in physics, chemistry, and mathematics to prepare him for the challenge ahead, and he eventually earned a doctorate.

Several members of the VfR were offered positions with the German Army to continue their work. Although some balked at the notion of working for the Army with the imminent rise to power of the Nazis, von Braun's youthful enthusiasm and vision for the future of space travel tended to obscure the political implications.

The Raketenflugplatz continued to operate briefly under the direction of Nebel and Klaus Riedel, who achieved motors with a thrust of up to 1700 pounds by March of 1933. However, the departure of von Braun to the Army and a general decline in membership and political disputes within the VfR caused several of its wealthiest backers to withdraw their funding. When Nebel received a water bill in September 1933 from the city for 1,600 Marks, the result of several leaky faucets, the VfR was forced to vacate the Raketenflugplatz.

Wernher von Braun

While most of the pioneers of rocketry are not household names, one who was known to most Americans for a period of 20 years was the German, Wernher von Braun. Few individuals have had a life that has spanned the most productive period of man's conquest of space. Von Braun's involvement was one that could be easily partitioned into several phases, beginning with his association with the VfR, *Verein fur Raumschiffarht* (Society for Space Travel), and Hermann Oberth in 1930, and culminating in the building of the Saturn V moon rocket in the 1960's and the redesign of the Space Shuttle, in the early 1970's. The details of von Braun's involvement with the first large liquid fuel rocket (the German V-2) and his later contributions to American rocketry are revealed in subsequent chapters.

Born in Wirsitz, Germany, on March 23, 1912, to Baron and Baroness Magnus von Braun, Wernher was an inquisitive child with subjects almost always outside the formal curriculum of his schoolwork. Oddly enough, physics and mathematics were not his strong points. His father was well placed in the Weimar Republic that preceded Hitler's Third Reich, becoming a member of the Cabinet and Minister of Agriculture and one of the founders of the German Savings Bank. Wernher's early life was comfortable.

As a youngster, he was captivated by the exploits of Max Valier, who was setting speed records with autos, gliders, and ice-boats. In 1925 the 13-year-old von Braun was entranced by an advertisement for Hermann Oberth's book *The Road to Interplanetary Space*. However, when the book arrived, it was filled with mysterious symbols and mathematical formulas that he could not understand. While he was certain that man would one day fly into outer space and that day would be within his lifetime, he was somewhat inhibited by the obvious need to learn more about the two subjects that had resisted his half-hearted attempts to understand. His desire to become a part of the solution to the challenge of space travel overcame his inhibitions, and he set his goal to master these subjects. He was an early member of the VfR and worked tirelessly on the experiments. Von Braun's efforts, knowledge, and enthusiasm were duly noted by Captain Walter Dornberger when the German Army paid its first visit to the VfR.

In the spring of 1932, von Braun graduated from the Charlottenburg Institute of Technology with a bachelor's degree in Aeronautical Engineering. On November 1, 1932, von Braun became employed by the German Army and was charged with the development of liquid-fueled rockets. Two years later, the 22-year-old von Braun earned a Ph.D. (doctorate) in physics from the University of Berlin. His thesis related to theoretical and experimental aspects of liquid propellant rocket motors but bore the simplistic title *About Combustion Tests*. The desire for the military to obscure this type of information was the reason behind the unassuming title. The paper earned the university's highest rating, *exmium,* but it remained unpublished until after the war for security reasons.

In 1931, at the age of nineteen, he began to take glider lessons at Granau, with Wolf Hirth as his instructor. It was at this time that he met Hanna Reitsch (latter a famous female test pilot for the Third Reich), and they became lifelong friends. A year later, von Braun took up powered flight. In the summer of 1933, he became a licensed pilot for single engine aircraft. In 1936 and 1938, he served briefly as a pilot in the prewar Luftwaffe. He received military flying experience in aerobatics and formation flight and became certified in multi-engine aircraft. (Von Braun continued his love affair with aviation, receiving an FAA Airline Transport Rating along with a Gulfstream Type rating in the 1960's. Between 1964 and 1970 von Braun logged more than 2,500 hours in the cockpit of NASA 3).

Von Braun's boyish enthusiasm that Oberth noted at their first meeting and that attracted the attention of Walter Dornberger of the German Army in the 1932 could be seen 30 years later in Senate hearings and presidential interviews. While he exhibited a cool confidence in his endeavors and rarely boasted of his achievements, he did reveal a temperamental attitude that Dornberger was able to subdue. Those who came to know him immediately recognized von Braun as a genius with a healthy sense of humor and quick with a joke. Most who met him were captivated by his charm and gracious manner that were an important part of his personality. He was quick to smile and often exhibited a wide grin, especially to those whom he sought to convert to his way of thinking. He was a good listener, especially using his eyes, and this attribute often won over those in opposition because he gave the impression that he had carefully evaluated all that they had to say.

Dornberger respected von Braun's imagination but occasionally had to demand he produce factual evidence to support his wide span of grand dreams. In this respect, Dornberger, von Braun's superior in the organization that created the V-2, helped create "von Braun the visionary." Dornberger was quoted as saying: *"He was erratic at first and not completely persistent. He would go from one thing to another, but not until they had a clear idea of what he wanted to achieve. Then he would grow stubborn, and would not tolerate any impediments or deviations."*

While von Braun appeared to live a life free of vices that often corrupt those who achieve high positions of influence and power, his desire to avoid compromising his integrity may have been hardened by his brief association with the Nazi Party. While many in the VfR refused to follow von Braun as civilian employees in the pay of the Third Reich, von Braun notes: *"I was still a youngster in my 20's and frankly didn't realize the significance of the changes in political leadership... I was too wrapped up in rockets."* In defense of his joining the Nazi party in 1938 and the SS in 1943, his supporters note that it was simply a wartime expediency to help ensure the A-4 (V-2) project held favor with those at the highest level of government.

While most who met him were favorably impressed, there were those who could not forget that here was a man who was able to quickly and conveniently switch loyalties from the Nazi regime of Adolph Hitler to the Constitution of the United States. To these people von Braun represented the ultimate opportunist who could adopt the good ideas of others and profess them as his original thoughts. This concept of duality of political thought and professional opportunism, which some believe is what kept President Eisenhower from embracing von Braun's proposal to orbit the first satellite in 1955, will be examined in a later chapter. Author Erik Bergaust, who knew von Braun quite well, notes that von Braun himself commented, *"I have to be a two headed monster—-scientist and public relations man."*

Walter Hohmann

The VfR brought into the mainstream of rocket science (as it was evolving in Germany in the 1920's) many individuals whose contributions would not be applicable for several decades into the future. This was the case with Walter Hohmann (1880-1945), a German engineer, whose analysis of orbital mechanics demonstrated a fuel-efficient technique to move a spacecraft between two different orbits. Hohmann, who had a degree in civil engineering, became interested in interplanetary flight. He calculated the requirements of sending a spacecraft from earth to other planets in the solar system. He understood the importance of minimizing the amount of fuel a spacecraft had to carry and analyzed a variety of orbits to accomplish the task.

The technique that resulted now bears his name, The Hohmann Transfer. It is applicable to all orbiting bodies including earth satellites, lunar probes, and interplanetary flights which use the basic laws of physicis that Hohmann sought to exploit. He wrote his findings in *Die Erreichbarkeit der Himmelskörper* (The Attainability of the Celestial Bodies) in 1916, but the manuscript was not actually published until 1925.

Athough Hohmann never worked in rocketry as a profession, he was an important member of VfR. He despised the Nazis regime that came to power in Germany in the early 1930's and remained outside the mainstream of rocketry to avoid any involvement with weapons.

British Interplanetary Society - BIS

The British Interplanetary Society (BIS) was founded by space travel enthusiasts in October 1933 in Liverpool, England. The BIS focused more heavily on theoretical research than actual development of rockets because a British law enacted in 1875 banned the private use of explosives and rockets. One of their first significant efforts was to design a large, solid-propellant moon rocket based on existing technologies of 1938. In the 1940's they proposed a more advanced moon rocket that included a small module for descending to the moon's surface. The American Apollo lunar landing program used a similar technique 20 years later. Perhaps the most enduring of the societies, the BIS made its mark with its periodical, *Journal of the British Interplanetary Society,* that was introduced in January 1934.

Guggenheim Aeronautical Laboratory - GALCIT

In addition to funding Robert Goddard's work on rockets, The Daniel Guggenheim Fund for the Promotion of Aeronautics provided $300,000 to the California Institute of Technology for the construction of a laboratory and the establishment of a graduate school of aeronautics. Dr. Robert A. Milliken, Chairman of the California Institute's Executive Council, was the driving force in obtaining the grant. The eminent Hungarian scientist, Dr. Theodore von Karman, was brought from Germany in 1930 to oversee the activities of this organization, which became known as the Guggenheim Aeronautical Laboratory California Institute of Technology (GALCIT).

A graduate student, Frank Malina, whose interest in space exploration was first aroused by his readings of Jules Verne, convinced von Karman to let him do his graduate research in the rocket field. From that beginning, in 1934, research at GALCIT on both liquid and solid propellant types began.

In March 1935, William Bollay, a graduate assistant to von Karman, became interested in rocket-powered aircraft because of a paper by the German, Eugen Sänger, who was working in Vienna, Austria, at the time. Bollay became a proponent of rocket-powered aircraft as a promising method of reaching extreme altitudes and velocities but noted: *"It seems improbable that the rocket plane will be a very hopeful contender with the airplane in ordinary air passenger transportation. For this purpose, the stratosphere plane seems eminently more suitable."*

A local newspaper account of a Bollay lecture resulted in two rocket enthusiasts, John W. Parsons and Edward S. Forman, joining GALCIT. Under the direction of Malina, Bollay, Parsons, and Forman began a program in February 1936 to design a high-altitude sounding rocket. Yet another addition to the group was Hsue Shen Tsien, who would eventually return to his native China to become the father of the Chinese space program.

In the summer of 1936, Goddard visited Caltech as Milliken was a member of a committee appointed by the Guggenheim Foundation to advise on the support given by the Foundation to Goddard. A reciprocal visit to Goddard's facilities was made by Malina, who recalled that Goddard did not offer any technical details of his work beyond that which had been previously published. Most of Goddard's papers were very general in nature and not useful to other experimenters—Goddard's apparent objective. Malina recalled Goddard's bitterness towards the press and the distinct impression that he felt *"rockets were his private preserve."* Because of Goddard's reluctance to share information, GALCIT independently proceeded in the development of liquid- and solid-propellant rockets.

The GALCIT team reviewed the existing literature from Tsiolkovsky, Goddard, Esnault-Pelterie, and Oberth. Their initial program provided for theoretical studies in an effort to save both time and money. Soon experiments began with the construction of a liquid rocket motor similar to that tried by the *American Rocket Society*, which used gaseous oxygen and methyl alcohol.

In October 1936, the first attempt to fire a gaseous-oxygen—methyl-alcohol rocket motor was made in Arroyo Seco on the western edge of Pasadena, California. The combustion chamber liner and exhaust nozzle were made of electrode graphite. The motor delivered a thrust of 5 pounds for only a few seconds before the oxygen hose broke. Subsequent tests were more successful with one lasting 44 seconds in January 1937.

A member of the U.S. Army Ordnance Division visited Caltech in 1938 and disclosed to GALCIT that the Army had determined that there was little possibility of using rockets for military purposes. Nevertheless, interest in rockets was growing in other areas. In May 1938 von Karman received inquiries from the U.S. Army Air Corps about rocket propulsion. A subsequent visit by General Henry (Hap) Arnold, the new Chief of Staff of the U.S. Army Air Corp, resulted in some funding through the National Academy of Sciences.

In August Ruben Fleet, of Consolidated Aircraft Co., began discussions with GALCIT on the use of rockets for assisting the take-off of large flying boats. These big seaplanes were becoming a popular means of commercial travel as well as having obvious military applications. However, getting a heavily laden boat into the air required lots of power.

The Air Corps issued a small contract to GALCIT in 1939 for the development of a jet assist take-off unit (JATO). Malina addressed the problems associated with using solid-fuel rockets for this application by using a slower burning propellant to extend the thrust impulse. One specific challenge to be overcome was that any cracks in the solid-fuel allowed the flame front to propagate into and expand the crack to the point of producing an occasional explosion. John Parsons, with a background in chemistry, decided to try flexible binding materials such as paving tar and asphalt. Parsons experimented by mixing these with potassium perchlorate, an oxygen-rich compound. The result gave good thrust and was not prone to cracking. Malina and his cohorts had achieved a solid propellant rocket that was more reliable than any previously constructed. This was a critical factor in using solid-fuel rockets in applications that involved man-carrying aircraft.

In August 1941, a small solid-fuel JATO unit attached to a 2 place, single engine Ercoupe demonstrated a jet-assisted take-off, and the proof of the concept was successfully established. However, it would be two more years before a liquid-propellant rocket engine, constructed by the Aerojet General Corporation, was tested in a Consolidated Aircraft Co. flying boat on San Diego Bay.

As rockets were still considered science fiction by the elite scientific establishment, von Karman decided to avoid the use of the word and substituted *jet* in its place. Thus in 1939, GALCIT's Rocket Research Project became the Air Corps' Jet Propulsion Research Project, and von Karman and Malina went on to found the Jet Propulsion Laboratory (JPL) a few years later.

American Interplanetary Society - AIS

Science fiction gave birth to yet another influential organization. David Lasser, a war veteran and MIT graduate, along with G. Edward Pendray, Fletcher Pratt, Laurence Manning, and Hugh Pierce formed the American Interplanetary Society (AIS) in 1930. All of them had earned their writing laurels working for science fiction impresario Hugo Gernsback. Their stated objective was the "promotion of interest in and experimentation toward interplanetary expeditions and travel." Meeting bi-weekly in Manhattan's Museum of Natural History, they soon moved to the point where experimentation was the next step.

Not realizing that Goddard was successfully experimenting with liquid-fuel rockets only a few hundred miles away, Pendray vacationed to Europe to witness some of the work being done by the VfR and began correspondence with Willy Ley. In 1932, the AIS tested its first liquid fueled rocket at Stockton, N.J. Based on the German Repulsor rocket, it produced 60 pounds of thrust for about 30 seconds but was damaged during the test and never flown. On May 4, 1933, the AIS achieved its first successful flight to an altitude of 250 feet with the six-foot-long Rocket No. 2 at Marine Park, Staten Island, New York. Rocket No. 4, designed by John Shesta, reached 400 feet in 1934.

Two years later the group conducted tests of sending mail by rocket. F. W. Kessler, Willy Ley, who had left Germany for the United States, and N. Carver launched two small, mail-carrying, unmanned "rocket airplanes" at Greenwood Lake, N.Y. The experiment was an embarrassing failure as the rockets traveled only 1,000 feet before crashing while the newsreel cameras recorded the non-event.

Because some considered the word *interplanetary* in the title of their society as being too presumptuous, the name was changed the American Rocket Society (ARS) in 1934.

In December of 1938, the ARS tested a 90-pound thrust regeneratively-cooled liquid rocket motor designed by Princeton University student James H. Wyld. With war on the horizon, society member Lovell Lawrence, with Pierce, Wyld, and Shesta, formed Reaction Motors to secure a government contract for providing a liquid-propellant rocket engine based on Wyld's design. With a meager start-up capital of $5,000, the group set up in the garage of Shesta's brother-in-law but quickly outgrew it and moved to Pompton Plains, New Jersey. They tested their creations in an isolated area in Franklin Lakes. The foundation had been laid for the development of several important rocket engines as the war began.

The Group for Investigation of Reaction Motion - GIRD

In Russia in 1931, another group of space enthusiasts, inspired by Tsiolkovsky, formed *Gruppa Isutcheniya Reaktivnovo Dvisheniya (GIRD)* under the leadership of Dr. Perelman and Professor Nikolai A. Rynin. The organization's title translated to The Group for Investigation of Reaction Motion. Like the many other rocket societies that formed during the period, GIRD was to build and test rocket vehicles and promote space exploration. There were two locations, one in Moscow (MosGIRD), and the other in Leningrad (LenGIRD). Among MosGIRD's founding members was Sergey P. Korolev (1906-1966), who had trained as an aeronautical engineer at the Kiev Polytechnic, and Fridrikh Arturovich Tsander, both of whom were employed by the Tupolev Aircraft Design Bureau. Mikhail Tikhonravov, who was to play a pivotal role in Soviet rocketry, soon joined GIRD.

By March of 1933, the group, after evaluating several proposals from its members, began work on Tikhonravov's design GIRD-09. This was a hybrid

Sergei Korolev

combination of a thrust chamber filled with jellied petroleum into which liquid oxygen was introduced. The hybrid system, perhaps the first of its kind, produced 115 lbs. of thrust. The eight-foot-long rocket was successfully flown on August 17, 1933, and reached an altitude of 1,300 feet before a failure in the combustion chamber seal sent it out of control, and it crashed after an 18 second flight. Six more GIRD-09 rockets flew with varying degrees of success over the next few months.

A follow-on (GIRD-X) was the first all-liquid-propellant rocket in the Soviet Union and was slightly more than five feet in length. It produced 160 lbs. of thrust, burned for 12 seconds, and flew on November 25, 1933. During this maiden flight the combustion chamber burned through (a common problem in those days), and the rocket reached an altitude of only 200 feet.

A parallel organization to GIRD was the Gas Dynamics Laboratory (GDL), which was likewise doing early research into the construction of rocket engines. A leading member, Valentin V. Glushko, had achieved several promising designs, and he too would play a significant role in Soviet space exploration. When the military determined that there was potential in the liquid-fuel rocket, GIRD and GDL were incorporated into the Reaction Propulsion Scientific Research Institute (RNII). RNII developed a series of small rocket-propelled missiles and gliders during this period.

RNII was well funded by the Armaments minister Mikhail N. Tukhachevski. However, he came under suspicion as promoting a military coupe against Soviet Premier Joseph Stalin in 1937 and was executed. Much of his staff, including Korolev, were imprisoned for "crimes against the state." Korolev was in the process of completing a design for a rocket-propelled aircraft (the RP-318), but it would never make its first flight.

Korolev's journey into oblivion began with an extended train ride on the Trans-Siberian railway and a period of time on a prison vessel at Magadan, in far eastern Siberia. He then spent a year in the Kolyma gold mines, perhaps the most tortuous part of the Gulag prison system. However, Korolev was just one of some 20 million political prisoners. As war with Germany loomed larger, Stalin recognized that there were many talented people in the Gulags who were far more valuable to the war effort than to gold mining. Korolev and selected others were "graduated" to a *Sharashkas* (prison design bureau). Senior aircraft designer Sergei Tupolev, who was also a member of TsKB-39 Sharashka, specifically requested that Korolev be assigned to his organization. Korolev's activities were recognized as being of high quality and innovative. The world would eventually be greatly influenced by the future work of Sergey P. Korolev and indirectly reshaped by it.

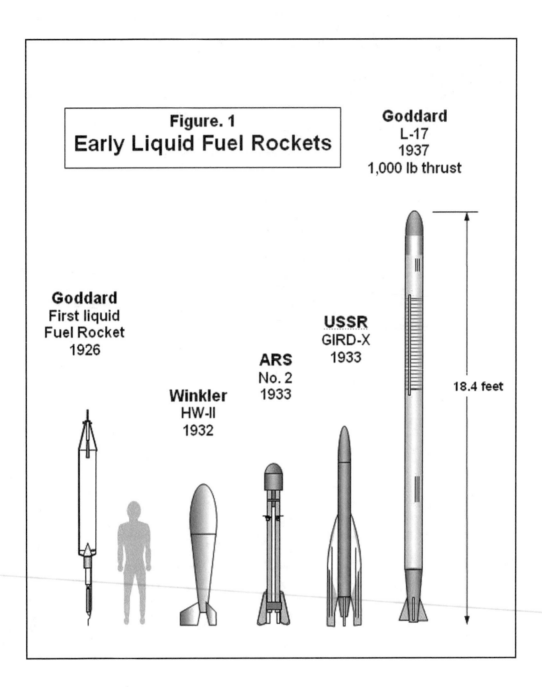

Figure. 1
Early Liquid Fuel Rockets

Goddard
L-17
1937
1,000 lb thrust

Goddard
First liquid
Fuel Rocket
1926

USSR
GIRD-X
1933

ARS
No. 2
1933

Winkler
HW-II
1932

18.4 feet

A Weapons Program

Interest in rockets as weapons of war was not lost to the German military of the 1930's as they contemplated their options under the arms limitations imposed by the treaty of Versailles that ended the World War. They looked for technologies not covered by the treaty that restricted the German Army to 100,000 men and only to specific defensive weapons. The German Army Weapons Ballistics and Munitions Branch, under Professor Colonel Karl Becker, examined the possibilities that new technological advances in rockets might offer. Although solid-fuel rockets had been used as weapons for several centuries, they had fallen out of favor with the military as the development of rifled artillery (helical groves in the gun barrel to impart a spin to the projectile) had proven extremely accurate and reliable.

One of Becker's staff officers, 35 years-old Captain Walter Dornberger (soon to be colonel) was tasked to review the progress being made by the various rocket societies in Germany. Dornberger, the son of a pharmacist, had enlisted in the Army in 1914 and had risen rapidly because of his abilities. He was one of a select few who remained in the small *Reichswehr* following Germany's defeat in the First World War.

In the spring of 1932, Captain Dornberger, dressed in civilian clothes, paid a visit to the VfR's *Raketenflugplatz* and witnessed their slow but potentially useful progress. A demonstration for Dornberger used a Mirak II that rose 200 feet vertically and flew several hundred yards horizontally. Although impressed by what he saw, he was disappointed by the lack academic discipline—specifically documentation and critical analysis of their activities. However, he quickly grasped the possibility that large artillery rockets might play a role in a future battlefield scenario, and he appreciated the talents of the youthful Wernher von Braun whom he observed on this occasion.

Dornberger recognized that many if not most of the VfR were primarily interested in space travel, and he made clear to them that the Army was interested only in producing a usable weapon.

Acting on Dornberger's report of the demonstration, Colonel Becker directed him to develop a military capability using the liquid-fuel rocket with a range that should surpass any existing artillery, and production of which was to be carried out in the greatest secrecy by industry. The conquest of space was about to receive a significant advance forward by a military application.

By mid-1932 Dornberger had established a test facility for the Army at Versuchsstelle (Experimental Station), Kummersdorf-West, 17 miles south of Berlin. He recognized that he could not depend on the research and development being conducted by undisciplined organizations such as the VfR. Thus, on October 1, 1932, he set about hiring into his command civilians, the first of these being the nineteen-year-old Wernher von Braun. By then von Braun had a degree in engineering. Dornberger arranged for him to attend Friedrich Wilhelm University where he obtained a Ph.D. in physics in 1934.

On December 21, 1932, less than three months after hiring von Braun, Colonel Dornberger observed the first test of a small pear-shaped, aluminum rocket motor about 20 inches in length conducted by von Braun, Walter Riedel, and Heinrich Grünow at the Kummersdorf proving ground. The test facilities at this point consisted of a few concrete walls and thin metal doors and were only slightly better than those enjoyed by the VfR. These block houses, as they were to be called, provided a degree of safety for the experimenters.

Valves to control the flow of fuel and oxidizer were operated manually, based on values observed from a few large pressure gauges, allowing the liquid oxygen and alcohol to flow into the combustion chamber. Von Braun ignited the mixture using a long wooden pole with a can of flaming gasoline at the far end. On this inaugural occasion, the result was an explosion rather than the test of the rocket engine. It was an ignominious beginning to what would become an amazing and infamous, history-making project.

Under Dornberger's management a program was established to create a rocket that would send a one-ton warhead 200 miles. In the technology of the time, this was an order of magnitude greater than existing artillery, which at best could hurl a ton no more than 20 miles. Accuracy of the impact of the rocket was required to be within three mils (3 feet for every 1000 feet of flight). Assuming a 200-mile flight, 50% of the warheads must impact within 3/5th of a mile of the target point. Thus began an extensive and expensive research and development effort for a weapon whose value would be questioned several times over the course of its development and for years afterward.

However, a rocket of the size needed to accomplish the mission as defined was not within the grasp of Dornberger's fledgling organization of the mid-1930s. Under Von Braun's technical direction, the team laid out a plan for a progressive series of rockets that would culminate in the ability to meet the goal established by Dornberger. These rockets were identified by the prefix A for the word aggregate—the sum total of their effort represented by the progressive number that followed. Their first liquid-propellant rocket in the series was designated A-1 and took the new team six months to build. Several thrust chambers and injector-nozzle designs were tested but, more often than not, failure seemed to dominate. Gradually, frequently by trial and error, despair was displaced by optimism as successes began to accumulate. Finally, a 650 lb. thrust chamber was constructed that gave consistent, although rather poor, performance. The ratio between fuel consumption and thrust performance proved a difficult stumbling block. Measurements of flame temperature, analysis of exhaust gas samples, and changes in propellant mixture ratio, gradually increased the exhaust velocity from 5,600 ft. per second to 6,200 ft. per second.

To stabilize the small, 4.6 ft.-long, 330-pound rocket, a rotating mass that weighed about 85 lbs. was placed in the nose and was spun-up to a high speed by an electric motor prior to launch. This mass acted as a gyroscope to stabilize the rocket on take-off in much the same way as a child's spinning top can remain balanced on its sharp pointed base.

Before the rocket could be flight-tested, it was destroyed as a result of a "hard start." A hard start occurs when an incorrect propellant mixture ratio accumulates in the combustion chamber and explodes when the ignition source is introduced. Even before a second rocket was ready to be flown, it was realized that the A-1, as constructed, was dynamically unstable.

The team was learning the physics of the interaction of the center of pressure on a rocket and how it related to its center of gravity and moved on to the A-2, incorporating what they learned. In December 1934 two A-2s, named Max and Moritz (the German names of the Katzenjammer Kids, two fictional misbehaving comic strip characters of the time), were successfully launched from the island of Borkum in the North Sea. Unlike their namesakes, the rockets behaved perfectly and climbed to over 6,500 feet. They were crudely controlled by the spinning mass that had been relocated between the propellant tanks to provide a gyroscopic effect to maintain a vertical ascent.

In anticipation of the larger rockets that were to come, test stands for motors as large as 3,300 pounds thrust were constructed. These structures allowed the rocket motors to be fired while being held down and allowed measurements of temperatures and pressures as well as observations that could not be done with the rocket in flight. These tests are referred to as *static tests* and are the basis for initial firing of virtually all rockets.

Following quickly, the A-3 design was the next step in evaluating and testing significantly larger liquid-fuel rocket propulsion and guidance systems. Attempts simply to scale-up the motor and various other components did not always work.

While the A-3 was still under construction, the final specifications of the deliverable weapon, the A-4, were defined. The two most basic requirements were somewhat arbitrary in that it had to have twice the range of the Paris gun (80 miles) used in World War I and ten times the explosive power of its warhead (200 pounds). Thus, the A-4 required a range of at least 160 miles and a warhead weighing 2,000 pounds. With these goals in mind, it was quickly realized that the size and range of rockets greater than the A-3 would require a new testing facility. Von Braun suggested an area on the Baltic Coast of Germany, where his father had gone duck hunting, named Peenemünde—a name that would forever be linked with the development of terror weapons of World War II as well as the conquest of space. Following Dornberger's inspection of the location and procurement of the land, the team moved to this desolate area that afforded seclusion, security, and the ability to fire the big rocket out over the sea. Under Dornberger's watchful eye progress continued, and a visit in March 1936 by high-level army officers, who were suitably impressed, assured a steady flow of funding.

As Goddard had understood, the control of a rocket by external aerodynamic fins could only be achieved after that rocket had developed sufficient velocity—typically several hundred feet in the air.

Some form of control had to be exerted that was independent of fin stabilization, and this control capability would also be needed in the later stages of powered flight when the rocket would be above 100,000 feet where fins would be of little value. There was a need for stabilizing and controlling the rocket during these two phases of its flight. Tsiolkovsky had suggested flat plates pressing against the exhaust gases, and Goddard had implemented such a scheme.

The A-3 employed a guidance and control system that used three gyroscopes: one for each axis of flight—pitch, yaw, and roll. It also had two integrating accelerometers, devices that integrated time with acceleration to calculate velocity. The guidance system sensed the attitude of the rocket, compared it against a desired track, and sent appropriate signals that turned vanes in the exhaust plume to correct for any deviations, much like the rudder of a boat. The vanes, made of molybdenum, had to withstand temperatures in excess of 5,000 degrees.

On December 4, 1937, the A-3 was the first rocket launched from Greifswalder oie, a tiny island in the Baltic near Peenemünde. The 3,300 lb. thrust motor lifted the 21-foot rocket briskly into the air. It exhibited instability as the gyro-stabilized control system initially proved inadequate, and it was quickly realized that larger control surfaces (vanes) and a more rapid feedback loop were needed to compensate for the dynamically changing flight conditions. The team learned that it was not enough to provide a control mechanism. Many specific parameters had to be determined such as the range of motion for the controls, angular velocity of the vanes to maintain a stable flight path, and the optimum pitch for a standard multiple range trajectory (one in which the only variable is the cut-off velocity).

A flight dynamics computation office was established under Dr. Herman Steuding, who developed analog computers and electronic simulators to address these problems. Dr. Rudolph Hermann, assistant professor of aerodynamics at the University of Aachen, supervised the building of a supersonic wind tunnel having a cross section of almost five feet.

A revised A-3 motor produced 3,300 pounds of thrust for 45 seconds. The original A-3 design called for supersonic velocities, but this was not to be achieved due to the excessive weight of the recording equipment. With the A-3 the magnitude of Goddard's work had been eclipsed.

Recruiting Technologists

As Technical Director of Dornberger's elite scientific team, von Braun's challenge lay in developing the technology for the first truly large, liquid-fuel rocket. He was both a scientist and manager, and to this end he was more fortunate than Goddard, who, aside from some talented assistants, worked each problem alone. With a much larger funding base, von Braun was able to seek out and hire a wide variety of technical specialists, many with advanced degrees in physics, mathematics, and chemistry. Thus, the myriad of problem solutions that Goddard had tackled virtually alone von Braun's team quickly reinvented one by one.

Recruiting for Peenemünde was done in a variety of ways. A typical example was the three-day conference held in September of 1939. As the war had just recently been declared, scientists and engineers were galvanized to support their country. This particular day was known among the team as the Day of Wisdom and was essentially a gathering of some of the top scientists from various institutions around Germany. Thirty-six men who represented a wide variety of scientific areas were brought together for a round of technical briefings and problem presentations. Included were mathematicians, physicists, chemists, electrical engineers, mechanical engineers, and gyroscopic theoreticians.

Von Braun, Professor Walter Thiel (a propulsion specialist), Klaus Riedel (head of ground support equipment), and other members of the team made presentations on the types of problems facing the development of the A-4. They stressed that they were looking for short-term solutions that could be implemented within two years. This was to discourage those in academia who had their sights set too far into the future. Von Braun was concerned with solving problems in these technologies that could directly move the A-4 project ahead—space travel would have to wait until after the war. The results of these recruiting sessions were excellent, and the Peenemünde team continued to grow with the best minds that Germany had to offer.

Another method of recruiting specific individuals who had talents needed by the A-4 program was to use "the old boy's network." Friends and colleagues identified those they may have known in school or from former jobs. In many of these cases, these people had already been conscripted into the Army and were fighting on the front lines. There were several stories of individuals being tapped on the shoulder in a foxhole and told to report to a place called Peenemünde. Such was the case with tank commander Kurt Debus and Dr. Ernst Stuhlinger.

Although progress through a myriad of problems was rapid, by the time the team finished their work with the A-3, it was clear that one more step was needed before a weapon of the capability of the A-4 could be developed. Thus, the A-5 designation was inserted between the A-3 and A-4. The first flawless launch of the complicated 21-ft. A-5 missile that weighed 1,650 lbs. quickly followed the last A-3 in the summer of 1938, and it reached an altitude of 5 miles (8 km) before being recovered by parachute. This recovery method was so successful that several A-5s were refurbished and re-launched.

By the fall of 1939, a new guidance system was installed that incorporated all that had been learned, and it worked without any problems. Twenty-five A-5s were launched over the next two years. Several different types of control systems were used, and radio guidance was also tried for the first time. The A-5 was ultimately capable of reaching an altitude of 8 miles (13 km.) in vertical firings, and these were followed by tilted trajectories to provide experience with establishing firing for range.

Third Reich Politics

On his 27th birthday, March 23, 1939, von Braun met Reich Chancellor Adolph Hitler for the first time, at Kummersdorf. Von Braun gave Hitler a briefing on the performance specifications of the A-5. Hitler seemed disappointed when von Braun indicated that the A5 did not carry a warhead but was strictly a research vehicle. He was then given a briefing on the proposed A4 and viewed some static firings of various rocket motors. On the whole, Hitler appeared unimpressed, and his questions to von Braun were simplistic, indicating his lack of knowledge of basic physics.

By the time work had progressed to the A-4 in 1941, Hitler's war machine was doing well. His *blitzkrieg* (lightening war) had conquered virtually all of Europe. The German Army was on the verge of completing its conquest of the Soviet Union, and England was an isolated island since the United States had yet to enter the war. Many in the Nazi high command saw no need for the A-4, and keeping the funding for the project was a constant challenge for Dornberger.

The budget for 1941 was set at $12 million marks but then was cut in half. Eventually the money was restored as the priority for various weapons moved with the whims of the military hierarchy. In March of 1942, Hitler shifted priorities, and the A-4 was again in jeopardy.

A year later, March 1943, Hitler reportedly reduced the priority on the A-4 program as a result of a dream, saying, *"I have dreamed that the rocket will never be operational against England. I can rely on my inspirations. It is therefore pointless to get more support to the project."* However, Dornberger's influence prevailed, and the money continued to flow into the A-4 program.

While there was an obvious gap between the intellect of von Braun and Hitler, an interesting interchange occurred during one rare meeting between the two in 1943. At this stage of the war, Hitler was described by many as having the appearance of a corpse, with trembling hands and waxy face. In the course of the interview, Hitler listened intently while von Braun described the impact of a ton of explosives hitting the earth at a speed of over 3,000 mph, explaining that this impact would multiply the effect of the explosives it carried. Hitler interrupted and, perhaps in a moment of rare insight, stated, *"I don't accept that thesis. It seems to me that the sole consequence of that high impact velocity is that you will need an extraordinarily sensitive fuse so that the warhead explodes at the precise instant of impact. Otherwise, the warhead will bury itself in the ground, and the explosive force will merely throw up a lot of dirt."* Upon further study von Braun conceded that Hitler was correct. In fact, the warhead of the competing V-1 cruise missile, which was essentially the same size as the A-4, had more explosive effect because it had a surface detonation whereas the A-4 tended to explode slightly below the surface.

Von Braun's intellectual abilities, coupled with a charm and enthusiasm that he could readily communicate, made him a likable personality. These traits would not only enable him to make converts to his views on the A-4 but years later would help advance his visions of space flight. However, his magnetic personality would not deter some of his detractors from trying to interfere with his plans. In March of 1944, the SS arrested von Braun and charged him with making statements that the A-4 was not intended as a weapon of war, but that his primary intent was space travel. Also arrested were A-4 scientists Klaus Riedel, Helmut Gröttrup, Magnus von Braun (Wernher's younger brother), and Hannes

Lührsen. Apparently, an informal, after-dinner discussion among the group had been overheard by an SS informant. In the allegations the SS said it had strong indications that von Braun regretted the imminent operational use of the A-4 as a weapon. Some historians have questioned this latter observation as a self-serving statement by von Braun. There is no corroborating evidence except that the arrests were made because it was felt these members of the von Braun team were not "giving their all" for the fatherland.

Dornberger defended his team members and insisted that without these scientists, the development of the A-4 would grind to a halt. As the repercussions of the arrest began to make their effect known at higher levels, the SS back peddled a bit by insisting that these men had not been arrested but had simply been taken into protective custody. Between the efforts of Dornberger and Albert Speer (the Minister of Armaments), the pride of the German rocket team were finally released to resume their work.

Adding a level of inter-service rivalry to the Army's A-4 was the presence of the Luftwaffe's pulse-jet powered Fi103 cruise missile (later identified as the V-1) of about the same range and warhead weight capability. While the Fi103 could be produced for about one-tenth the cost of an A-4 (about $1,500), its speed was only 350 mph which meant it was capable of being shot down by aircraft or anti-aircraft guns. The A-4 arrived at its target at the end of a ballistic trajectory traveling at almost 3,000 mph. The Fi103 could be heard as well as seen before impact. The A-4 arrived unannounced. Thus, Dornberger claimed that the A-4 was the ultimate weapon—it could not be stopped. Compared with the 10 year period required to develop the A-4, the V-1 flew only nine months after conception. Although initial production estimates of the V-1 were for 5,000 per month, actual numbers were only a fraction of that, with June 1944 achieving the highest with 2,500 completed.

The debate between the cost effectiveness of the Fi103 versus the A-4 raged in several quarters. In May of 1944, a demonstration comparing the A-4 and the Fi103 was staged. An A-4 was first launched to a range of 173 miles but missed its target by three miles. A second rocket, fired a few hours later, failed within sight of the evaluating assembly of German officers. Both V-1s subsequently launched for the evaluation failed. No definitive decision came about as a result of these trials; therefore, both weapons continued in development. As the fortunes of war began to turn against Germany, Dornberger began to sell the program to Hitler as an unstoppable weapon. Without the benefit of an operational effectiveness study on the weapon and without realistic estimates of when it could be ready and in what quantities, Hitler awarded the A-4 the coveted top priority. Spear played an important role in the development of the A-4 but, not unlike Hitler, had his doubts as to the effectiveness of this new weapon.

A-4 Technology Development

Dr. Walter Thiel was von Braun's primary propulsion specialist. When Thiel joined the Peenemünde staff in 1936, the largest liquid propellant rocket motors of the time produced 3,000 pounds of thrust. To achieve greater thrust required higher propellant delivery rates. Smaller rockets typically had their tanks pressurized to deliver the fuel and oxidizer. The 56,000 pound thrust of the A-4 dictated delivery rates that could only be achieved with a high performance pump. Thiel's innovative approach quickly advanced the state of the art by developing the turbine-driven pump.

In an effort to avoid reinventing pump technology, a bit of searching determined that fire trucks require high flow-rates and stabilized pressures for delivering their water to the fire hose. Manufacturers of those types of pumps were engaged, but two more problems remained. The first was that LOX is a liquid at -297 degrees below zero, and this requires special materials and seals for the pumps. Significant research by chemists and metallurgists was required to resolve these challenges.

The final problem was how to power a pump of this size and capacity. It must come up to operating pressure within a few seconds and be driven at high speed to provide the delivery rates and consistent pressure stability. The answer came with an understanding of hydrogen peroxide (H_2O_2). In levels of high purity, such as 80%, hydrogen peroxide will decompose rapidly when exposed to a catalyst such as sodium permanganate—producing water as a byproduct. The decomposition releases a high amount of energy in the form of heat which turns the water into steam—the power to drive a turbine. The steam turbine (operated at 5,000 rpm) drove two pumps that injected the alcohol, at 128 lbs./sec and LOX at 158 lbs./sec into the combustion chamber—rates in excess of 50 gallons per second to produce a phenomenal 56,000 pounds of thrust in the new engine for the A-4.

Fuel and oxidizer mixing posed another interesting problem. Simply impinging streams of liquid oxygen (LOX) and alcohol against each other, as had been done in the first engines developed by Goddard and the VfR, was neither efficient nor effective. The liquids had to be reduced to a fine mist and mixed so that the correct proportion of each was available for complete combustion. Creating a propellant injector for the 56,000 lb. thrust engine of the A-4 proved difficult. After several disappointing attempts, Dr. Thiel decided to try an injector head composed of an intricate set of 18 smaller injectors developed for the A-5 engine. Referred to as *rose cups* because of their distinctive, circular, rose-like shape, the new injector showed noticeably higher efficiencies than achieved with any of the other injector heads. While the *rose cup* injector created a plumber's nightmare because of all the separate LOX and alcohol lines that had to be routed and connected to the combustion chamber, it went into production. Later, a more efficient single head was developed, but by then it was too late to incorporate it into the production line.

While the A-4 may have functioned using the same principles as Goddard's rockets, it produced about 100 times the power of his most advanced designs. Temperatures of 3,000 degrees Fahrenheit and pressures of 225 PSI were encountered, which Goddard could only dream about. The intense flame in the combustion chamber was kept from melting the engine through the use of regenerative cooling— routing the alcohol fuel around passages in the thrust chamber.

However, as there were still occasional failures in the chamber walls due to the heat, an additional cooling method was incorporated. Thousands of small holes were drilled in the combustion chamber wall penetrating the alcohol 'jacket' to allow a small quantity of the fuel to actually enter the throat of the chamber. As there was little or no oxygen available for combustion of this fuel at these points, the insulating affect of the alcohol flowing along the inside of the walls of the thrust chamber provided additional cooling. When this unburned alcohol finally exited the combustion chamber, oxygen in the atmosphere ignited the fuel resulting in a long orange-red sheath of flame that tended to reduce the brilliance of the white-hot plume of the working exhaust.

From the beginning, two methods of guidance were considered. The first of these was an electronic guide beam being, perhaps, the most accurate. However, this would require more extensive ground facilities and could be subject to jamming by the allies. The other method was to use a completely inertial (internal) system. A series of gyroscopes and accelerometers fed data to an analog computer that sent signals to the control vanes during powered flight to establish the trajectory as demonstrated with the A-5. It was this second method that was used for the A-4.

Working with new technologies and volatile liquids (especially cryogenics) created many hazards. In one particular situation the low temperatures of LOX caused the steel in the pipes in which it flowed to become brittle, causing cracks that subsequently burst explosively, killing ten people.

Its mission essentially determined the A-4's size. However, its ability to be transported via railways and roads to firing sites determined the shape. The dimensions of the standard railway tunnel, for example, dictated the span of the tail fins.

Early tests

The first launch of an A-4 was scheduled for February 1942 with production to follow two months later, but this proved to be a very ambitious schedule, and the first launch did not occur until June 13, 1942. The rocket lifted off normally amid a roar greater than any man-made sound yet produced. The missile rose from the launch pad but then began a slight roll as it disappeared into the cloudy sky. Following the sound of a muffled explosion, the roar of the engine ceased. The A-4 shortly reappeared out of the clouds without its fins, crashing into the Baltic a few miles away.

The second launch on August 16 was more successful, with the rocket climbing to an altitude of about 7 miles and passing through the sound barrier without incident—the first such airframe to successfully make the penetration. Some had feared that the aerodynamic pressures of the transonic region would destroy the relatively fragile structure. Many a heart hesitated as the flight proceeded through the lower level of the stratosphere where the water vapor in the exhaust plume condensed to form a white trail. Although these vapor trails had been a common occurrence with high-flying aircraft, the apparent erratic shape the A-4 created by its rapid vertical advance through the horizontal high-altitude winds caused many on the ground to believe that it had gone out of control. However, the telemetry indicated all was well. The phenomena was termed "frozen lightening."

As the flight progressed through its 45th second and was still within view of the observers on the ground, the missile broke up and fell into the Baltic 5 miles from the site. A structural flaw in the instrument compartment accounted for the failure at the point of maximum dynamic pressure (referred to as Max-Q), and this area was reinforced for the third flight.

On October 3, 1942, the third rocket rose into a clear sky, and observers noted good stability about all three axes. The missile had a series of black and white geometric markings, allowing observers (and film) to determine if and when any roll or pitch occurred. The end-of-burn (brennschluss as it was called by the Germans) occurred at 63 seconds, and the Doppler shift of the radio tracking continued to indicate the vehicle was following its prescribed path. It continued its upward (and now un-powered) path, reached a height of 53 miles, and went 120 miles down range. By all standards it was a successful test. Hermann Oberth was present at the launch and said with self-depreciation, *That is something only the Germans could achieve. I would never have been able to do it.*" (Oberth was from Transylvania). Von Braun remarked, *"Today the spaceship is born."*

Numerous technical problems continued to plague the program, but one-by-one they were tediously resolved. The ability to monitor what was going on in the rocket during flight was very limited, and the capability to send back to earth (by radio transmission) various readings from sensors of temperature and pressure was just being developed as a process called *telemetry*. Within the A-4 there were only four channels of telemetry being returned. As a comparison, the Saturn V moon rocket in 1969 transmitted more than 3,500 measurements. This lack of information about what was actually occurring onboard the rocket required large numbers of tests in order to make changes by trial and error. Monitoring propellant shut off valves, for example, required more than 20 launches.

Large scale operational testing of the A-4 began in the summer of 1944 when an average of ten missiles was launched per day. Only about 20% of the missiles fired were actually impacting on or near their targets after flight times of slightly over five minutes at distances up to 200 miles. The primary problems related to several areas, most prominently in the launch phase. Vibration tended to break electronic components such as relays and to cause motors to shut down, with the rocket returning to earth after only a few seconds of flight.

During this period, an estimated 60% of those rockets that completed the powered phase came apart several thousand feet over the target in the final stage of re-entry when the structure experienced aerodynamic heating and pressures. The phenomenon, known as an "airburst", was visually observable if a person knew where and when to look. These visual observations and examinations of the wreckage of the rocket revealed aerodynamic panel flutter in the forward part of the center section. More insulation was added to the rocket body, and the warhead was encased in plywood in an early attempt at handling the high temperatures of reentry. The airburst problem was never effectively resolved because of the requirement of depending on visual observation rather than sophisticated telemetry.

Allied Intelligence and Defensive Measures

When Willy Ley, one of the original members of the VfR, departed from Germany in 1934 for the United States, his contacts with the VfR and its news decreased rapidly. In part this was because the rocket society itself was in decline. It was also because the German government put a lid on all references to rockets in the media to tighten up on security.

However, in November 1939, just two months into the war, an article appeared in the British publication *Astronautics* that quoted Adolph Hitler as saying that, if the war continues for four or five years, Germany will have access to a weapon now under development *"that will not be available to other nations."* The magazine noted that Germany had been in the forefront of experimentation with rocketry primarily through the VfR and that the society had since broken up. The speculation was that Hitler was indeed referring to large ballistic rockets.

During the first few years of the Second World War that began in September 1939, pieces of information began accumulating from a variety of intelligence sources, indicating that Germany was perfecting a super weapon. There was great debate in various political and military circles about the viability of a large ballistic rocket and one outspoken critic was Frederick Alexander Lindemann (1886-1957). Known as Lord Cherwell, this former university professor and personal assistant to Prime

Minister Churchill refused to accept increasing evidence that a large rocket program did exist and might pose a danger to Britain. Lord Cherwell said that it was not that the Germans could not develop such a weapon but that he did not believe it would be militarily practical, and, therefore, the Germans would not be so foolish. He also refused to believe that German scientists had pre-empted British and American scientists by such a large margin. Another report flatly stated, *"We are of the opinion that the possibility of such a rocket development in Germany can be ruled out."* Several American scientists were also skeptical of a rocket carrying a one-ton warhead. They could accept the possibility that the threat was bacterial and germ warfare delivered by rocket.

The British War Office Intelligence Branch assigned the task of investigating German long-range rocket development to 35 year-old Duncan Sandys, who reported directly to Churchill. His appointment was somewhat controversial in that he was married to Churchill's daughter and there were others with more seniority. Nevertheless, he did have reasonable credentials and proved to be a good choice. By the spring of 1943, allied intelligence from a variety of sources gave strong indications that important research was occurring at a Baltic Sea site and that it was probably rocket technology.

Peenemünde was first photographed in May of 1942, and at first glance the photos were inconclusive. There were unusual structures and what appeared to be circular dirt embankments. The fact that there were actually several separate weapon systems being developed that were significantly different in their technologies led to some of the confusion. Often reconnaissance photographs cannot be properly interpreted without a knowledge base that allows the interpreter to recognize what they are looking for. For a time the photos were simply stored, but they were revisited later when further information gave the interpreters something more tangible to seek.

A reconnaissance flight on June 12, 1943, provided more insight into the A-4 when the photographs revealed a white cylinder, approximately 45 feet long and 5 feet in diameter, that appeared to have fins on one end. The report was immediately sent to Cherwell, and a subsequent flight several days later revealed the profile of two more rockets. Reexamination of old existing photos subsequently found four tailless aircraft, ultimately defined as the Messerschmitt Me163 rocket powered fighter, yet another secret weapon. The Fi103 (V-1) being a more conventional airplane configuration was not identified until November of 1943.

Complicating the analysis of possible German progress was the fact that there was virtually no one in England who had any significant experience with liquid-fuel rockets. Thus, the ability to analyze effectively what the Germans might be doing was notably hindered. In an effort to become more knowledgeable of rocket technology, the Shell International Petroleum Company, conducted several experiments with rocket motors fueled by gasoline and liquid oxygen under the direction of Isaac Lubbock. That their experiments were on such a simplistic scale, however, delayed a proper understanding of the rapid advances that had been made in Germany. As late as April 1943, the official position was that *"it is clear that a heavy long range rocket is not an immediate threat."*

Based on this faulty understanding of technology, some estimates indicated that a large rocket could weigh up to 100 tons and possess a warhead of 8 tons. These prognostications produced great concern. Other guesses put the size of such a missile at 60 tons with a range of 200 miles and a warhead up to 12 tons.

Despite a lack of accurate intelligence, the British knew that Peenemünde was the site of some form of advanced weapons development. On the night of August 17, 1943, almost 600 British bombers attacked the research and development facility. The attack was well planned, although, for variety of reasons, it was not carried out to perfection (several groups missed their intended targets). Key scientists including von Braun, Steinoff, Kurt Debus, and the famous female test pilot, Hanna Reitsch, as well as Dornberger (who had recently been promoted to Major General) were all present when the bombs began to fall at about midnight. While there was serious damage, only two primary scientists were lost—Walter Thiel and Erich Walther.

Following the raid, productivity at the facility dropped noticeably. However, as most of the development of the weapon had been completed, the bombing itself was not a major impact to the A-4 program. German estimates indicated development of other projects may have been delayed by two months, but the A-4 had progressed to the production and operational testing phase.

The primary effect of the bombing, from the German perspective, was to reinforce the idea that production of the A-4 and other weapons developed at Peenemünde would have to go underground. Four days after the bombing, the decision was made to move most of the flight-testing activities from the Baltic to a remote site in southern Poland. In November 1943 Peenemünde's supersonic wind tunnel was dismantled and shipped to the Bavarian Alps. While the raid may not have had the destructive impact that the RAF had hoped for, it is obvious that it did noticeably redirect production. Peenemünde began its evacuation in March of 1944.

The RAF did not follow up with a subsequent raid for about a year as a result of reconnaissance photos that showed heavy destruction and, in some instances, no effort to rebuild. Of course, the Germans were careful not to fill in many of the bomb craters or to rebuild some of the buildings but to leave them in a destroyed condition to give the impression that the site might be abandoned (which it ultimately would be).

The U.S. Eighth Air Force attacked Peenemünde three more times in the summer of 1944. These subsequent raids were not so much inspired by concern for weapon development but because it was suspected of being a manufacturing facility for hydrogen peroxide, a chemical believed to be in demand in Germany but whose use the allies did not really understand. Allied scientists had indicated that they did not believe that liquid oxygen could be a component of the German rocket because they did not believe that large quantities of this cryogenic could be pumped into a missile in field conditions. Hydrogen peroxide, of course, was used to power the turbo-pump for the A-4 propellants, but it was also the primary propellant of the Me-163 rocket-propelled interceptor. It was only after the British were able to examine the wreckage of an A-4 that fell in Sweden that it was accepted that liquid oxygen was indeed the oxidizer.

As for launch sites themselves, reconnaissance photos of various areas of France, Holland, and Belgian, in late 1943, revealed strange structures that seemed to resemble a pair of large skis. What the photo interpreters were looking at was the launching sites for the V-1. In an effort to gauge the effect of these mysterious weapons, a worst-case scenario by the British Air Ministry intelligence made a prediction that 2,000 tons of explosives falling on London every 24 hours could be achieved from 100 launch sites. This was the equivalent of a 1,000-plane raid every day, something the Luftwaffe had never even approached in its most aggressive days of the Blitz. With 96 ski sites identified, it was important to destroy these facilities before they could be put into operation.

Now that the threat was recognized, the British turned to planning and executing some form of countermeasure. The initial V-1 launch sites were rather elaborate structures that took several months to construct and were easily destroyed by allied bombing. As a result, a more portable structure was developed that actually could be installed in less than 24 hours.

As for the A-4 launch sites, initially envisioned were elaborate underground production sites that had the missiles actually launched at the end of their fabrication. Von Braun in particular championed this scheme as he felt that the weapon was too complex to launch by a field unit of non-scientist soldiers. Dornberger vetoed the idea, and, in the end, relatively inexpensive and easily transported field units were established that did not give the British any real opportunity to interfere with launch operation. In fact, the Allies were at a loss to determine where the A-4 was being launched from as they had no understanding of how simple its launch requirements were.

For a time the British felt that offensive operations against suspected V-1 and the A-4 launch sites had been effective. This was to change on June 13, 1944.

Slave Labor Production

Development of the revolutionary A-4 in such a short time was another miracle of wartime expediency, but production was another matter. Three factories were planned for A-4 quantity production: Weiner Neustad, south of Vienna, a facility on the northern outskirts of Berlin, and the Zeppelin works in Friedrichschafen, on Lake Constance. With the British discovery of the secret development facility at Peenemünde and the subsequent raid in August of 1943, the decision was made to move all production underground to the Harz Mountains, in Southern Germany. This factory was known as *Mittelwerk* (central work).

Near the town of Niedersachswerfen, in the Harz Mountains, was a network of tunnels constructed during World War I for the production of ammonia and gypsum resources. These were readily enlarged and converted into the required manufacturing environment, despite the cold wet dampness that prevailed. The facility consisted of two parallel tunnels, each two miles in length, and approximately fifty feet in width and height, and connected laterally by forty-eight smaller tunnels.

While it was claimed that production planning was not seriously affected by the August 1943 British attack, other more insidious problems began to plague production. Severe labor shortages in attempting to manufacture 1,000 missiles a month required that concentration camp workers be assigned to the task. While some of the manufacturing and assembly of the A-4 demanded skilled labor, POWs and political prisoners did a significant amount of the missile construction. Albert Speer stated that, by 1944, approximately 40% of all POWs in Germany were employed in some form of war work.

The Dora SS concentration camp was founded to provide slave labor to enlarge the Mittelwerk tunnels and build the rockets. A personnel director of the Mittelwerk was appointed to coordinate with the commandant of Dora to provide the labor. Beginning in September 1943, the first of 60,000 prisoners from more than a dozen countries began arriving. An African-American American flier named Johnny Nicholas, captured in France, was among them. He was notable in that there were few Negroes in Europe at the time, and his presence was obvious.

Conditions in the tunnels were deplorable to say the least, especially since many of the laborers actually slept there, and the death rate from pneumonia was very high. Unlike some of the more well known concentration camps, Dora had few Jews. The conditions were exacerbated because the prisoners included not just POWs but criminals and homosexuals. The criminals were often used as "straw bosses" who exacted their own private revenge on the others.

There is some divergent opinion as to just how badly the prisoners were treated. Because of the need for the laborers, there were reports that an effort was made not to mistreat them. However, these reports are in sharp contrast to the German's own records that show, of the more than 60,000 prisoners who were employed, about one third died over the course of an 18-month period from harsh treatment and malnourishment. A case could be made that more people were killed building the V-weapons than were killed from their use. In addition, because the Mittelwerk was a top-secret operation, once a laborer arrived, he was there for the duration, whatever his fate may be. Arthur Rudolph, an engineer from Peenemünde, was in charge of A-4 production. His first task was to dismantle a pilot production plant and move it to the Mittelwerk. Rudolph said he walked through the tunnels once or twice a day and even visited Dora's SS camp commandant. *"I knew that people were dying,"* he told an interrogator after the war. Following a December 1943 visit to Mittelwerk, Albert Speer described conditions as *"barbarous"*.

Dora prisoners found subtle ways to exact revenge for their plight. Many A-4s produced at Mittelwerk and test-fired at the proving ground in Poland failed. The SS maintained control over POWs and were always on the alert for sabotage. Yet the prisoners managed to sabotage rockets by urinating on wiring or simply loosening screws. About 200 prisoners were hanged at Dora or from overhead cranes in Mittelwerk's tunnels for suspicion of sabotage.

Wernher von Braun occasionally visited the facility and was quoted in a 1971 interview as saying, *"I saw Mittelwerk several times, once while these prisoners were blasting tunnels in there, and it was really a pretty hellish environment…The conditions there were absolutely horrible."* While numerous scientists (including von Braun) were members of the Nazi party, and aware of the situation at Dora, charges were brought by the United States government only against Rudolph (who had been a key player in the Saturn V moon rocket). The allegations were for his knowledge of, and possible involvement in, the alleged atrocities at Mittelwerk. As an elderly man in ill health, he elected to return to Germany rather than face the charges in 1984. A subsequent investigation revealed no creditable evidence against Rudolph.

As for von Braun's possible involvement, whether it was his credible denial, political connections, the aura created around him and his work that help America beat the Russians to the Moon, or his untimely death in 1977 from cancer, he was never indicted for his role—whatever that might have been. Even today, many wonder how a man, who seemed to encompass numerous personal moral attributes, could engage in, or contribute to, the horrors of the Dora concentration camp.

While labor problems were being addressed, securing reliable components and materials and attempting to minimize design changes became a constant challenge. When specified materials were not available, the engineers redesigned components to take advantage of what was available. An estimated 50,000 changes were made to the A-4 between the time it first flew in October of 1942 and its first operational use in September of 1944. For a time, the only aluminum came from downed Allied bombers. Another example was the jet vanes that controlled the rocket. Relatively expensive molybdenum was originally specified but was in critical supply. Vanes made from carbon were substituted, but these often exhibited excessive erosion and occasionally broke loose in flight.

While the use of alcohol was preferred to petroleum-based fuels, its availability was dependent primarily upon the yield of the Polish potato harvest. Thus even though A-4 production rates as high as 2,000 per month were occasionally talked about, there was no guarantee that there would be enough alcohol to launch even 900 a month.

Those who had some knowledge of the A-4 occasionally felt that it was still a research and development project and that to attempt to mass-produce it, in a cost effective manner, was unrealistic (von Braun himself felt the missile needed perhaps as much as a year more development). The German Air Ministry arms chief, Field Marshal Milch, was reported to be outraged that A-4 production priorities were having a negative affect on fighter production, as increased bomber attacks by the British and the Americans continued to devastate German industries. Armaments Minister, Albert Spear soon became convinced that, important as the A-4 was, it should not interfere with aircraft production.

In quantity production, the A-4 required about 15,000 man-hours and cost 38,000 Reich Marks. It was expected that this figure would eventually decline to approximately 7,500 man-hours and possibly as low as 3,500 for the higher production runs, and each A-4 would cost about 14,000 Marks. Of course, these prices are exclusive of the forced labor used for the production.

The Me-262, Germany's top jet fighter, at the end of the war cost about one-half million Marks each. Many believed that this fighter was needed to keep the American B-17s from demolishing Germany. Although the A-4 was unstoppable, it could be used only once, and then only about 70% fell within a mile of the target. The debate raged as to whether Germany should concentrate its dwindling resources on an offensive weapon such as the A-4 or the defensive Me-262. An interesting dilemma, it had many military people on both sides of the issue. The V-1 required only a fraction of labor and materials consumed by the A-4, but within two months of its introduction into combat, only a small percentage were making it to the target.

Although the Germans never reached their manufacturing target of 1,000 A-4s per month, they did produce about 600 missiles each month between September 1944 and February 1945. Approximately 6,000 missiles in total were produced at Mittelwerk before the collapse of Germany in April 1945.

It is remarkable that Germany, faced with deteriorating conditions both militarily and economically, was able to field a new weapon in a period of less than two years from its first successful test to its introduction in combat. The final end of A-4 production, though hindered by spot shortages of sheet metal and a variety of other components, was ultimately due to the inability to transport raw materials to, and completed missiles from, the production environment. Rocket production at the Mittelwerk facility continued until an Allied attack on April 10, 1945, which caused most of the 4,500 German factory workers to abandon the site.

Operational Use

Many of the scientific staff, including Von Braun, felt that the A-4 could not be fired successfully in combat unless there were elaborate testing facilities and professional technicians. One plan called for a large, bombproof, concrete installation in the Calais area of France, with the capability to launch 36 A-4s per day. Dornberger, again providing realistic leadership, knew that mobile firing batteries by well-trained military personnel were the key. He believed that, regardless of the thickness of the roof of such sites, as envisioned by von Braun, they would be the target of intense allied bombing. Dornberger predicted that carefully trained men who were relatively unfamiliar with the technical aspects of the rocket would achieve greater operational success. Dornberger's view persisted and his approach was highly successful.

The testing and modifications needed to make the rocket reliable continually delayed its introduction into combat. The German High Command, and Hitler in particular, were anxious to see the effect that thousands of these rockets raining on London might have. While it was originally envisioned that hundreds of both the V-1 and A-4 would be launched almost simultaneously, by the early morning hours of June 13, 1944, only few launch sites were ready, and these were for the V-1, not the A-4. Thus, instead of a rain of destruction, only two V-1s made it into the general vicinity of London. However, the destructive effect of the V-1 was considerable, often devastating an entire block of buildings. Damage for up to a quarter of a mile was noted. In the first three weeks, 370 V-1 missiles hit London, and by July 5th the death toll was almost 2,500.

By the third of August, one flying bomb arrived every six minutes in the ten-mile by ten-mile area of London proper. Initially, Hitler decided to increase the production of the V-1 over the V-2, but soon it was evident that the Allies were stopping upwards of 80% of the slower weapons. It took two months for appropriate tactics to be established by the British, using anti-aircraft, barrage balloon, and fighter interceptors. On August 28th, 90 of 101 V-1s that crossed the Channel were downed before they could reach London. French based V-1 attacks ceased on September 1st when the last of the launch sites was captured by the Allies.

Three V-2 rockets on display
(note stabilizing wires attached to the nose)

Just when England's defenses began to severely limit the number of "doodlebugs" (as V-1s were occasionally called) getting through the defensive network, the A-4 began its assault on England and the Belgian port of Antwerp, on September 8, 1944. The first tactical A-4 launched from the outskirts of The Hague, Netherlands. At 6:43 P.M. the rocket detonated at Cheswick-on-Thames near London. The first application of a large liquid-fuel rocket, born from the dreams of countless visionaries and created with theory and work by persistent experimentalists, had arrived—but it was probably not what any of them had imagined. It was a destructive weapon of war, not a scientific leap into the cosmos.

Twenty-six A-4s hit London over the ensuing ten days, and Goebbels announced in a radio broadcast that retaliation weapon number two, as promised by the Fuehrer, was operational against England. It was through this announcement that von Braun learned his rocket would become known to the world as the V-2 (the launch crews continued to call it the A-4).

As the guidance of the rocket was rather crude by today's standards, it had an actual circular error probability (CEP) of about 3/5 of a mile over a range of 200 miles. This meant that it could not target specific locations but only fall within the general limits of large cities such as London—indiscriminately killing non-combatants. Determining just where a specific A-4 impacted was also difficult for the Germans to determine. This information was important to adjust target location information, as the maps of the day were not very accurate. The British intentionally gave out disinformation as to where the weapons landed and it was almost impossible for Germany to get photo-reconnaissance aircraft over London at that time of the war. Most of the credible information came to the Germans from neutral journalists (Swedish in particular) who accurately reported the impact sites, thus allowing the Germans to make adjustments. Even obituary notices often gave vital impact information. The British government made numerous attempts to influence the reporting and were somewhat successful.

While strategic facilities were not significantly damaged, production in London factories dropped 16% to 25%. Most of this was due to lost work time from the constant V-1 alerts, which kept many Londoners in bomb shelters. Over one million inhabitants left the city during the summer of 1944 in response to the onslaught.

Launch Procedure

For launching, the rocket was trucked to a desired site on a special trailer called a *Meilerwagen,* where it was raised into firing position on a simple launch table and fueled. The site was typically a thick wooded area to hide it from marauding allied fighters. Initially, setting the launcher on soft soil was a problem, but the crews soon learned to embed a set of logs in the ground and pour some of the super cold LOX over it to freeze it solid for the few hours needed to erect, fuel, and fire the V-2.

Once set up on the launcher, the orientation of the rocket's gyroscopic guidance system was achieved by manually rotating the firing table until fin III (3) was precisely aligned with the azimuth (direction) of the selected target. A timer within the rocket was then set that would shut the propellant valves after a predetermined time of firing. A range table had been established that showed how far the rocket would travel for each second of engine burn time. Thus, the location of the firing site as well as the target were critical to achieving any semblance of accuracy, and maps of Europe were not very accurate. Launch control was an armored car configured as a mobile Block House and stationed about 500 feet from the missile.

At ignition, the fuel and oxidizer valves opened and the volatile liquids flowed into the combustion chamber under the force of gravity. A simple fireworks-like pinwheel in the exhaust provided the ignition source. The engine developed only 8 tons of thrust by gravity feed at this point, not enough to lift its 13-ton weight. After the fire control officer saw that all was well (about 4 seconds), he issued the command that switched the rocket over to its internal electrical batteries and started the hydrogen-peroxide turbo pump that brought the thrust up to its full 28 tons needed to lift the rocket off the launcher. In little more than 20 seconds, the rocket accelerated through the sound barrier as it arched towards its target to an angle of about 45 degrees. By the half-minute mark, the projectile was six miles high and traveling twice the speed of sound, and the "frozen lightening" was making its distinctive signature across the sky. The fuel was shut off at a predetermined time for the range (average burn time was 63 seconds). Combustion was completed at an altitude of about 22 miles, and the missile's momentum continued to take it to its maximum height of over 60 miles. The velocity at thrust cut-off was about one mile-per-second (3,600 mph). The missile impacted at a speed of about 2,100 miles per hour, faster than a rifle bullet, after a total flight time of just five minutes. It is interesting to note that even the primitive radar being used by the allies during World War II was able to detect the launching of V-2s in their early stages. Nevertheless, there was no stopping the weapon.

Quality and performance of the early V-2s was poor. Of a batch of 1,000 inspected, 339 were defective and returned to the factory; of the remaining 661 launched, five percent either did not rise at all or tumbled after take off. However, after October 1944, 85 percent of the missiles received by tactical units were successfully launched, and all of these reached the general vicinity of the target, although a small percentage still disintegrated in midair on re-entry.

While the actual effect of the destruction was not a major factor in the war effort, General Eisenhower diverted over thirty percent of RAF and USAAF bomber strikes during mid-1944 to address the threat of these weapons. In fact, this was one of Hitler's objectives. Every bomb that fell on a suspected V-1 or V-2 site in France was one less bomb that fell on German soil.

Of the 6,000 V-2s launched, 4,300 were operationally directed against England. Of this number, 1,115 impacted on British soil, and over 2,100 against the allied port of Antwerp. The last V-2 was fired March 27, 1945, on Orpinton, Kent. Total V-2 casualties were 2,742 people killed and over 6,000 injured. Ultimately, more than 25,000 V-1s were launched with 2,419 impacting on British soil.

Stealing a V-2

When the bombing of Peenemünde caused the dispersion of A-4 activities, the flight-testing phase moved to an area of southern Poland called Blizna. The Polish underground became alerted because of reports of *tremendous rumbles and explosions, frost on torpedo shaped bodies,* and reports of *strange objects.* The Polish partisans began observing and reporting to England what they saw, both at the launch site and in the impact area. Sometimes they arrived at the impact site before the Germans and recovered small pieces that were ultimately returned to Britain. On occasion, larger pieces, including the rocket motor itself, were hidden by rolling them into nearby lakes so that the Germans would not find them.

In one particular recovery operation, the Poles had a virtually complete A-4 flown out of the area at night by a C-47 and subsequently delivered to England. This would have been a significant intelligence coup had yet another A-4 not erroneously strayed into neutral Sweden. It had been thoroughly examined several weeks before the Polish A-4 arrived.

The Allied technicians and scientists who examined the crushed pieces of the A-4 were amazed. The systems that they were looking at were so different from anything they had knowledge of that they occasionally had a difficult time interpreting functions. As one scientist put it, *"At times we felt we were putting together a huge three dimensional jigsaw puzzle, with only faint clues and hunches as to which pieces fitted where."*

By calling on information that had been published a decade earlier by Robert H. Goddard and that of the Jet Propulsion Lab at Caltech, in Pasadena, California, it was determined that the warhead weighed one ton and the range of the V-2 was about 200 miles. There was no doubt that the Germans had developed a very impressive artillery rocket that was capable of reaching London from the European continent.

Because the V-2 that fell in Sweden was being used as a test bed for the *Wasserfall* antiaircraft missile, it was being guided by radio control and not the inertial gyroscopic guidance of the operational A-4. Allied intelligence concluded that it might be possible to identify the frequencies used and jam them. This was not to be the case, of course; once launched, the weapon was on its own and unstoppable.

Advanced Projects

As the development of the A-4 reached the production stage, several of the primary scientists were redirected to other weapons systems. Von Braun in particular had his attention focused on what was referred to as a guided A-A (anti-aircraft) rocket in November 1942. This weapon was to be a liquid-propelled rocket using much of the technology of the A-4. The Wasserfall (Waterfall), as it became known, used a different set of propellants that were hypergolic; they ignited spontaneously on contact. Thus, no ignition was required, and, perhaps more importantly, the rocket might have its propellants loaded for days or weeks, unlike the super-cold LOX of the V-2, which continuously boiled off and had to be replenished if the rocket was not fired shortly after loading. Nitric acid provided the oxidizer and a petroleum byproduct called a Visol was used as the fuel in the Wasserfall. The first successful guided flight of the missile occurred on February 5, 1944.

By December 1944 the course of the war made it clear to both Dornberger and Speer that the Wasserfall could not become operational until 1946. It is ironic that a wonder-weapon that might have protected Germany from allied air attacks, which devastated its potential to wage war, was an afterthought of A-4 technology.

Towards the end of the war, as the allies pushed the German Army farther from the English Channel, the need to extend the 200-mile range of the A-4 became apparent. As early as 1940, wind tunnel data showed that the range of the A-4 could be extended significantly through the use of wings to permit it to become a hypersonic glider in the upper atmosphere, doubling its range. It was felt that this could be done with minimum cost and effort. It was not until October 1944 that serious design and development began on the project, when coastal launching sites for the A-4 in Holland were lost. Although several test flights of this rocket (designated A-4b) were flown, the problems of high temperatures and aerodynamic forces involved were beyond the crude adaptation being attempted. Years later, the U.S. would revive the hypersonic glider concept in the form of the X-20 Dyna-Soar program and the Space Shuttle.

While the major emphasis of the Army's research at Peenemünde was in developing the A-4 and Wasserfall, Ludwig Roth's small group of designers was considering advanced rockets that rivaled even the Saturn V and Space Shuttle. Using the A-4 as a baseline, the team worked between 1940 and

1945, although it was officially abolished in 1943 by Dornberger following the allied raid on Peenemünde. Roth would come to the United States in 1945 as a part of von Braun's elite team.

One project proposed in the fall of 1943 explored a submarine-towed, watertight compartment that could launch as many as five A-4s. These missiles would be transported across the Atlantic and fired on New York City or other major industrial areas in America.

The A-6 was a modified A-4, using storable propellants of nitric acid and kerosene but never reaching the flight test stage. The A-7 was a forerunner of the A-4b and the A-9 but was only half as long and half as great in diameter. Thrust was 3,600 pounds—essentially an A-5 with wings. It was to be a test bed for studying control systems for winged vehicles. Test drops from aircraft were made in 1943 to study its glide characteristics, but no powered flight tests were ever undertaken. The A-8 was again similar to the A-5 but with a thrust of 6,800 pounds. It was essentially an A-4 with reduced payload and range. It would have used storable propellants of nitric acid and diesel oil. It never left the drawing boards.

In an effort to demonstrate more futuristic projects, the von Braun team had made some studies towards missiles with intercontinental capabilities. A scaled up and winged A-4, designated as the A-9, was envisioned as a second stage to an all-new booster with a half million pounds of thrust. The combination, known as the A9/A-10, had a range of 3,000 miles that would have allowed attacks on the East Coast of United States. This represented the first specific engineering planning for a multi-stage rocket. The concept of staging, or discarding unwanted weight to increase the velocity of an ultimate stage, was pioneered by Hermann Oberth and developed in a mathematical proof in 1923. The design of the A-10, with a thrust of 440,000 pounds, was begun in 1943 but terminated in 1944 in an effort to apply resources to the A-4b. In order to use the priorities for the A-4 project, the A-9 was designated A-4b. Some drawings for the A-9 show a pressurized cockpit in place of a warhead, and tricycle landing gear.

The A-11, with a thrust of 3.5 million pounds, never progressed beyond preliminary design studies. It would have been a lower stage for the A-9/A-10. It was estimated that this three-stage rocket could reach orbit velocity of 25,000 feet per second—the first engineering assessments of an orbital vehicle. The winged A-9 essentially would be today's Space Shuttle. Roth's team also considered an A-12, which would have had a thrust of 12,800 tons. With visionary designs like these being considered by the von Braun team, perhaps the Gestapo's allegations about their commitment to the war was accurate.

At the end of the war, the Peenemünde agenda contained the following projects:

Automatic long-range single stage rocket (A-4)
Automatic long-range hypersonic glider (A-4b)
Automatic two-stage boost glide missile (A-9, A-10)
Manned long-range hypersonic glider (A-9b)
Hypersonic two-stage boost glide aircraft (A-9b, A-10)
Unmanned satellites
Manned satellites
Manned ferry rockets to satellites (A-10)
Automatic deep space vehicles
Manned deep space vehicles

As noted, not all the work being done at Peenemünde was by the von Braun team. An interesting study conducted by a pair of Austrians, Dr. Eugen Sänger and Dr. Irene Bredt, analyzed the intriguing possibility of using the upper atmosphere to skip-glide a rocket-propelled bomber across great distances (similar in concept to the A-9/A-10). The work was based on an earlier study Sänger had conducted in 1933 while at the Technische Hochschule in Vienna, which proposed a liquid-fuel-rocket-powered, high- altitude airplane.

As the design evolved over the next several years, Sänger added another innovation in an attempt to save weight. The vehicle (which Sänger referred to as the 'Silverbird') would be launched from a rocket-powered sled that would accelerate the craft to 1,200 mph before using on-board rocket power. It would then reach a velocity of 15,000 mph before exhausting its fuel and continue to coast upward to about 160 miles altitude. It would then proceed to skip off the upper layers of the atmosphere much like a flat stone skipping across a pond.

Known as the Sanger-Bredt Report, the highly classified document with its novel concept of launching was appropriated by the Germans during the war as a bomber. Called the *Racketenbomber,* it was to be driven by two A-4 engines. The work was put on hold in 1942, and Sänger spent the rest of the war working on ramjets for fighter aircraft.

It was envisioned that the *Racketenbomber* could bomb New York and land in Japan, where it would be refueled and re-launched on the reverse course for another mission. While schemes such as these were too far out for consideration during the closing year of the war, the report found its way to the United States and Russia following the war, where it would be studied closely in light of the dawn of the atomic age.

Surrendering the Team

At the beginning of 1945, the elite of von Braun's group had been evacuated from the area around Nordhausen, some 300 miles to the north, and relocated to Oberammergau, near the Austrian boarder, on orders from SS Obergruppenfuhrer, Hans Kammler, to allow him to exercise more direct control over the von Braun team.

In January of 1945, von Braun called a highly secret meeting of his most trusted cohorts. He expressed the rather obvious opinion that Germany had lost the war. Then he added, *"But let us not forget that it was our team that first succeeded in reaching outer space. We have never stopped believing in satellites, voyages to the moon, and interplanetary travel. We have suffered many hardships because of our faith in the great peacetime future of the rocket. Now we have an obligation. Each of the conquering powers will want our knowledge. The question we must answer is: to what country should we entrust our heritage?"*

The decision was obvious and unanimous. The French and British would not be in any position (economically or politically) to fund rocket research following the war, and neither country would look favorably on these German creators of the V-2 that had terrorized London and Antwerp. The notoriously savage Russians, while eager for the secrets, could not be trusted, as evidenced by the unstable paranoia Stalin had exhibited with his purges before the war. They all agreed to surrender to the American Army. Here was a country that had not suffered the ravages of war. Germany had treated its American POWs reasonably well, so the personal hostility might not be too intense. Most importantly, the United States would be economically strong enough to fund continued rocket research.

Dornberger and von Braun received a variety of conflicting and often vague orders from the Nazi hierarchy during the closing months, with communications channels in disarray. On one day they were ordered to defend Peenemünde and on the next to move to a more secure site. As a result, it was then relatively easy for them to select the order that best suited their plans and to defend that selection. They elected to obey the order that moved them to the town of Bleicherode in the Harz Mountains, where they had the most likely chance of capture by the Americans. Von Braun arranged for several tons of technical documents that supported the work at Peenemünde to be moved, as well as the families of key team members.

On the morning of May 2, 1945, Magnus von Braun, Wernher's younger brother, was chosen to attempt to make contact with the Americans because he spoke English better than any of the others. Riding a bicycle toward the American lines, he encountered Private First Class Frederick P. Schneikert. Magnus explained that the scientists responsible for the V-2 were only a few kilometers up the road at Schattwald (a small village on the Austrian border of Germany) and wished to surrender to the Americans.

Schneikert was somewhat skeptical. The rumors circulating among the American troops, who now occupied key areas in southern Germany, were that elements of the German Army were withdrawing into the impregnable Alps of southern Germany and Austria. In this region, called the National Redoubt, it was believed that the Germans could continue fighting for years, perhaps getting better terms of surrender. Schneikert wondered if this could be a trap for the Americans.

However, after further interrogation, Schneikert felt there was some credibility to the man's story and handed him over to First Lieutenant Charles L. Stewart of Counter-Intelligence Command (CIC).

Magnus von Braun told Stewart that his brother's group was in immediate danger from the SS of being shot or used as hostages, since Hitler had ordered that no scientists or engineers should be allowed to escape to the Allies. After further interrogation, Magnus was issued passes that allowed the German group to move safely through the American lines. By nightfall, they were all in American hands. Wernher von Braun and 118 of his most skilled team members were about to begin a second effort to conquer the cosmos in a desert far from the rubble of their devastated homeland.

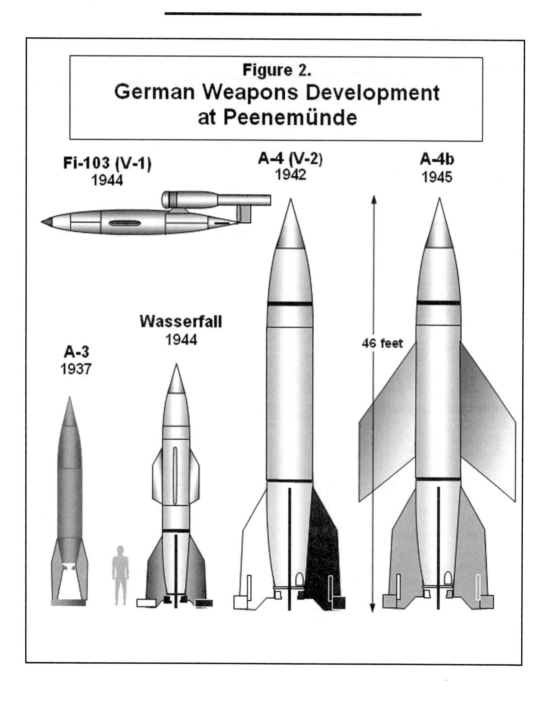

Figure 2.
German Weapons Development at Peenemünde

Fi-103 (V-1) 1944

A-4 (V-2) 1942

A-4b 1945

Wasserfall 1944

A-3 1937

46 feet

Planning Technology Transfer

The Second World War was unique for many reasons, not the least of which was the significant leaps in technology facilitated by developing the weapons of war. Had it not been for the British development of radar, the air battle that raged over that small island in the late summer of 1940 could have proved the downfall of the empire. Underwater detection was also a strong suit of the British, enabling them to turn-back the U-Boat menace in the Atlantic that threatened to strangle the island nation.

While the British excelled in early radar, sonar, and jet engines, America depended on its vast natural resources and industrial capacity to out-produce the Axis. With its depth of research facilities in academia such as the Massachusetts Institute of Technology (MIT), the University of California (UC), and the California Institute of Technology (CalTech), America was also able to take British innovation and improve upon it. Coupled with outstanding immigrant scientific talent, the harnessing of Atomic energy was one of the most notable American technological achievements during the conflict.

However, both Britain and America were continually surprised by the technology that Germany introduced into combat. Had it not been for nationalistic pride and complacency at the highest levels of their respective governments, this Germanic creativity could have readily been recognized many years before the war began. So much advanced science was being done in Germany between the wars that any student of physics or chemistry in other countries would have done well to learn the German language, as much of the happenings in those two disciplines (and several others) were being published in that language.

Not only were the innovative capabilities of Germany not fully appreciated within Britain but technology within Britain and America was not enthusiastically endorsed, either by the military or the government. Goddard's rocket activities in America and Frank Whittle's jet engine in Britain would have greatly benefited the military preparedness of both countries had they been even modestly funded. Of course, the economic times following the depression were difficult, but the fact remains that Germany (spurred by Hitler's desire for conquest) had managed to create a highly advanced research and development effort in many disciplines.

In trying to ferret out what Germany was doing after the war began, British intelligence excelled in several areas, but was myopic in others. Its tenacious work with ciphers (decrypting coded messages) allowed it to read many important German communiqués that permitted the British to head-off several critical German military advances. However, experts in British intelligence also predicted in 1943 that Germany could not sustain its war effort because of the allied stranglehold on strategic materials—including aluminum, chemicals, and petroleum. What these experts failed to recognize was that the German people understood that there was no turning back with Hitler. Nor was any peace likely to be offered that would be acceptable to Hitler, and thus it was a fight to the death for the nation (especially after Roosevelt's unilateral proclamation that only unconditional surrender would be accepted). Regardless of people's feeling towards their Nazi leader, the prospect of being overrun by the Russians in the East was enough to squeeze the most out of the German intellect. When this feeling of inevitable momentum was coupled with the German ability to innovate, many amazing advances were made in a wide variety of the sciences that allowed the German war machine and the economy to continue.

As early as 1942, the British recognized that they had a good chance of surviving the war now that America had been drawn in. They also realized they had been outclassed in several areas of weaponry. As the war moved towards a successful conclusion for the allies in late 1944, the dominant powers (America, Britain, and Russia) knew that there were technological prizes to be secured from the vanquished foe. The official stance, as dictated by previous agreements between the major leaders (Roosevelt, Churchill, and Stalin) and formalized at the postwar Potsdam Conference in July 1945, was that no reparations would be demanded from Germany. The allies would not remove non-military industrial or economic capacity from Germany so that it could sustain itself following the war. However, the technology and its products were fair game. Many recognized the fantastic breakthroughs in aviation, rocketry, submarine, and electronic technology that Germany had made. Thus, each victorious country put together teams that would immediately follow behind the combat troops in the closing year of the war and attempt to discover the many secrets that lay amid the ruins.

In Britain, Lieutenant General Sir Ronald Weeks devised a plan for advanced parties to travel with the frontline troops and secure promising technology for more thorough investigation at a later time. On the American side there were several, though often not well coordinated, activities aimed at understanding and acquiring German technology. Not the least of these was the *Alsos* mission, set up by General Leslie Groves, the military head of the Manhattan Project. His objective was to determine, and confiscate, progress the Germans had made with the Atomic bomb. However, Groves was to find that little significant research had been accomplished towards perfecting a nuclear weapon. Perhaps it was because the renowned German physicist, Werner Heisenberg, had not believed that a deliverable sized atomic bomb was practical.

As for the guided missiles and the V-2 in particular, Colonel Holger Toftoy was to play a significant role in altering the course of rocket history. As chief of the U.S. Army's Ordinance Technical Intelligence mission stationed in Paris, Toftoy was responsible for making inspections of captured equipment and frequently sent specimens of German weapons back to the United States. As the V-2 began its "rain of terror" on London in the fall of 1944, that weapon quickly rose to the top of Toftoy's wish list. A cable from Washington specifically requested that his command provide a shipment of operational V-2s to the United States.

Toftoy gave his subordinate, Major Robert Staver, the job of following the U.S. First Army into Nordhausen, Germany in April 1945. Nordhausen, the city closest to the Mittelwerk, had been identified as a major production center as early as August 1944, during the interrogation of a captured electrician who had worked there. However, the allies had never been able to effectively pinpoint the facility with enough assurance that a bombing raid would be effective until April of 1945.

Arriving at the V-2 production facilities during the last week of April, just days before the war ended, the 28 year-old Staver was amazed by what he saw, since he had some knowledge of rocketry from his work in the fledgling American effort in the Ordinance Department. In the underground tunnels was a production facility that the mechanical engineer held in awe. Scores of V-2s lay in various states of assembly. Reporting this back to Toftoy, he was directed to prepare 100 V-2s for shipment back to the United States as part of a project to capture and continue development of the V-2 under a program that was code named Project Hermes.

Staver and Toftoy were about to create one of the most morally controversial aspects of America's involvement in World War Two. Staver's contribution was one in which this young and somewhat naive officer (with the encouragement of his superiors) began a process that enabled hundreds of potential war criminals to be transported into the United States. He looked upon the von Braun rocket team as simply a group of men trying to do their best for their country. While he was keenly aware of the Dora Concentration Camp, he was unsure of the possible culpability of the rocket scientists.

Using the information supplied by a Dutch Army officer, William Aalman, who was acting as an interpreter, Staver was unable to make an immediate direct connection between those responsible for horrors being uncovered at Dora and the scientist who had so willingly surrendered to the U.S. Army. However, there was one individual, Albin Sawatski, an engineer identified as the Mittelwerk Technical Director, whose alleged arrogance and indifference towards the horrors of Dora so infuriated one of the U.S. soldiers that Sawatski was taken to an isolated location and executed.

Although the records of the interrogations have long since been lost or destroyed, there was little doubt that those scientists, including von Braun, had little moral reluctance to reject the use of POWs and others conscripted to serve as slave labor in the production facilities. The Darwinian concepts of higher and lower order human species that Hitler had ingrained in his subjects, or perhaps simple elitism, had a disastrous effect on the way many Germans behaved towards those who were considered *sub-human*. A generation of Germans, who had a highly developed culture, abandoned much of their moral codes as the war progressed.

However, many Germans pointed out that Hitler, through the SS and the Gestapo, had a tight hold on the population with the use of terror and intimidation. Even if von Braun or other members of his team had objected, would it have made any difference? They claimed that it might only have led to their being considered a threat to the Fatherland and sent to a camp themselves. Recognizing that one day they would have to account for their actions, many of the von Braun team began, months before the end of the war, to rationalize their participation in the development and production of the V-2.

While the Germans, and von Braun in particular, were friendly and helpful during their initial sur-render and internment, they were determined to get the best deal from the Americans. That meant a reluctance of full technical disclosure until some form of a contract with the American government for their services was concluded. *"We were interested in continuing our work not just being squeezed like a lemon and then discarded,"* von Braun was later quoted. He was walking a tightrope between some questionable aspects of his past and a nebulous future; for Dornberger the prospects were even more precarious. As a member of the German military he was a POW of the allies, and his closer associa-tion with the atrocities (which would ultimately earn him several years imprisonment by the British) left him little room to bargain, although he and von Braun maintained tight control over their team dur-ing these uneasy days.

In defense of their actions, German citizens contended that they were obligated to carry out the law-ful directions given to them by their government, or in Dornberger's case, the military. The expression that was to become infamous, especially for those involved with the concentration camp atrocities was *"We were just following orders."* The dilemma for Staver and others processing the technologists was determining if the person was an "ardent" Nazi. That descriptor became somewhat subjective and was applied to, and subsequently removed from, several members of the rocket team at various stages.

In the case of the rocket team, there were two moral issues: one was developing the V-2 that was allegedly used as a weapon of terror against a civilian population, and the other was the extent of their involvement with the Dora Concentration Camp. While no direct connection between the atrocities in the camp and the scientists themselves was unequivocally proven, Dornberger, von Braun, and Rudolph (among others) had enough knowledge to present a culpable image.

Staver tried to assess what had happened and who was responsible as quickly as possible for sev-eral reasons, one being that the area around Nordhausen was about to be turned over to the Russians for their occupation on June 1 (just three weeks away) as agreed to at the Yalta conference. In addition to identifying and locating the scientists themselves, there was also the matter of the supporting research and development documentation and facilities—the sum total of 13 years of the German rock-et program. The Germans had hidden much of the documentation in a mine near Goslar, not far from the Mittelwerk. Through Major Staver's initiative, the Germans revealed the location of this stash, and tons of paper began the trek back to the United States along with the V-2s. Much of the research facil-ities throughout Germany were destroyed, either intentionally as the Germans withdrew, or by Allied bombings.

Major William Bromley was sent by Toftoy to assist Staver in removing as much equipment from the caves as possible. The actual crating and shipment of the estimated 400 tons of "booty" did not begin until May 22. Local British officers found out about the activities and asked Eisenhower to put a stop to it. But Ike recognized (and commented) that technology was likely the only reparations America would get for the enormous expense and the lives that it had invested in the war effort. Sixteen Liberty cargo ships were needed to move the spoils of war from the Belgium port of Antwerp to New Orleans and then by rail to White Sands, New Mexico. The actual movement of the scientists was more difficult however.

Staver recommended to Toftoy that the Germans be offered short-term contracts to come to America for extensive debriefing and to assist in sorting out the tons of rocket parts and documenta-tion. There was speculation that the effort might be of some use in the continuing war against Japan, and the request was forwarded to the Chief of Ordinance in Washington. Staver himself did not believe that the scientists were war criminals and was instrumental in passing this belief on to others in high-er echelons.

Toftoy met personally with von Braun to size-up this "boy genius" and was, as most who met him were, impressed by his intellect and personality. Toftoy recommended that 300 of the V-2 development team be brought to the United States for interrogation by technical and scientific panels. He felt that the knowledge they had was far beyond the ability of a field team to properly record. Washington reduced the number to 100. Toftoy eventually brought 118 to the U.S. in the initial group.

Operation Paperclip

The plan to bring the German rocket scientists to the U.S. to work, called Operation Paperclip, was begun. The code name was arrived at quite simply: paperclips were affixed to the dossier of each scientist who was to be shipped to the United States. Ultimately 118 technologists were selected, with the help of von Braun, to quietly and allegedly illegally, enter the U.S. for extensive debriefing and possible future employment. The Americans would receive the cream of the Peenemünde rocket team.

The primary objective of Operation Paperclip was for defense contractors, such as General Electric, to interrogate the Peenemünde team in order to save time and money in guided missile development as well as to assist in assembly and firing of the newly acquired V-2s. To accomplish this, the War Department Joint Chiefs approved the *temporary* exploitation of the scientists on August 14, 1945, with an eye to a more suitable arrangement when there were less pressing wartime priorities. The end of the war with Japan, just a few days later, removed the obvious need for the Germans to assist with projects that might help defeat Japan, and other justifications would eventually be forthcoming.

With the wheels of the bureaucracy set in motion, the first group of Germans arrived in the United States in September 1945 and were soon transported to Ft. Bliss Texas. This was the closest military base to the White Sands Proving Grounds (near El Paso). For most of the German team, this would be their base of operations (and home) for the next five years.

On October 9, 1945, Major General Clayton Bissell (Assistant Chief of Staff) further legitimized the German presence in America. He issued a press release which stated that the German scientists were *"indispensable to the successful accomplishment of the most vital military research program"* and that *"certain outstanding German scientists and technicians are being brought to this country to ensure that we take full advantage of the significant developments which are deemed vital to our national security."*

But the Cold War was beginning, and the ultimate justification for the presence of the German colony in New Mexico was provided in 1948. The Joint Intelligence Committee released a report with the following observation: *"Unless the migration of important German scientists and technicians into the Soviet zone is immediately stopped, we believe that the Soviet Union within a relatively short time may equal United States developments in the fields of atomic research and guided missiles and may be ahead of the U.S. development in other fields of great military importance, including infrared, television and jet propulsion. In the field of atomic research for example we estimate that German assistance already has cut substantially, probably by several years, the time needed for the USSR to achieve practical results."* Keeping the German rocket team employed in the States was an obvious way to keep them out of Russian hands.

With the rise of the Communist threat to America, the need to move the Germans into a more permanent relationship with the American defense establishment dictated that some way be found to have them legally enter the country. To accomplish this, those who desired to make the United States their permanent residence were shuttled across the border from El Paso, Texas, to Juarez, Mexico, in 1948, where they were issued visas at the American Consulate, then re-crossed the border to officially enter the United States.

The activities of the Americans in sequestering the German rocket team were of great interest to the Soviet Union where a similar pursuit was taking place.

Operation Ost

As with the Americans and British, the Soviets had their covetous eye on the technology of the Third Reich. Even before the bombs stopped falling, the major research and development centers were marked for early entry by Soviet troops and subsequent intelligence teams. Stalin initially ordered the formation of a secret group of Soviet specialists to investigate the remains of the Polish firing site at Debica in late 1944. The ultimate objective was to determine if it was feasible to create a weapon similar to the German A-4. However, an assessment of that magnitude was not realistic at that point in time, given the fragments that could be gathered.

Although the Soviet leadership, as early as 1944, expressed no great interest in the development of large ballistic missiles for the war effort, Stalin was looking ahead to the period that followed the war and the role that the Soviets would play in world power politics. He was also acutely aware of the progress the Americans were making on the Atomic bomb project, although (at that time) the bomb had yet to be tested.

The initial Soviet effort to acquire missile technology and expertise in the closing days of the war could be categorized as disorganized (as it had been for the Allies). In the early months of Soviet occupation, these activities were not well coordinated. Many of the officers in charge had no knowledge of missile technology and relied heavily on the interpretations of aviation engineers.

Stalin diverted combat troops to Peenemünde. That facility was essentially deserted, and the Soviets faced no resistance. The expected cache of information and equipment was meager, and the Soviets intelligence teams were dismayed. One significant find was a set of 50 tested and certified combustion chambers at an underground depot and fifteen railway cars containing equipment.

It soon became apparent that the majority of engineers and scientists had not only evacuated the major R&D centers, but most had moved into the American and British occupation sectors. When it was realized that the principals, including von Braun, had willingly given themselves up to the Americans, Stalin was outraged.

In an effort to assure that German technology was fully discovered and exploited, a central command for rocketry operations was established in Berlin following the end of hostilities. A Special Technical Commission (Abbreviated OTK in Russian), referred to as Institute Rabe, was established. Soviet specialists in rocketry were flown to Germany and a joint Soviet-German commission provided for the collection of information about the various rocket programs that was being uncovered. In addition to the A-4 the Russians were also interested in weapons which held promise such as the Wasserfall anti-aircraft missile, the solid-fuel Rheintochter, and the Schmettering guided missiles.

Because the Americans had removed most of the completed A-4s from the Mittelwerk, the Soviets decided to re-open that production plant to produce several dozen A-4s for testing and evaluation. Test facilities for the rocket engines were restored as well. German engineers were indispensable in this effort, and within a year after taking over the Mittelwerk, the Soviets were ready to produce A-4s.

In the early fall of 1945, the Soviets formally began Operation Ost, the equivalent to the American's Operation Paperclip. It was an intense search for Germans who had the talents needed to exploit the A-4 technology for the OTK. In the desperate days that followed Germany's surrender, most Germans were willing to work at any job that offered the opportunity to feed and house their families. Leading the Soviet's recruiting effort was Boris Chertok, of Institute Rabe, who achieved some success with those who had been passed over for Operation Paperclip or those who preferred to remain in Germany with their families. Virtually all of the activities were held in strictest secrecy, and the NKVD secret police embedded informants at all levels.

The hunt for the more technically capable Germans, although intense, was kept as subtle as possible to avoid the possibility of alarming them. The Russians were well aware of their reputation and the behavior of their troops in the closing days of the war in looting and raping, which simply reinforced what most Germans had feared. One of those who responded to the benign but intense solicitation of the Soviets was Helmut Gröttrup, who made several secret trips into the Soviet sector to discuss his future. A former assistant to the director for guidance and control at Peenemünde, Gröttrup elected to participate in Institute Rabe in September. This decision was motivated more by his desire to keep his family together than by science or politics. While many voluntarily joined with the Soviets, there were others whose participation was coerced through intimidation or outright threat. When it became apparent that America had the core of the von Braun team, a proclamation was issued on October 11, 1945, directing all scientists to register with the Soviet occupation forces. At that point, a hard line was taken towards recruiting.

Eventually, 200 German engineers were a part of the Soviet effort to provide expertise in aerodynamics, guidance and control, propellant chemistry, propulsion, and ballistics. While few of these people had played a primary role in developing the A-4, they would supply important capabilities to the Soviets in mastering production and testing of liquid-fuel rockets. The Russians also let it be known that they would pay large sums of money for von Braun or Dr. Ernst Steinhoff, who had been the principle engineer in formulating the V-2 guidance. It was reported that they even initiated a kidnap plot for the latter. However, cooperation between the Germans and the Russians was generally amiable.

By October 1946 it was realized that Institute Rabe had achieved as much as it could, and it was decided to transfer all German and Soviet personnel to Soviet territory. This decision was concealed from the Germans, who were fearful of that possibility.

The surprise notification of transfer was made following a technical meeting that the Russians turned into an all night party, most likely to ensure that the Germans were well intoxicated before hearing the news. At about 4:00 A.M. the Germans were handed a document that essentially said that their employment was to be continued in the USSR under a five-year contract. They were directed to gather their families and take food and clothing for a trip that was expected to last up to four weeks. The 152 rocket engineers were only a small part of an estimated 6,000 Germans from a wide variety of technologies that were relocated.

The Germans arrived at the Science Research Institute No. 88 (NII-88) at Kaliningrad, about 10 miles north of Moscow, where they found deplorable living conditions. The Soviets themselves were not living any better. The post war conditions in the Soviet Union were about as bad as any civilization had experienced. Shortages of food, clothing, and housing were obvious. Medical treatment was almost nonexistent, and transportation was in great demand. Most everything had to be moved by rail since vehicles and gasoline was in short supply.

The basic hierarchy of who got the better accommodation was directly related to the educational standing of the worker. Dr. Gröttrup was housed in a large, six-room villa while most lived in single room dormitories, and some had to make do with tents. The workday was often 12 or more hours, six days a week. Some workers actually preferred to be on the job, as there was no real home to return to at the end of the day. Even the working conditions were primitive with a lack the basics such as tables and chairs. The buildings were in disrepair and often leaked and were cold. As for pay, most of the Germans received better compensation than an equivalent Soviet worker.

For most of the Germans and their families, Operation Ost was a frightening experience. The work being done was often obscured by a wall of separation between the real Soviet rocket program and the activities that the Germans were given to perform. As von Braun had prophesied, these Germans were being *"squeezed like lemons"* for their knowledge, and most would return to Germany after the five-year period with no knowledge of the current state of Soviet rocket technology that would be of any value to western intelligence. Additionally, they were removed from any possible assistance to the Americans or British during this period of being "on ice." Unlike the Americans, there was never any intent of incorporating the Germans into the mainstream industry or social life of the Soviet Union. Likewise there was no differentiation between "good Germans" or "bad Germans"; thus there was no moral concern over using any and all Germans who were available. Helmut Gröttrup was among the last group of Germans to be repatriated in November, 1953.

Creating a Super-Power

As World War II came to a close, three key technologies literally fell into Soviet hands. The first of these was the atomic bomb. The Soviets had skillfully seeded a wide spectrum of spies into the fabric of American, Canadian, and British industry and government in the 1930's and 40's. However, several individuals, outside the mainstream of Soviet influence, surfaced during the war to provide far more damaging information to the Soviets than cipher clerks, bureaucrats, and machinists could possibly pass on. At the super-secret Los Alamos atomic weapons laboratory, sequestered in the northern mountains of New Mexico, scientists Klaus Fuchs and Ted Hall, acting unilaterally on repressed political beliefs, supplied detailed research and design data to the Soviets. Neither Hall nor Fuchs was aware that the other was sending this highly classified information. Soviet dictator Joseph Stalin and Lavrenti Pavolvich Beriya, his head of the NKVD Secret Police, were incredulous at the level of detail and significance of the information.

So taken were they that any one would volunteer such information (money was not a factor in the exchange) that both remained skeptical for some time and suspected that it might be an American plot to send the Russians off on an erroneous scientific tangent. However, Soviet scientists were able to validate the information against that which was coming in from other more reliable sources such as Julius and Ethel Rosenberg's son-in-law, David Greenglass. So comprehensive was the data, that the first Soviet atomic bomb detonated in 1949 was a virtual copy of the Mark I "Fat Man" dropped on Nagasaki in 1945.

The second technology windfall was the unexpected arrival of three B-29s that diverted to Vladivostok on the eastern edge of the Soviet Union during the last year of WWII. The pilots of these planes experienced a variety of difficulties that prohibited a return to their base in the Marianas Islands of the Pacific, following bombing missions over Japan. The U.S. had a loose understanding with the Russians that American planes would find safe haven in such circumstances. In fact, the first such intrusion was a B-25 of the Doolittle Raiders in April of 1942.

Over the course of three years following the war, Soviet technologists of the Tupolov Design Bureau managed to reverse engineer the B-29 and produced several hundred identical copies. This was no easy feat for soviet industry, for not only had it been severely crippled by the war but it had also suffered for years under Stalin's brutal purges and ambiguous economic policies. The required skills and technology in metallurgy, plastics, electronics, rubber, and machining were woefully behind western standards. At one point when their industry could not provide for the B-29's tires, they were forced to purchase quantities from the U.S. surplus market. Their engineering perseverance and skillful ability to duplicate the B-29 into their TU-4 pushed the Russian aircraft industry ahead 10 years.

By 1950 the Soviets had a comparable level of mass destruction (the atomic bomb) and the means to deliver it effectively over long distances (the TU-4). With the appearance of a large number of these bombers in the skies over the Kremlin's 1950 May Day celebration, the debate in the United States over military parity began to take on significance.

This debate was heightened when the Soviets encouraged the North Koreans to invade South Korea in June of 1950. Air power, exerted by the Americans, turned the tide and quickly pushed the North Koreans back into the Chinese mainland of Manchuria, but the surprise appearance of the Soviet MiG-15 jet fighter into combat created a new environment for the U.S. and its allies. The MiG clearly outclassed American fighters and threatened to upset the air superiority of the United Nation's forces. Again, U.S. technology apologists pointed out that the MiG derived its swept wing design from German research, and its engine was an import from Rolls Royce. As with the Japanese fighters of a few years earlier, the MiG owed its superior performance to its light weight, achieved at the expense of more durable construction, and pilot protection and safety. With these factors in mind, and with the F-86 Sabrejet arriving on the scene, the MiG threat was suppressed, and America could move forward with its image of technological superiority.

The third technology handed to the Russians was that of the large liquid-propellant, surface-to-surface rocket—the German V-2. While the Russians were not on the receiving end of the V-2, they were keenly aware that it represented a technology that could provide a dominant military capability in the future. Following hard on the heels of their advancing troops, a small staff of knowledgeable Russian engineers scoured the German countryside seeking the rocket scientists responsible for the development of the V-2 as well as equipment and supporting documentation. Although the results of their efforts were not as qualitative as those of the U.S. technologists, who were moving through Germany from the west, they did provide enough resources to move Russian rocket technology forward many years.

The spoils of war in the form of theV-2 had accelerated soviet rocket science ahead a generation. Unbelievable luck with the unexpected arrival of the B-29s had given them a first rate strategic bomber. Moreover, the patience in planting a spy network and allowing it 20 years to germinate and mature provided for an atomic and diplomatic espionage network that was used very effectively.

The Soviet Union had moved from a backward, war-ravaged nation to a position of nuclear parity with the United States in just five short years.

White Sands: America's First Rocket Test Facility

A desolate stretch of land covering 1.2 million acres in the Tularosa basin of southern New Mexico was acquired by the United States government early in 1942 to provide a training area for the beleaguered British Royal Air Force. Some permanence was given to the Alamogordo Bombing and Gunnery Range and the supporting Alamogordo Army Air Field when the GALCIT Caltech group drafted a proposal for the Army to fund the development of missiles in response to Germany's V-2 rocket. Because of this proposal, the chief of the Research and Development Service, Major General G.M. Barnes, directed the Corps of Engineers, in late 1944, to conduct a survey of open areas in the United States to find a place suitable for testing missiles. The results of these studies indicated that the Alamogordo Bombing and Gunnery Range fulfilled most of the specifications, although, at 40 by 90 miles, it was not quite as large as desired. Nevertheless, it was isolated, and initial development work could be easily facilitated.

The White Sands National Monument was situated within the new missile range, as was a public access highway, but wartime expediencies set those concerns aside for the duration. The climate in this region was almost ideal as the low humidity and abundant year round sunshine allowed for almost continuous operations except for the occasional sand storms in the spring and the general lack of water. An area of the southern part of the range, approximately 40 miles north of El Paso Texas, was selected for erecting the required facilities. The site for actual firing was located approximately 6 miles east from the assembly area with an anticipated line of fire due north.

Named for the natural silicon deposit that brought many to visit the region, White Sands Proving Grounds was formally activated on July 13, 1945. Three days later, the Alamogordo Bombing and Gunnery Range extension received notoriety when a brilliant flash in the northwestern corner heralded the birth of the atomic age. At the time, it was announced that a large ammunition depot had blown-up and that there were no casualties. Only after the first bomb was dropped in combat a month later, was it revealed that the giant explosion in the remote New Mexico desert was a test of the first atomic bomb. No further tests of nuclear weapons were conducted in any portion of the range however.

The initial effort at the missile range during 1945 was directed towards the tests scheduled for mid-September of the rockets being developed at Caltech. However, towards the end of August, the range commander was notified that 300 railroad cars containing 100 German V-2 missiles and material would be arriving at the camp in less than a month. These were the results of Colonel Toftoy's and Major Staver's efforts in Germany to bring to the United States the German rocket team and the fruits of its labors. The size of the contingent of troops and civilians in the middle of the desert was about to increase. America began building what some considered its equivalent of Peenemünde. However, White Sands was to be primarily a firing range, with the actual technical developments taking place in other, more diverse industrial locations around the country.

As the first Caltech rockets were being launched, new construction and modification of facilities began to provide for the V-2 and follow-on projects. A large gantry crane was constructed to improve the efficiency of preparing large missiles for launch. Additional facilities were provided for radar tracking, instrumentation, and photography. Enclosed hangers permitted the missiles to be thoroughly inspected and assembled. A static-test stand allowed the V-2 propulsion system to be erected and fired without flight. Additional static-test stands with a capacity of up to 500,000 lb. of thrust were planned.

In an effort to avoid duplication and to bring the Navy into full partnership on the missile activities, the Army Chief of Ordnance invited the Navy to participate at the White Sands Proving Ground. In part, the Army was looking for additional funding and also recognized that the Navy had expertise in atmospheric research. The Naval Research Laboratory (NRL) provided telemetry for the missiles and created instruments designed to gather data in the upper air and transmit this information by telemetry to the ground stations.

WAC Corporal: Hypergolic Propellants

Caltech had begun work on an 8-foot-long solid-fuel rocket called the *Private A* which was test-flown in December 1944. Its rather crude propulsion system was based on the jet assisted take-off units (JATO) being developed for the Navy. The rocket was not an end to itself but simply a vehicle for gaining experience in building and launching larger rockets.

The *Private A* was a 530 pound unguided ballistic rocket designed to test fin-stabilization in free flight. It used a standard 1,000 lb. thrust JATO unit for propulsion and had four rectangular tail fins. Because the JATO rocket, which burned for 30 seconds, was developed for manned aircraft and did not provide for rapid acceleration, four 5,500 lb. thrust T-22 artillery rockets were used as a cluster to accelerate the rocket rapidly (burning duration was two-tenths of a second) to a speed where the fins were effective. A short, 36-foot launch tower with rails guided the rocket until enough velocity was reached for the fins to perform aerodynamically. The booster assembly was jettisoned when the rocket left the launcher.

GALCIT fired 24 *Private As* over a two-week period in early December 1944 at Leach Springs, Camp Irwin, near

An American V-2 with the WAC-Corporal second stage

Barstow, California, with some success and proved the effectiveness of fin-stabilization. The *Private A* achieved a maximum velocity of 1,200 fps and could fly a distance of 12 miles after a flight of about 90 seconds.

The *Private F* rocket was the next series and consisted of a *Private A* with the cruciform tail fins replaced by a single vertical fin and two five-foot horizontal wings. Two small canards (forward horizontal fins that spanned three feet) were added for aerodynamic trimming. The objective was to see if an unguided "cruise missile" was practical. Seventeen *Private F* rockets were fired in April of 1945 from Hueco Range at Fort Bliss, Texas, and were generally unstable in free flight. The tests appeared to prove that a winged missile would need an active control system, *i.e.,* an autopilot.

On the strength of the progress made with the *Private*, and the steady bombardment of England by the V-2s, the Army officially began funding GALCIT's efforts in January 1945, at which time the project was referred to as ORDCIT (ORDinance California Institute of Technology). Of particular interest were rocket propulsion systems and supersonic aerodynamics. At about the same time, GALCIT's rocket group adopted the name Jet Propulsion Laboratory (JPL).

JPL had decided to use military rank to name their advances in the project which had the ability to achieve higher performance. The first operational missile was to be called *Corporal*, a relatively sophisticated liquid-fuel, surface-to-surface-ballistic-artillery rocket for the Army, with half the warhead weight as the V-2 and half the range. However, as the Germans had discovered, there had to be small steps when technology needed to be proved. Thus, JPL's first liquid-fuel creation stood 16 feet tall (without its booster), had a one foot diameter, and was called the *WAC Corporal*. This name was a play on words, as officially the acronym WAC stood for "without attitude control." (Some sources use the term Without Any Control.) The missile was guided only by the aerodynamic flow around its fins. However, the acronym was also an obvious reference to the Women's Army Corps. One popular story relates that a general once asked Dr von Kármán how far up in rank JPL would go in naming their missiles. *"Certainly not over colonel,"* replied the academic. *"That's the highest rank that works."*

In the process of developing the *Corporal* technology, JPL would use the *WAC Corporal* in much the same manner as the Germans had used the A-3 and A-5 in developing the A-4. It was also realized that the missile could also be a useful tool as a small sounding (high-altitude research) rocket which could loft 25 lb of instrumentation to an altitude of at least 20 miles.

The *WAC Corporal* was boosted into the air by a solid-fuel rocket that developed 50,000 lbs. of thrust for six-tenths of a second. This booster was originally developed for the Navy as the *Tiny Tim* air-to-surface rocket that was used during the final months of World War II. By using high acceleration and a launching rail for initial guidance, the JPL program could proceed rapidly in developing the propulsion technology without the need for a complex guidance system, which could take a parallel course and merge at a later stage. This approach reduced the initial cost of the project while allowing the rocket to scientifically probe the upper atmosphere.

The main engine of the *WAC Corporal* was to be fueled by gasoline, but an effort was made to find an alternative oxidizer for the difficult-to-handle, super-cold LOX which readily evaporated. GALCIT team member Martin Summerfield wanted propellants that could be stored at room temperature. Parsons suggested nitric acid (HNO_3) diluted with about 10% nitrogen dioxide (NO_2) to produce "red fuming" nitric acid (RFNA). Summerfield experimented with RFNA as the oxidizer and gasoline or kerosene as fuels, but his rocket engines sputtered violently and often blew up or ceased burning.

Dr. Frank Malina visited Annapolis to confer with two other groups who were developing a liquid-fuel-rocket-assist take-off system (similar to the solid-fuel JATO units). Formerly of GALCIT and now a member of the JPL propulsion team, Malina discovered that a group headed by Navy Commander Robert Truax had established that certain chemicals ignite spontaneously on contact and thus needed no ignition system. This knowledge was immediately communicated back to the *WAC Corporal* team, and aniline, with a mixture of 20% furfural alcohol, was subsequently used for the *WAC Corporal*— the first free flight missile to use the combination. It is of historical interest to note that the second group visited by Malina during his east coast visit was headed by Dr. Robert Goddard.

Although performance of RFNA was not up to that of LOX (theoretical exhaust velocities of 11,800 ft/sec compared to 14,400 ft/sec), it was sufficient, and RFNA was easier to handle and store under combat conditions. Nitric acid had been used by the Peenemünde team in the Wasserfall and the Taifun missiles. Von Kármán was to note: *"Nitric acid and aniline took to each other beautifully. The flame in the engine was absolutely steady."* With this, the GALCIT team had another important innovation: a liquid-fuel rocket that used storable propellants.

Compressed air (rather than nitrogen) pressurized the fuel and oxidizer tanks for delivery of the propellants to the combustion chamber—flow rates did not require a turbo-pump. The liquid-fuel rocket motor was started by an inertia valve, which sensed acceleration of the booster and opened the propellant valves, allowing their tanks to receive pressure from the compressed air tank.

After the *WAC Corporal* had left the 100 foot-high launch tower, the missile was stabilized in flight by three tailfins. A small series of experiments at the GALCIT Goldstone firing range in California in July had confirmed their adequacy, as well as the design of the booster arrangement, with a one-fifth-size model known as the *Baby-WAC*.

Following preliminary booster tests with dummy *WACs*, the first flight of a complete *WAC Corporal* occurred on September 26, 1945. It reached an altitude of 43 miles (as determined by radar), but the nose-cone recovery mechanism, which was designed to separate and allow the instrumentation to parachute back to earth, failed, as it did on most flights. However, much of the data was radioed back to the ground station, so instrumentation recovery typically was not critical. Additional flights followed until mid-1946, mostly with meteorological payloads like radiosondes (electronics packages that record meteorological data and transmit it back to ground stations).

Eight improved rockets, called *WAC B*, with a more advanced engine (that developed 1,500 lbs. of thrust) as well as a new telemetry system, were fired from December 1946 through mid-1947 before the program was completed. With a burn time of 45 seconds, velocities of 3,000 fps were achieved, and the highest altitude reached was 250,000 feet (50 miles) for the rocket that weighed 690 lbs. when fully fueled (without the booster). Several of the *B* models were used to create an experimental, two-stage research rocket, when mounted on top of the V-2 missile. Eight of these vehicles, known as RTV-G-4 *Bumper*, were tested between 1948 and 1950.

The *WAC Corporal* represented the highest level of technology reached by the United States before the influx of German influence. It essentially established America as being about 8 years behind German progress at the war's end. Although overshadowed by the V-2 as a high-altitude research rocket, the ability of the *WAC Corporal* to quickly and inexpensively loft scientific packages to impressive heights spawned a second generation general-purpose sounding rocket, the *Aerobee*.

With the completion of the *WAC Corporal* firings, enough knowledge had been gained to permit JPL to move forward with developing a combat weapon. Although the war had ended, the Army decided that it needed to be able to compete in any future conflict with the most up-to-date technology, and work continued on the full-scale version of the Corporal. However, before it arrived at White Sands, the V-2 would provide spectacular displays for the next few years.

Project Hermes: Blueprint for the Future

In November 1944 the U.S. Army awarded a contract to General Electric for research, development, and engineering work on a family of test vehicles and operational guided missiles. The project was named *Hermes* after the Greek god of science. It was not realized at the time the extent of the German rocket program, except that it was causing some alarm on the British home front as well as the battlefront.

With the acquisition of the V-2 (it was rarely referred to as the A-4 at White Sands), the *Hermes* program was broadened to include its use as research and test vehicles. While the Hermes programs tended to be pursued in parallel, they were revised over the succeeding years to reflect the results of research as well as the availability of funding and the world's political situation.

As the spoils of German research began to arrive in the form of a vast amount of hardware, documentation, and the von Braun rocket team, a wide spectrum of activities was planned by General Electric and the U.S. Army. Under the umbrella of Project Hermes, these programs were designated by the prefixes A, B, and C (not to be confused with the German A series designations).

The Hermes A-1 surface-to-air (SAM) missile was based on the German Wasserfall. Much of the component became a test vehicle for guidance and control systems, but the role of the operational SAM was taken over by yet another program called Nike, a further offshoot of Hermes. Much of the component flight-testing used the V-2, as problems with the rocket engine development delayed the first launch of the A-1 until 1950. Five Hermes A-1 tests were made before the program was concluded in 1951.

A wingless surface-to-surface derivative of the Hermes A-1, the A-2, was a 75-mile- range, tactical missile designed for the Mark 7 (20kT) nuclear warhead. The design reverted to the use of solid-fuels, and the project became a test vehicle for improving solid-fuel performance and guidance techniques. The first firing, in February of 1953, used a large (for its time) rocket motor developed by the Thiokol Chemical Company. The project was cancelled when the development contract for the tactical missile went to JPL for what became known as the *Sergeant*.

The U.S. Army established specifications for the A-3, in 1947, as a liquid-fueled surface-to-surface missile that would deliver a 1,000 lb. warhead 150 miles with an accuracy of 100 yards. The program, plagued by lack of funding and numerous requirement changes, was to have carried the Mark 5 nuclear warhead (47 kT). The first flight of the A-3 failed in March 1953, but the second, in June, was a success. By January 1954, of seven launches five failed. The high-performance rocket engine and improved inertial guidance system were noticeable advances in technology. The *Hermes A* program ended in 1954 and did not result in an operational missile; the follow-on *Redstone* and the *Sergeant* benefited from the research performed.

Perhaps the most advanced work of Hermes to come directly from the German team was the *Hermes B* program that began in 1946. This ramjet-powered surface-to-surface cruise missile (sometimes referred to as *Hermes II*) was to carry a 1,000 lb. warhead over a range of 1,000 miles at a speed of Mach 4.

The basic V-2 was modified to place a winged vehicle (named *Ram*) in the nose and was launched to position the *Ram* at an altitude of 100,000 feet with a speed of Mach 4, flying parallel to the earth. Here the *Ram* would separate and continue to the target with the power of the ramjet engines. The *Ram* employed wedge shaped wings with no sweep. The power plants were fed through small rectangular air intakes. To provide more stability to the first stage V-2 booster, its fins were enlarged.

The first flight occurred in May 1947, with a mockup *Ram* to measure dynamic pressure in the ramjet ducts. This is the flight that veered off course and crashed in Juarez, Mexico. Three additional flights occurred, the last in November 1950, but no live ramjet power was employed. The objectives of the program were well ahead of the technology and the project languished, although some exciting and interesting experiments were performed.

The program was concluded in 1954 without producing a proof-of-concept vehicle. By this time North American Aviation was well into developing its Navaho cruise missile, which had similar requirements but which also did not produce an operational missile. By that time the Atlas ICBM was progressing well enough that the cruise missile concept was redundant and ultimately capable of being intercepted.

The *Hermes C* program, begun in July 1946, employed the Eugen Sanger boost-glide concept (of Peenemünde fame) in a three-stage configuration. This huge, 250,000 lb. vehicle was to have six engines with a total thrust of 600,000 lbs. and a single 100,000 lb. thrust engine for the second stage. The glider would deliver a 1,000 lb. warhead after a flight of 2,000 miles. No hardware was ever built.

Exploiting the V-2

The V-2 research program in America did not proceed as rapidly as the von Braun team had antic-ipated, as the first static firing of a V-2 did not occur until March 15, 1946. The first launch on April 16, 1946, of V-2 No. 2 failed after 19 seconds of flight when the graphite vane on fin IV disintegrated causing the fin to tear itself from the rocket body. (Each missile was assigned a number for adminis-trative tracking, although V-2s were not necessarily fired in their numerical order.)

Because of the close proximity of inhabited areas to the west and south of the launch area, a radio-controlled fuel shut-off capability was provided as a precaution against a rocket going out of control. The third V-2 flight was more successful, reaching some 70 miles into space.

Each rocket was carefully assembled from the parts that Major Staver and his troops had packed-up at the Nordhausen Mittelwerk. Of course, Staver didn't have a parts book, so there were many of some items and few of others. Occasionally, an item had to be manufactured locally, but for the most part Staver did, in fact, provide enough parts for almost 100 complete missiles. As the Germans already understood, several of the components had very short shelf lives, and the highest missile reliability was from those most recently assembled.

Each launch often served several purposes. Various aspects of rocket technology were examined, and the payload typically consisted of instruments to measure performance parameters such as temper-atures and pressures. However, scientists were eager to probe the upper limits of the atmosphere, and scientific objectives included measurement of the ionosphere, and solar and cosmic radiation, and the detection of micrometeorites. Live animals were flown for biological research and Earth photography was conducted (V-2 No. 40 photographed over 800,000 square miles of Earth's surface on July 26, 1948).

Flights occasionally had some equipment or data that required recovery. It was obvious from the first few launchings that the impact of the missile was so traumatic (even without a warhead) that lit-tle was left that was useful. Beginning with V-2 number 5, a small amount of explosive was attached to the primary structural members, just forward of the fuel tank. At about 20 miles altitude on the way back down, a signal was sent to explode the charges in what was referred to as an "airburst." With the aerodynamics of the missile destroyed, both the forward and aft sections tumbled to the ground at a much slower speed. No. 9 had the first successful airburst. There were also payloads that required a softer descent, and Project Blossom provided parachute deployment.

By January of 1947, seventeen V-2s were fired at a rate of about two per month. With the realiza-tion that the number of V-2s was limited, an Upper Atmosphere Research Panel was organized. The V-2 Panel (as it was sometimes called) reviewed requests for experiments for each rocket, identified regions of particular scientific interest in the upper atmosphere, and assigned payload space to specif-ic academic research institutes.

Assembly and launch of the rockets was accomplished through coordination between the von Braun team, Army Ordnance missile technicians, officers and men of the First Guided Missile Battalion sta-tioned at Fort Bliss, and the General Electric Company. Each missile was assembled and thoroughly tested before undergoing a static firing of its rocket motor while held to the launch pad. No. 13 provided motion pictures showing Earth's curvature on October 24, 1946, while No. 19 employed a different form of guidance. V-2 No. TF-1 (a *Hermes* Test Flight) achieved the highest altitude of 132.6 miles on August 22, 1951.

By 1947, the telemetry capabilities were significantly improved, and over 50 different channels of information were sent back from the missile, including the first heat transfer data from supersonic flight at over 3,400 mph. More than 70% of all V-2 launches were above the original design weight, and 50% had contour modifications. Performance enhancements were introduced as well as changes to the payload compartment on some flights. These compartments were lengthened about five feet, to increase in payload space from 16 cubic feet to 80 cubic feet.

The Navy was interested in how the launch of a large liquid-fuel rocket could be achieved from a moving ship. Operation Sandy accomplished this on September 6, 1946, from the carrier USS Midway. The V-2 was successfully launched but almost immediately tilted about 45 degrees. It made an attempt to right itself but was destroyed by the range safety officer. Although the missile misbehaved, the experiment fully completed the objectives, in that the handling, fueling and launching were accomplished with minimal disruption to normal flight operations.

By the time the V-2s were expended, 67 had been launched from White Sands, two from the new Joint Long Range Proving Grounds at Cape Canaveral in Florida, and one from the USS Midway in operation Sandy. The last V-2 launch occurred on September 19, 1952.

Project Bumper: First Multi-Stage Liquid Fuel Rocket

In February 1946, JPL initiated a study for a two-stage 28,300 pound, 57-foot-long research missile that used a V-2 first stage and a *WAC Corporal* second stage. The project was approved and given the code name *Bumper* as a part of Hermes. As the first two-stage liquid-fueled rocket ever assembled, it explored problems related to stage separation and rocket ignition at high altitude. It provided the ability to double the maximum altitude attainable by a research rocket. As an indication of how important this experiment was, eight V-2s were allocated to it—almost 15% of the total launched by the United States.

The V-2 engine and structure remained unchanged, but the missile required an interface in the nose to secure and ignite the *WAC Corporal* (which was referred to as the *Bumper WAC*). Because the *WAC* would begin its flight 20 miles up, the thin atmosphere dictated that the three fins be changed to four and enlarged. Two small solid-fuel "spin" rockets were attached to induce a stabilizing rotation for flight beyond the atmosphere. Staging was achieved by shutting down the V-2 propellant turbo-pump, which reduced the V-2 engine's thrust to about 16,000 lbs. This allowed enough G forces to keep the *Bumper* WAC fuel firmly seated in its tanks. A signal was then issued to the *Bumper WAC* to open the propellant valves, and the hypergolic action of the propellants ignited on contact in the WAC's combustion chamber. The exhaust from the WAC burned through a wire, signaling the V-2 propellant valves to shut. As the V-2 decelerated, the *Bumper WAC* began its flight.

In May 1948, the first *Bumper* was launched with a dummy upper stage and a small solid-propellant charge to test stage separation. The first flight with a live *Bumper WAC* on September 30, 1948, failed when the *WAC*'s engine exploded on ignition. The problem was identified as a "hard start" caused by the propellants vaporizing as they entered the combustion chamber that had little atmospheric pressure. It had been anticipated that this could be a major problem in achieving a high altitude ignition.

For the next flight, a frangible diaphragm seal was placed over the exhaust nozzle to retained sea level pressure. When ignition occurred, the diaphragm was blown out, and *Bumper* No. 5 (following a V-2 failure in November) achieved full success on February 24, 1949, reaching an altitude of 244 miles. It is interesting to note the physics involved here. Had the *WAC* been fired at the high point of the V-2's flight (at best 115 miles), the capability of the *WAC* would have simply added another 40 miles to the altitude (perhaps resulting in a 155-mile altitude). But by igniting the second stage at the end of the first stage burn (where it had achieved its greatest velocity), the cumulative velocity provided a record for a man-made object, which would hold for almost 8 years in the United States.

Following the failure of Bumper No. 6, it was decided that the remaining two *Bumpers* would fly trajectories for range (rather than altitude), which would result in higher velocities and attain additional aerodynamic heating data. The V-2 would fly a trajectory that would release the *Bumper Wac* at an angle of about 20 degrees to the earth's surface. These tests required a firing range with a longer distance capability than available at White Sands. The area of Cape Canaveral at the Joint Long-Range Proving Ground (JLRPG) in Florida was chosen as it provided for firing southeast across the Atlantic Ocean. *Bumper* No. 8 and then 7 became the first flights for a new launch facility whose name would eclipse both White Sands and Peenemünde. On the first launch of Bumper No. 8 at the Cape, the upper stage failed to ignite, but the second and final attempt on July 29, 1950, was successful, and the *Bumper No. 7* reached a speed of 3,270 mph.

JPL Corporal: America's First Combat Ballistic Missile

In 1947 the first *Corporal E* missile, a follow-on to the *WAC Corporal*, was successfully launched. It provided for evaluation of the structure and guidance for the operational version. However, the second and third tests failed, and JPL decided to redesign the rocket motor for better cooling.

The 40-foot, 30-inch diameter *Corporal* used red fuming nitric acid (RFNA) as an oxidizer. This chemical is a highly toxic and corrosive substance, and the fueling process was a high risk adventure. The guidance system used radar to track the trajectory and velocity from the ground, and to compute and transmit course corrections to the rocket. The motor produced 20,000 pounds of thrust (about one-third that of the V-2) for 60 seconds, giving a maximum range of 64 miles. As with the V-2, the missile was steered in flight by jet vanes in the exhaust and aerodynamic rudders. In 1949 the *Corporal* was the first missile approved to carry nuclear warheads.

JPL and Firestone contracted to produce 200 units per year by 1952. Production numbers like this were typical for nuclear tipped missiles, as the days of 900 missiles per month were left in the German hinterlands. The first flight of a complete combat configuration of the *Corporal* came in August 1952, and by April 1954, the Army units began to receive the first operational SSM missile in the United States.

Poor reliability of the Guidance system, a long nine-hour launch preparation time, and a large launch contingent (35 vehicles) assured that the 9,250 pound missile would see many upgrades and be replaced as soon as the solid-fuel *Sergeant* could be proved. The last *Corporal* was retired in 1964.

The expertise JPL gained in systems engineering in meshing various tasks of a complex project served it well in coordinating the development, manufacture, and support systems for a next-generation guided missile, called *Sergeant,* and for space-exploration projects that replaced missiles as JPL's mainstay by 1960.

Aerojet Aerobee: Workhorse of the Upper Atmosphere

The Navy Research Lab (NRL) played an important role at White Sands. A 1946 study looked at the requirements for future upper atmospheric sounding rockets and quickly determined that a follow-on would be required for the V-2 and *WAC Corporal*. A young member of the Applied Physics Lab at Johns Hopkins University, named Dr. James Van Allen, was a key member of that study team. The *WAC* had proved that a small, inexpensive rocket was often sufficient for the lighter experiments weighing about 100 pounds, while there would still be a need to occasionally lift a half ton. Thus, NRL embarked on a program to develop two distinct types of rockets, initially called Venus and Neptune but ultimately changed to Viking and Aerobee respectively.

The U.S. Navy Bureau of Ordnance awarded a development contract in May 1946 for the *Aerobee* to Aerojet General. The single-stage liquid-fueled rocket used an 18,000 lb. thrust solid-propellant rocket motor as a booster and a tall launch tower to guide the rocket until the fins became effective, a technique similar to that used in the *WAC Corporal*. The booster burnout and separation occurred at 2.5 seconds, and the 2,600 lb. thrust liquid-fuel sustainer continued to power the rocket for 45 seconds. Helium was used to pressurize the fuel tanks. The nose cone with the scientific payload was recoverable by parachute.

The first launch occurred in November 1947, and by 1950 the *Aerobee* was well established as the standard lightweight (1,200 lb.) research rocket capable of lofting a 150-pound payload to over 75 miles. There were many versions of the *Aerobee* that resulted in lengthening it from 20 to 25 feet, while the fifteen-inch diameter remained. Engines of up to 4,100 lb. of thrust were used with larger fins for improved stability. Velocities of up to 4,600 mph were achieved.

More than 800 *Aerobee* versions were flown by the U.S. military services and NASA between 1947 and 1985, although several solid-fuel sounding rockets using Nike boosters made their appearance in the intervening years. The technology of the Aerobee would find itself as the second stage of America's first rocket designed specifically to launch an earth satellite and for subsequent moon shots.

Martin Viking: Advancing Rocket Technology

The Glenn L. Martin Company was awarded the contract for twelve Viking missiles that would ultimately replace the V-2 for heavy payloads. Reaction Motors of New Jersey built the 21,000-pound thrust XLR-10 engine (only one-third as powerful as the V-2). Under a 1944 Navy contract, Reaction Motors had built larger versions of Wyld's engine, capable of producing 1,000 pounds and later 3,000 pounds of thrust.

The primary objective in the design of the Viking was to produce a rocket with state-of-the-art technology that could provide a high degree of efficiency for its size. To accomplish this, it was decided to control the rocket by swiveling (gimbaling) the rocket motor to provide for pitch and yaw. This would eliminate the troublesome vanes used by the V2 that occasionally failed, but more importantly, reduced the exhaust velocity and impeded the performance of the engine. Roll control was achieved by routing the exhaust steam from the propellant turbo-pump to the tips of the fins and valving the flow to provide jet action.

The first seven Vikings had a length of 48 feet (slightly more than the 46-foot V-2, but a diameter of only 30 inches. This size was determined by the largest width of aluminum sheets available at the time (100 inches), as it was desirable to have only a single welded seam. To assure the highest mass ratio, the structure provided for integral fuel and oxidizer tanks. The body of the missile itself was used as the propellant tanks with the top and bottom tank bulkheads welded into the body. This reduced the gross weight (10,800 lbs.) and assured that any leakage would not go undetected and cause an explosion. Eight thousand pounds of propellant allowed for 70 seconds of burn time.

The first launch occurred in May 1949, but it achieved an altitude of only 50 miles when the engine shut down prematurely. This occurred again with Viking II, which reached only 32 miles. The problem was traced to the turbo-pump housing's not retaining its seal under the vibration in flight. Beginning with Viking III, these units were welded rather than bolted together, and there were no more failures of this type. Unfortunately, Viking III drifted towards the boundary of the range, the engine had to be shutdown by the Range Safety Officer, and another disappointing 50-mile altitude was recorded.

Viking IV launched May 11, 1950, from the USS Norton Sound, a converted seaplane tender, near Christmas Island in the Pacific. It carried almost 1,000 pounds of instrumentation to 105 miles in an effort to gather information about the magnetic field of the earth. Viking V went to 108 miles, but Viking VI achieved only 40 miles altitude when a control failure caused it to literally fly a loop.

Beginning with Viking VIII, a new configuration appeared with smaller triangular fins. The diameter was widened to 45 inches and the length reduced to 41.5 feet. The propellant load was increased by 3,450 pounds (about 1/3 more), while the overall empty weight of the rocket shed some 100 pounds. This resulted in a noticeable improvement in the mass ratio (M_R) to 3.6:1 (the V-2 was 3.1:1) and an increase in burn time to103 seconds. Unfortunately, during the static test of the rocket on June 6, 1952, the missile broke loose from its launch pad and climbed 50 miles into the night air—without returning any data.

The remaining four Vikings performed perfectly with Viking XI reaching an altitude of 158 miles and a velocity of 4,300 mph. The scientific payloads included temperature, density, and composition measurements of the upper atmosphere, measurements of solar and cosmic radiation, and ionospheric experiments. The later Vikings also used small gas jets to permit orienting the rocket after burnout, as it was discovered that any combustion instability in the last impulse of powered flight often caused the rocket to tumble as it continued upward (the fins became useless above 20 miles).

The Viking is significant for several reasons. It was the most technologically advanced missile of its time, and it was the basis for the first stage of the Vanguard satellite launch vehicle. Two additional Vikings (Numbers 13 and 14) were produced and flown in 1956 and 1957 in support of the Vanguard project.

Viking also provided the Glenn L. Martin Company with experience in building and launching liquid-fuel rockets, which enabled them, in October of 1955, to bid successfully for the back-up ICBM that became known as the Titan. The Titan was pivotal in the manned Gemini program that moved the United States ahead of the Soviets in space technology by the mid-1960's.

While the White Sands Missile Range continued to be an important facility, by the time the Viking XII thundered aloft in 1955, the larger rockets being developed needed an entirely new range with the ability to launch over several thousand miles. The Joint Long-Range Proving Grounds at Cape Canaveral in Florida, initiated by the *V-2/WAC Corporal* flights in 1952, is where the action would be for the next generation of missiles.

Kapustin Yar: First Soviet Rocket Test Facility

In July 1947 State Range No. 4 was established 50 miles South East of the town of Volgograd in a barren stretch of desert in the Soviet Union. The first launch from Kapustin Yar, as it was commonly called, of German and Soviet built V-2 missiles was set for September 1947. The living conditions were primitive, and engineers used special trains that served the dual role of work place and living quarters. Others had to live in tents or in the village with the local inhabitants, who were none too pleased to share their meager housing. The temperatures in the summer often exceeded 100 F degrees, while winter could see -30 F.

Authorization to test missiles did not come until July 26, 1947, with the first launch of a V-2 from the Soviet Union occurring on October 18, 1947 (almost 18 months after the first American V-2 test). A total of eleven were fired, with the last two on November 13th. Five had been assembled at the Mittelwerk and six at a factory on the outskirts of Moscow called the M.I. Kalinin Plant No. 88. It would become famous by its acronym NII-88 (pronounced "nee 88"). The Germans under Gröttrup assisted in the launchings, of which only five were successful. A second series of 10 rockets proved no better. However, the objective was to gain experience in handling and firing large liquid-fuel rockets and to measure a variety of points on the missile, in order to expand the knowledge of this new and often volatile environment.

Because the Soviets were only able to fashion less than a dozen V-2s from the parts left at Mittelwerk by Major Staver, they were forced to manufacture copies (which they would probably have done in any case). The Americans, with their 100 V-2s, and tight post-war budgets, were in no hurry to manufacture an improved version (although von Braun's team certainly was). The United States used its V-2s for experiments and research, and design on an improved version would not take place until 1950.

Sergey Korolev, an original member of the Soviet GIRD society of the 1930's, had survived his imprisonment in the Siberian Gulag and had joined the Tupolev Design Bureau in 1944, although he had yet to be exonerated of his "crimes against the state." The lack of available V-2s with which to experiment, benefited Soviet technology as it prompted Korolev's NII-88 team to move forward more rapidly in improving the basic design of the V-2 rather than just replicating it, as was being done with the American B-29. Thus, the apparent success of America's Operation Paperclip had an unintended effect on the direction of immediate postwar rocketry in both the Soviet Union and the United States.

VR-190: A Vision of the Future

While the missile development at NII-88 (and other design bureaus) would reflect the military focus that dominated the Soviet government in the post war years, the dreams of Korolev and others continued to focus on the stars. Mikhail Tikhonravov was a former member of GIRD who had designed the first liquid-fuel rocket in the Soviet Union. He convened an ad hoc group of engineers at the institute in 1945 to investigate the use of existing technology (the V-2 or its immediate derivative) to develop a rocket for carrying two passengers to an altitude of 190 km.

The proposal, termed VR-190, was the first in the Soviet Union (and the world) for launching humans into space. The plan envisioned the use of a modified A-4 with a recoverable nose cone containing a pressurized cockpit for carrying two "stratonauts." The passengers would remain in a semi-reclined position on custom-made couches in the capsule from launch until touchdown. There were several objectives of the plan including research on the effects of weightlessness and the separation and stabilization of the cockpit.

Much of the design was remarkably advanced for its time. It included a parachute system for the return to earth, which was coupled with a breaking rocket for softening the landing of the cabin on the ground. The pressurized cockpit provided life support systems and attitude control jets, allowing the vehicle to orient itself during its trajectory beyond the atmosphere. A heat shield on the bottom of the cabin was similar to that used on the Soyuz spacecraft. Several other design elements of the VR-190 would be incorporated into the first Soviet spacecraft in the early 1960's.

Recognizing that the hierarchy above him would never approve such a plan, Tikhonravov appealed directly to Stalin. Although the response to his letter was somewhat positive in tone, the project never received funding. Korolev did not support the VR-190 project as he felt that a piloted rocket plane, an outgrowth of his interests and experiments of the 1930's, was the preferred method of exploring the upper atmosphere, at this point in his thinking.

R-1: Cloning the V-2

Although the plan to replicate the V-2 was conceived in 1946, with Korolev appointed the chief designer, the R-1 (as the missile would be known) was delayed until April 1948 while Stalin apparently pondered the best course of action with his spoils of technology and German specialists. The R-1 borrowed heavily from the V-2 but with improvements that included structural modifications and a guidance and control system based on Russian components.

Valentin Petrovich Glushko, a compatriot of Korolev's at GIRD in the 1930's, was assigned the task of producing a copy of the V-2 engine (designated the RD-100) for the R-1. (The two space enthusiasts would collaborate and clash over the next two decades.) The RD-100 provided 61,000 pounds of thrust to lift the 29,000 lb. missile. With a diameter of 5.4 feet and a length of 46 feet, it was a virtual copy of the V-2 externally. The missile could accelerate an 1,800 lb. payload to a velocity of 3,300 mph. resulting in a range of about 200 miles.

By September of 1948, the first of two series of tests were prepared at Kapustin Yar that involved the launch of twelve R-1 missiles to verify the basic design. None of the German contingent was present when the first missile thundered aloft on September 17th but failed within seconds. A second, launched on October 10th, was considered a success. Although nine of the twelve flew the distance, only one landed in the target area. The second series of 20 missiles was launched in 1949 following improvements, dictated by the first series, to increase reliability, although accuracy was little better than the V-2.

The R-1 was an important tool for high altitude scientific research for the Soviet Union as well as for the development of the technologies that made it work. In the fall of 1947, a modest, upper-atmospheric-research program was prepared. Beginning with the fifth flight in November 1948, the R-1 was equipped with instrumentation containers, a separable nose cone for recovery of biological specimens (dogs and rabbits), and other apparatus to study cosmic rays, the upper atmosphere, solar radiation, high altitude winds and the ionosphere. The total scientific payload was about 1,000 pounds. The missiles reached altitudes of up to 50 miles. These early excursions into space were unfortunately marred by frequent equipment failure, and little scientific results were achieved.

Although the R-1 program was overseen (and psychologically driven) by Stalin's secret police chief, Beriya, progress was hampered by the backward state of Soviet industry. While Korolev could produce working missiles from the NII-88 facility, it was asking too much too soon of Factory 586 at Dnepropetrovsk, which had been assigned the manufacturing responsibility. Factory 586 was not able to complete it first missile until November 1952. As Gröttrup had estimated in 1946, Russia was fifteen years behind Germany, but that gap would shrink rapidly in the years to come.

The R-1 was accepted into the Soviet arsenal in November 1950. With respect to its military capabilities, it was as awkward to launch in combat conditions as the American Corporal missile. It required twenty vehicles and used the super cold LOX. About six hours were needed to prepare for launch, and perhaps worst of all, it used the Soviet soldier's favorite beverage, ethyl alcohol, as its fuel. With the availability of the R1, the Soviets had beaten the U.S. to an initial deployment of a short-range ballistic missile by about one year, although the Corporal had a nuclear capability. The R-1 was designated SS-1 by the U.S. military and its NATO allies. There were 184 R-1 launches recorded between

September 1948 and 1964.

While Stalin was initially pacified by the availability of the V-2 copy, it was obvious that this new weapon of war was still in its infancy and of limited value. Accuracy, reliability and greater mobility were needed to make it effective. Korolev, who had never been enthusiastic about the R-1, was soon advocating a more advanced design, the R-2.

R-2: Expanding Technology

Korolev met with Stalin for the first time in the Kremlin in April of 1948. Stalin was a strong supporter of the development of German technology, especially long-range rockets that might one-day span the distance to America. He seemed to understand that the technology as it currently existed was only a stepping-stone to greater things to come. Although Stalin was reluctant to approve the follow-on R-2, he did want continued research and technology testing to move forward at the fastest possible pace.

Designed by Korolev over the two-year period of 1947-1948, the R-2 had twice the range of the R-1. The length increased by about 8 feet while the empty weight grew by almost 1,000 pounds. Several major innovations including integral fuel tanks, and a separable warhead provided for a considerable savings in weight of the rocket body. Improved guidance was critical, as the wide dispersion of V-2 hits on London had proved. A quicker and more efficient means of preparing the missile for launch was also a prime consideration.

Moving forward with work to improve the V-2 engine (begun by the German engineers at Mittelwerk) Glushko replaced the 75% ethyl alcohol/water mixture with 96% methyl alcohol. The change to a lethal liquid when ingested also reduced the tendency of the launch crew to drink the rocket fuel. Along with several other relatively minor changes, he was able to produce the RD-101: an engine with 74,000 pounds of thrust. The increased thrust allowed the rocket to absorb the 8-foot stretch, resulting in a lift-off mass almost 50% greater than the R-1. With a weight of 44,000 pounds, a length of 58 feet and diameter of 5.4 feet, the missile retained its family resemblance to the V-2. The rocket had a range of 350 miles with a 1,000 lb. warhead.

The German team under Gröttrup had proposed a competing missile designated the G-1, which had several interesting and innovative ideas. The German design avoided using the hydrogen peroxide steam generator by diverting gas from the combustion chamber to power the propellant turbo pumps. It also provided improved targeting precision, a shorter launch preparation time, and twice the range of the V-2.

However, Korolev and others in the design bureaus resisted any effort to allow the Germans to proceed. While the Soviets would examine and emulate good ideas, they would not allow the Germans to take a leading role in Soviet rocket development. It was important to insure that emerging technology was *Soviet* from top to bottom. Korolev was not about to be upstaged by his German contractors. To achieve this without revealing their intent, Gröttrup's team was requested to present detailed development plans for the G-1. Constant requests for further clarification of relatively minor points ultimately made it obvious to the Germans that they were not to play a major role in Soviet rocketry. By May 1948, there were only a few Germans at the main institute offices, and pay rates began to shrink.

This philosophy was in direct contrast to how the Germans were viewed in the melting pot of America. Within a few years after their emigration to America, most of the Germans could see that their integration into the mainstream of the various U.S. missile projects almost insured that they would be allowed to stay, and that citizenship would also be available.

When a State Commission evaluated the G-1 and R-2 in December 1948, the G-1 was found to be a superior design. Korolev fought the decision, which was finally reversed after some changes were made to the R-2. The R-2 (NATO designation SS-2) had a high explosive warhead, but a "Geran" warhead was developed that dispersed a deadly 'radioactive rain' in a wide area around the target. In combat use, a crew of 11 and a period of six hours were required for launch preparations. Although the rocket did not use storable propellants, after fueling it could be held in launch condition for 24 hours before it had to be de-fueled and recycled.

The first full-scale launch of an R-2 took place on October 21, 1950, and ended in failure. The second attempt on October 26 was a partial success. All twelve missiles failed to achieve their primary objectives, with problems occurring in the engine and guidance systems. Subsequent improvements finally resulted in a usable weapon, and the R-2 was accepted by the military in November 1951.

As with the R-1, several factors precluded rapid development and production of the R-2 such as the lack of raw materials and the relatively poor state of supporting technologies (such as metallurgy) in the Soviet Union. The same factors had impeded the Soviet copy of the American B-29. The first production rocket was delivered in June 1953 (just six months after the first Factory 586-produced R-1). The R-2A version was primarily used for scientific work, and reached altitudes of up to 125 miles. Korolev considered the R-2 for manned suborbital flights in 1956, but the plan was never approved.

R-3: Beyond the State-of-the-Art

The R-3 was a part of the comprehensive plan, authorized in April 1947, to develop a family of missiles that included the R-1 and R-2. The R-3 was to provide the ability to reach virtually any point in Europe from deep within the Soviet Union. The basic specifications called for a range of 2,000 miles, and for the first time, the ability to deliver an early model atomic bomb that weighed over three tons.

Several alternative configurations were evaluated including a winged rocket similar to the German A-9/A-10; a two-stage, rocket-boosted, Mach 3 ramjet cruise missile; and a more conventional large, single-stage ballistic missile that could be grown into the ICBM as a follow-on. As the technologies for the first two alternatives were still beyond the immediate capabilities of Soviet science, the third and more conventional approach was chosen.

A very impressive missile was designed that had an initial thrust of 300,000 pounds (more than five times the V-2, which was still considered a big rocket) and a gross weight of 150,000 pounds. The missile would stand about 100 feet tall and have a diameter of about nine feet. The studies accomplished in support of the R-3 were on the leading edge of rocket technology. These included the theoretical limitations of the mass ratio of single-stage rockets as well as optimization of multiple stages. Lightweight aluminum structure, integral propellant tanks, separable warhead, and the elimination of aerodynamic fins would provide considerable weight savings, but the latter would require the development of the gimbaled (movable) engine for control.

As the design advanced, it was obvious that it was too big a step for the Korolev team. The mass ratio of the rocket and the Specific Impulse of the engine, would have to improve dramatically. It was determined that this was not possible with the state-of-the art.

The German team under Gröttrup was tasked to submit a competing design, the G-4, which had even more impressive numbers. The R-3 proposal was completed in June of 1949 and was not well received. While the German design was again deemed superior to the R-3, it was felt that, with so much new technology being required, neither could be produced without another intermediate step. It was decided to proceed with the R-3A, using less ambitious specifications, with the anticipation of the availability of lighter weight nuclear warheads.

As the design continued to evolve through the early 1950's, the Germans were frequently queried for specific information on selected technical problems. No feedback was ever provided, so the Germans had virtually no knowledge of how their answers might be integrated into any design or engineering studies. The R-3A itself was cancelled in June of 1951 as the Soviets moved on to the R-5 and R-11.

R-5 and R-11: Building a Creditable Capability

The R-5 (SS-3 Shyster) was designed for delivery of radiological weapons, but this version was not fielded in deference to the nuclear-capable R-5M that could send a 3,100 lb. warhead over a range of 750 miles, doubling the performance of the R-2. Significant improvement in guidance provided for a dispersion of about 3 miles from the target, and the R-5 could be prepared for launch in two hours. It was the first independent design to emerge from the Soviet technologists and went from design to first firing in less than two years.

Liftoff thrust of the R-5 was 97,000 lbs., using the RD-103M that had a specific impulse of 244 sec. [in a vacuum] and a firing duration of 112 seconds. The 60,000 lb. 69-foot long rocket was the last

Soviet design to use the V-2 diameter of 5.4 feet and alcohol as a fuel.

The first Launch in March 1953 was a failure, but six of the remaining seven, fired over a period of two months, were termed successful. The R-5 was accepted by the military in 1955, although it was deployed in small numbers. It was used as a sub-orbital carrier of scientific instruments through the 1960's.

The short-range R-11 is notable as being the first Russian ballistic missile to use storable propellants as pioneered in the German Wasserfall antiaircraft (SAM) rocket. The Americans had extensive experience with the RFNA and Aniline combination that powered the WAC Corporal, Corporal, and Aerobee designs. The Russian military was anxious to get storable propellants, as the LOX/Alcohol combination was difficult to make and keep in field operations. The Russian Navy was even more desirous of the storable concept for use in submarine launched missiles. Korolev was not motivated to work with storable propellants, as they did not provide the high energy levels he would need for rockets capable of orbital flight. The design of the R-11 was eventually assumed by Viktor Petrovich Makeyev.

The R-11 adapted the rocket engine used in the Russian version of the Wasserfall, called the V-300 / R-101, that used kerosene as a fuel. Work began in October 1951 on the 160-mile range missile with the first flight in April 1953. Significant problems were encountered with handling and leakage of the propellants, and with engine reliability. The R-11 was redesigned to include a more reliable and higher performance propellant combination. About 2,500 R-11s were produced in the Soviet Union where it was also used as a sounding rocket. A refined model, the R-17, became an export version more infamously known as the Scud-B.

The work done by the Soviets in the late 1940s closely paralleled the effort of the United States, at least as far as paper studies on cruise missiles and ballistic missiles. By the early 1950's the Soviets had achieved remarkable success despite their lagging technology base and poor supporting infrastructure. They had managed to compete successfully with their German counter-parts (albeit a second string team). They had initiated advanced studies not only towards the ICBM but also for instrumented satellites and manned sub-orbital flight. As the Cold War grew hotter in the mid 1950's, the pace of Russian technology continued to quicken. While the Americans were concerned about Soviet developments, the Iron Curtain precluded any peeking into the progress made by these "backward barbarians."

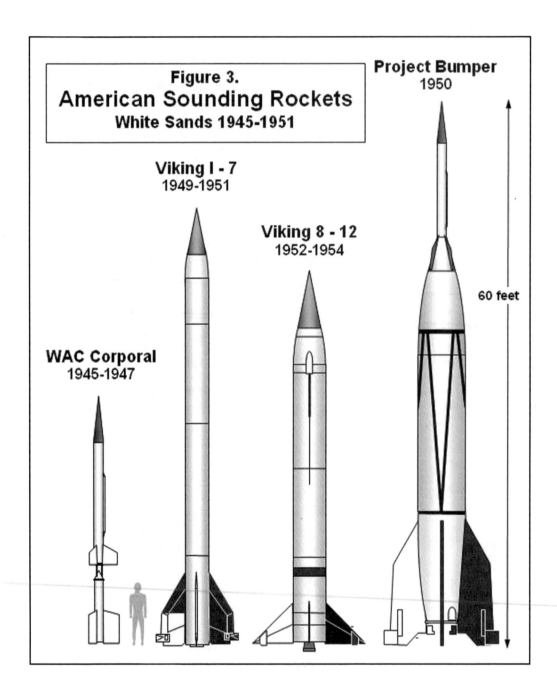

Figure 3.
American Sounding Rockets
White Sands 1945-1951

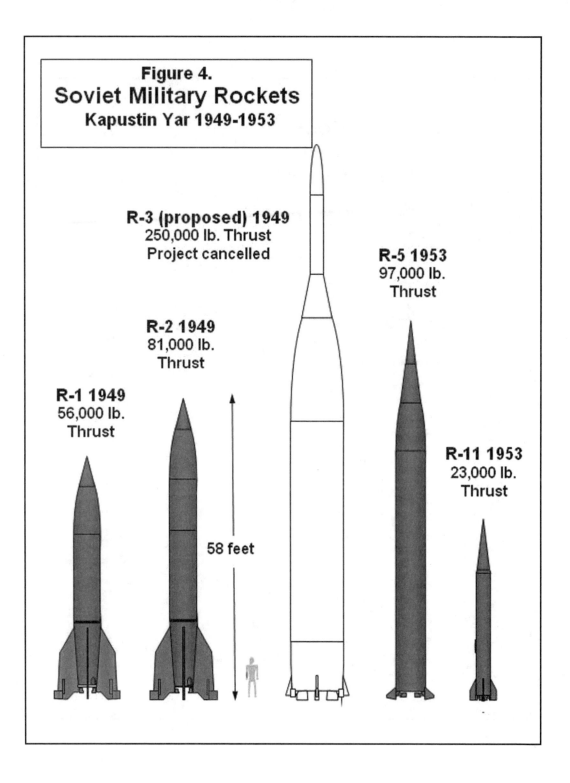

Figure 4.
Soviet Military Rockets
Kapustin Yar 1949-1953

R-3 (proposed) 1949
250,000 lb. Thrust
Project cancelled

R-5 1953
97,000 lb.
Thrust

R-2 1949
81,000 lb.
Thrust

R-1 1949
56,000 lb.
Thrust

R-11 1953
23,000 lb.
Thrust

58 feet

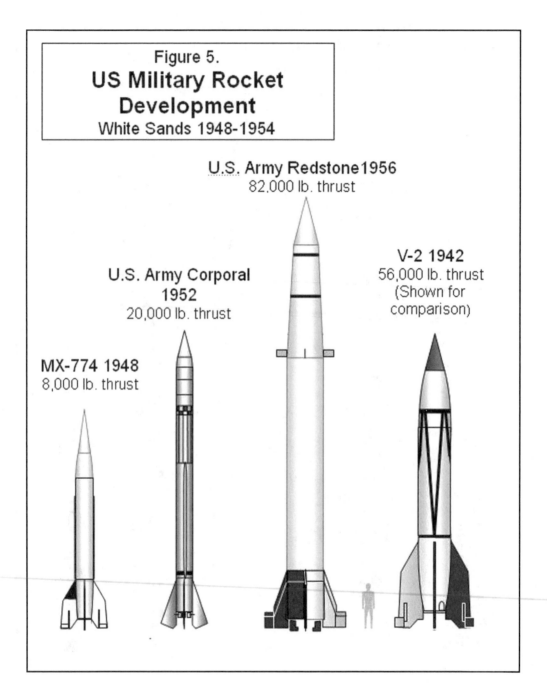

Figure 5.
US Military Rocket Development
White Sands 1948-1954

U.S. Army Redstone 1956
82,000 lb. thrust

V-2 1942
56,000 lb. thrust
(Shown for comparison)

U.S. Army Corporal
1952
20,000 lb. thrust

MX-774 1948
8,000 lb. thrust

Fritz von Opel: Rak1

While rocket propulsion was envisioned as a "Method of Reaching Extreme Altitudes" (as defined in Dr. Robert H. Goddard's 1919 publication) it was also seen as a means of propelling aircraft. During the 1920's, aviation was experiencing exciting growth after being spurred on by Charles Lindbergh's dramatic flight to Paris. Germany was again in the forefront of this activity. Fritz von Opel (heir to the Opel Car Company) was the first to successfully combine the airplane and the rocket engine while working with Max Valier (of the VfR) and Friedrich Sander, a pyrotechnics specialist.

In 1928, von Opel purchased a tailless sailplane from Alexander Lippisch, a well-known builder of gliders in Germany. To the rear of its short, canoe-like fuselage, he attached two 44-pound thrust Sanders solid-fuel rocket motors, creating the world's first rocket plane. Fritz Stamer, one of Lippisch's test pilots, was selected to fly it.

With a wingspan of 40 feet and a length of 14 feet, the aircraft had canard surfaces, and rudders mounted outboard on a straight wing. As was typical for launching gliders, an elastic rope was used to sling the plane into the air off the side of a hill. The rockets were then ignited sequentially by the pilot, and an automatic counterweight system adjusted the aircraft's center of gravity as the rocket fuel was consumed.

The first test on June 11, 1928, was a failure when rockets failed to provide the specified thrust. A second launch later that day was successful. After being slung into the air, Stamer reportedly flew a complete circle that covered almost a mile. On the third flight of the day, when one of the rockets exploded, Stamer quickly landed the airplane, which was then consumed by fire.

Nevertheless, von Opel was enthused and had Julius Hatry build a custom rocket plane that had a wingspan of 36 feet and length of 16 feet. The craft had a conventional high wing with twin rudders mounted on booms that assured that the empennage would be clear of the rocket exhaust.

Sanders installed 16 of his 50-pound thrust black powder rockets on Rak 1, as Opel had named the craft. The first flight on Sept. 30, 1929, was made in the presence of a large group of spectators near Frankfurt. With von Opel as the pilot, the rocket plane made a successful flight of almost two miles in 75 seconds, reaching an estimated top speed of 90 mph. Rak 1 made a hard landing and was destroyed. While von Opel's experiments did not advance the basic technology, they were highly publicized and resulted in much speculation on the future of rocket-propelled aircraft. The concept of using simple and inexpensive solid-fuel rockets to provide a jet-assisted takeoff for heavy aircraft was quickly applied. The first test occurring in August 1929, when a set of solid rockets attached to a Junkers Ju-33 seaplane allowed it to get airborne after an exceptionally short take-off run.

Alexander Lippisch: Messerschmitt Me-163

While the von Opel experiments were essentially a dead-end, they did influence Lippisch to continue work on a much more advanced concept. Lippisch's interest in aviation began when he witnessed a flight by Orville Wright in Germany in September 1909 at the age of 14. His motivation to create a tailless aircraft began when, in the early 1920's, a friend sent him a tropical plant seed that could glide long distances in the air currents. This seed had an arrow shaped wing, and Lippisch based the design of his tailless aircraft on this example from nature. Without the added weight and drag of a conventional fuselage and horizontal tail, a "flying wing" would perform much more efficiently.

Lippisch was not the first to successfully demonstrate a flying wing as the D-1 Burgess-Dunne seaplane tested by the U.S. Navy before World War I was the first truly successful example of a tailless airplane. He effectively addressed the problems of controlling the craft having studied the work of Dunne and the problems of inherent instability. Lippisch recognized the need to use the wing itself to enclose all of the engines, fuel, and passengers and was drawn to the delta shape as providing more internal room.

Lippisch's first powered delta wing, the DFS 39, flew in 1931 using a traditional gasoline engine housed in a small fuselage along with the pilot, and later he built an improved version called the DFS 40. His innovative work led to his being named the director of the Aviation Research Institute in

Vienna, Austria. In 1937 Lippisch was informed that the German Reich Air Ministry (Germany had recently annexed Austria) wanted a more advanced design with a slightly changed fuselage to allow the installation of a special power plant. He began design on the DFS 194, a single-seater, similar in layout to the DFS 40, with a larger vertical fin and rudder. In 1939 Lippisch's team was placed under the Messerschmitt Aircraft Company, and work started on converting it to rocket power, with much of the effort being performed at Peenemünde in October of that year.

Hellmuth Walter was a German engineer who, during the 1920s, recognized the limitations of the internal combustion engine. He examined a variety of possible propellants for a reaction motor (rocket) and determined that hydrogen peroxide in concentrated amounts, when used with an appropriate catalyst, decomposed rapidly and could produce high pressures and heat that would result in reactive forces. The von Braun team in Peenumunde used the process to power the turbo-pump for the propellants in the V-2. Further experiments into hydrogen peroxide-based engines had produced light weight and simple rockets that were suited for aircraft.

In the late 1930s, Wernher von Braun's rocket team in Peenemünde had installed liquid-fuelled rockets in a Heinkel He 72 and an He 112, which retained their piston engines and simply tested the rocket in flight. A new design, the He 176, was built around one of the new Walter engines and was the first aircraft propelled solely by a liquid-fueled rocket during a 50 second flight on July 20, 1939. But its performance was poor, thus the interest by Reich Air Ministry in Lippisch's DFS 40 creation.

An outgrowth of the DSF-40, the DFS-194, was flown as an un-powered glider and then with an 882-pound thrust Walter rocket motor at Peenemünde-West in August 1940, with test pilot Heini Dittmar at the controls. It achieved a maximum speed of 342 mph in level flight, and the success of the flight test program led directly to the creation of the Me-163, a rocket powered interceptor.

The first prototype Me-163 was tested as a glider and towed aloft behind a Messerschmitt Bf 110; its flying qualities were good and progress continued rapidly. In the summer of 1941, powered flight tests began with the 1,650 lb. thrust Walter HWK R11-203b rocket motor, and extraordinary performance was achieved with speeds of up to 550 mph. Because the Me-163 carried a rather small quantity of fuel, it was difficult to explore the full performance envelope. To avoid using fuel for the basic take-off and climb to altitude, one Me-163 piloted by Dittmar was towed aloft as a glider, and then the engine was ignited. A speed of 623 mph was attained before the aircraft encountered stability problems, most likely the result of shock wave compressibility effects as the aircraft approached the speed of sound. Dittmar was able to land successfully, and some changes were made to the design of the wing.

Of greater danger to the pilot was the instability of the fuel, which was a mixture of 80 per cent hydrogen peroxide with oxyquinoline of phosphate (called T-Stoff) and 30 per cent hydrazine hydrate solution in methanol (called C-Stoff). These chemicals were highly toxic and corrosive and had to be carefully handled. The engine had enough propellant for only 7 minutes and 30 seconds of firing. This typically limited the range to little more than 20 miles.

The aircraft was small, having a length of 19 feet and a wingspan of 31 feet. It weighed 9,500 pounds at gross and 4,200 pounds empty. To keep the Me-163 as light as possible and to achieve maximum performance from its small fuel supply, the landing gear was eliminated. Take-off was made using a two-wheel dolly jettisoned by the pilot at a height of about 20 feet, and a steep climb was initiated which took only four minutes to reach 40,000 ft. The landing used a retractable skid beneath the forward fuselage and a small tail wheel.

The Me-163 was designed as a point defense interceptor to shoot down bombers. On reaching 40,000 feet, the aircraft would accelerate to 500 mph within seconds, at which time its fuel supply would typically be exhausted. If the intercept was made close enough to the Me-163's base, it was possible to turn the engine off before the propellants were completely consumed. With the advantage of altitude and high speed, it would then dive un-powered into the bomber formations. Using two 20mm cannons, the attacks were usually made head-on (later versions had two 30 mm). If fuel remained, the engine could be re-ignited to use the last few seconds of power to climb for another attack or to evade enemy fighters. Two minutes had to elapse between shutting down and relighting the engine.

The Me-163 interceptor was used in combat for the first time on July 28, 1944, when five attacked a formation of Boeing B-17s, but they failed to down any of the bombers. It is believed that, throughout the war, only nine bombers ever fell to the guns of the Me-163, while 14 Me-163s were lost in combat, mostly due to accidents, as landing a high speed glider was a difficult task.

Production ended in February 1945 after 400 were built. Recognizing the impractical and hazardous nature of the landing gear, the designers gave the Me-163D a retractable, tricycle landing gear and a larger engine, but it was never put into production. A license-built version, the Mitsubishi Ki-200 (J8M1), was to be built in Japan. It flew in July 1945 but was destroyed when the motor failed.

Vertical Launch Natter

The Ba.349 Natter (Viper) was built by Dr. Erich Bachem of the Fiesler company to a different specification than the Me 163 in that it was launched vertically through a launch tower using four solid-fuel boosters that provided power for about ten seconds. The booster then dropped off, and the 3,800 lb.-thrust Walter 109-509A-2 liquid-fueled rocket motor provided thrust for 70 seconds. With a length of only 20 feet and a wing spread of 12 feet, the small, 4,850-pound aircraft could reach a speed of almost 500 mph.

Able to climb to 35,000 feet by the time the fuel was exhausted, this point defense surface-to-air, manned missile was guided initially by radio control to the intercept point where the pilot would take over and attack the bomber formations. The armament was twenty-four 2.87-inch diameter unguided solid-fuel rockets in the nose. Covered with a plastic fairing, all 24 were to be fired in a single salvo at a bomber. The pilot would then dive away and direct the craft back towards a recovery area. The nose would separate, allowing the pilot to parachute out while the reminder of the craft came down by parachute so that it could be salvaged for reuse. As ten rocket planes were being prepared for combat in early 1945, the launch area was overrun by the allies.

Northrop MX-324 Rocket Wing

In America the first manned flight of an aircraft propelled by rocket thrust alone was in California on Aug. 23, 1941. Using a small general aviation airplane (the Ercoupe) whose propeller had been removed, 12 small solid-fuel rockets had been installed under the wing. The plane was first towed by a truck to a speed of about 25 mph where the pilot, Army Captain Homer Boushey, then released the tow rope and fired the rockets. It was a modest beginning as the plane climbed to 20 feet, and then landed straight ahead on the runway. The Ercoupe had previously demonstrated a successful rocket-assisted takeoff, with its piston power plant using with three rockets attached beneath each wing.

In September 1942 Jack Northrop, who had recently left Douglas Aircraft to start his own company, began a feasibility study for a rocket-powered interceptor. His radical flying wing design was interesting enough for the U.S. Army Air Service to issue him a contract for three gliders, two of which were designated MX-334, and a powered version, the MX-324.

The MX-334 was designed to test the control and stability characteristics of the flying wing. On a low budget, these were constructed of a metal tubing and plywood. The pilot lay on his stomach, allowing a true all-wing aircraft, with no significantly protruding cockpit and permitting the pilot to withstand higher g-forces during maneuvering than the traditional sitting position. Although initially designed with no vertical surfaces, it was later determined that a dorsal fin was needed at higher speeds for directional stability.

The first flight tests of the MX-334 began on October 2, 1943. After a series of flights determined the handling characteristics were acceptable, it was time to move on to the rocket-powered MX-324. The Aerojet Corporation XCAL-200 liquid-fuel rocket engine used monoethylaniline and red fuming nitric acid and produced a mere 200 pounds of thrust. The proof-of-concept aircraft had provisions for only three minutes worth of fuel at which time it would glide back for a landing.

Test pilot Harry Crosby flew the first American rocket powered aircraft on the morning of July 4, 1944. The MX-324 was towed to 8,000 feet by a P-38, and the Aerojet motor was ignited. The flight of the small, 32-foot-wing-span craft lasted just four minutes and ended safely. While the project continued and ultimately produced the Northrop XP-79, the inadequacy of available rocket engines led to the installation of two Westinghouse J-30 turbojet engines. American rocket engine technology had not been able to produce a truly flight worthy product that could compare with the German Me-163, which had flown three years earlier and was in the process of being introduced into combat. The XP-79 had a short career because the only example became uncontrollable 15 minutes into its maiden flight on September 12, 1945, and the pilot, who had bailed out, was struck by the rotating aircraft and killed.

The Bell X-1: First Supersonic Manned Aircraft

General "Hap" Arnold, who headed the U.S. Army's air force during World War II, had essentially stated in 1943 that the war would be won with conventional and existing aircraft such as the P-51 and B-17, both of which had flown before the U.S. entry into the conflict. Perhaps this statement was more for public consumption, as Arnold had been blindsided by the advances in jet engine technology made by the British. However, while more advanced aircraft and concepts (including rocket and jet aircraft) continued to be developed by the Americans, these technologies were initially several years behind the British and Germans.

As for performance, some existing aircraft, such as the Lockheed P-38, were capable of approaching the speed of sound (763 mph at sea level) in a dive, but unknown forces were at work that would cause an aircraft to become uncontrollable at those speeds. Stick forces rose dramatically in the transonic region, and Lockheed eventually fitted a dive flap on the P-38 that could be extended at high speed to slow the plane so that the pilot could regain control. The center of lift appeared to move aft along the camber of the wing causing the nose to pitch down. What was known about these strange affects was that a shock wave was apparently building up on portions of the airfoil and upsetting its aerodynamic qualities and controllability—the phenomena was known as "compressibility." Even though the aircraft might be flying at Mach .8 (eight-tenths the speed of sound), the airflow over the curved surface of the wing was accelerated by the *Bernoulli effect* and could be as high as Mach 1.2.

John Griffith, who would later pilot the Douglass Skyrocket, noted a flight he had in New Guinea during the WWII: *"I got a P-40 up to 32,000 feet and came straight down, and I first experienced a stick that felt like it was cast in two feet of concrete. It just didn't move back until you get a little denser air, and the drag increases, and the temperature goes up a little bit, and the Mach number comes back. If you throttle back its easy enough to pull the airplane out."*

Because the speed of sound decreases with altitude, the term "Mach" is used to define that speed regardless of the flight condition. The term was derived from name of the Austrian physicist, Ernst Mach, who had theorized the effects of high speed on an object almost 100 years earlier.

Piston engine aircraft lacked the power to be capable of supersonic flight. While jet engines of that era were able to exceed the speed of piston powered aircraft by a considerable margin, as aircraft approached Mach 1 (specifically beyond Mach .8), the parasitic and induced drag increased dramatically. In an effort to explore this unknown region that had already killed several pilots, the U.S. Army and Navy, in collaboration with the National Advisory Committee for Aeronautics (NACA), began a program to explore high-speed flight. The rocket engine was selected as the power source for a series of experimental aircraft that would probe the mysterious region that became known as the "sound barrier."

The British, French, and Soviets also had programs to investigate the problems of supersonic flight. The British Miles M.52 project, initiated in 1943, showed great promise but was discontinued after the war for economic reasons. The myth has grown that the design information was transferred to Bell Aircraft and aided in the engineering of the X-1. But at the time the M.52 was cancelled in February 1946, the X-1 had already flown. The de Havilland D.H. 108 Swallow, which had a close resemblance to the Me-163, had set the world speed record of 605 mph in April of 1946. In September 1946, while preparing for a new speed record, Geoffrey de Havilland Jr. was killed when his D.H.108, flying at about Mach .875, experienced several longitudinal excursions and broke up in the air. This and other highly publicized accidents had given the sonic barrier a reputation of being a deadly, invincible wall. The director of British Scientific Research for Air had already declared that probing the sonic barrier was too risky for piloted operations and indicated a move to unmanned instrumented radio controlled models.

The Soviets had developed an experimental rocket plane, the Samolet 346, using one of German Lippisch's DFS designs. It was carried aloft under the wing of one of the American B-29s (Ramp Tramp) that had landed in the Soviet Union during WWII. Piloted by Wolfgang Ziese, it had attained a speed of Mach .93 by May of 1947.

None of the European projects, however, approached the sound barrier with the same degree of engineering innovation and planning as did the coordinated effort of NACA and its director, Walter C. Williams, who was then head of the High-Speed Flight Research Station (HSFRS) at Muroc,

California. Williams, who had come to NACA in 1939, would play an ever expanding role in America's quest for supersonic flight development and its future space program.

Wind tunnel technology of the era suffered from the inability to understand supersonic flow and to adequately recreate it in a laboratory environment. So it was left to building an experimental aircraft to explore the transonic region (Mach .8 to 1.2)—the first of these was designated as the XS-1. The *X* prefix defined that its purpose was solely experimental and that it was not intended to become an operational aircraft, and the *S* represented its goal of supersonic flight (the S was soon dropped). The Bell

The Bell X-1 in flight The world's first supersonic aircraft.

Aircraft Company of Buffalo, N.Y., received a contract in 1944 to build three aircraft, each with slightly different wing airfoils, under the technical direction of NACA and funded by the Army Air Force. The fuselage was to present the most streamlined shape possible, to reduce the amount of air resistance, and was patterned after a .50 caliber bullet—known to be stable at speeds up to 2,000 mph. Even the cockpit was blended into the fuselage to reduce drag, although that also reduced visibility and the ability to exit the aircraft quickly, since ejection seats had not yet been perfected.

The engine chosen to power the X-1 was really the only engine that was suitable, the Reaction Motors Inc. XLR-11. This small New Jersey company was founded by several members of the American Rocket Society in the late 1930's. The XLR-11 was composed of four combustion chambers that each produced 1,500 pounds of thrust. This was ideal for the X-1 in that the flight test profile could be ramped up by firing any combination of chambers to produce incremental thrust levels (throttleable rocket engines were still years away from being practical). The engine used ethyl alcohol and liquid oxygen (LOX) as propellants, and the X-1 carried enough fuel for 150 seconds of powered flight. Best of all, the XLR-11 weighed only 200 pounds.

Because rocket motors devoured their propellant rapidly to generate high thrust levels, it was decided from the beginning that the aircraft would be air launched from the largest plane then available—the B-29. This procedure would allow more time at altitude for the test flight. The first of a series of ten un-powered X-1 flights took place on January 19, 1946, at Pinecastle Field (now McCoy AFB), Florida, with Bell test pilot Jack Woolams at the controls. However, Woolams was killed, preparing for the Cleveland Air Races, before the powered test segment of the program could begin. Chalmers "Slick" Goodlin, then only 23 years old, became the primary Bell test pilot.

For the powered tests, the program shifted to Muroc Army Air Field on the west shore of Roger's Dry Lake in the Mojave Desert of southern California. The word Muroc is the name of one of the original settlers—spelled backwards. The facility was later renamed Edwards Air Force Base in honor of Glen Edwards, a B-49 pilot who lost his life testing that advanced aircraft. Carrying almost 500 pounds of test equipment to monitor a variety of test points for aerodynamic pressures and temperatures, the X-1's first powered flight was achieved on December 9, 1946.

The X-1 had a conventional tail with elevators for pitch control. However, at high speeds a shockwave formed on the tail surfaces near the elevator hinge and rendered them ineffective. But the X-1 also had a trim system that adjusted the angle of incidence of the horizontal stabilizer, which could also control pitch. It was proposed that this system be used during flight in the transonic region. The technique worked so well that an "all flying" horizontal stabilizer/elevator was secretly incorporated into the next generation of American fighter planes, beginning with the F-86. (The "all-moving" tail had long been used in a variety of aircraft, including the XP-42, but not for its implications in addressing the compressibility problems.)

One of the eighteen pilots who flew the three X-1s from 1946 until they were retired in 1951, Goodlin was an American who had joined the Royal Canadian Air Force in 1941. Following Peal Harbor, he transferred to the U.S. Navy as a test pilot before he had a chance to see action. He then joined Bell Aircraft in 1944, testing the first American jet fighter, the P-59A, and the follow-on XP-83. Goodlin had carefully put the X-1 through its early test program, achieving Mach .82 and demonstrating its structural integrity by pulling 8.7 Gs. Although the initial data acquired during the first flights had some believing that the X-1 would never break the sound barrier, the Army/NACA test team was convinced that it could. Goodlin's memoirs include an interesting report by ground observers of a double thunder-clap during one of his high G flights. Was it possible that Goodlin was the first to generate the phenomena known as the "sonic boom?" He maintained that this was the case until his death, in 2005, at the age of 82.

By the summer of 1947, the preliminary testing had been completed and negotiations between Bell and the Army were undertaken regarding the cost of the follow-on contract to attempt to go supersonic. Because of the hazards involved in testing the X-1, Goodlin had received a $10,000 bonus for the initial sub-sonic test flights (a portion of which he gave to Woolam's widow). He was also to receive $150,000 for his participation in the supersonic portion of the program (what has been reported as a standard industry contract). However, the Army, in an effort to reduce costs, decided that they would take over the next stage of the X-1 tests. As Goodlin's contract with Bell had been a "handshake deal," Bell decided that they were under no legal obligation to honor their commitment, and an unhappy Goodlin resigned. With the notoriety that Chuck Yeager received as a result of Tom Wolfe's book *The Right Stuff*, Goodlin has often been characterized as having lost out on the fame of being the first to break the sound barrier to greed. A good natured man, Goodlin attempted to counter this perception for the reminder of his life.

The Air Force (which became a separate branch of the military that year) assumed responsibility for the X-1 flight test program on July 27, 1947, and Captain Charles "Chuck" Yeager became the prime test pilot. The drop sequence required the B-29 to climb to 25,000 feet and pitch over to a shallow dive, to achieve an indicated airspeed of 250 mph. The airspeed had to be that high because the fully loaded X-1 stalled at 240 mph.

Tuesday, October 14, 1947, was Yeager's thirteenth flight in the X-1; the last nine had been powered. During the preceding flights, he had carefully evaluated the handling characteristics, while the engineers recorded the data from the sensors mounted on the aircraft. The program had steadily progressed in small increments of 10-15 mph toward the objective of Mach 1. That morning, Yeager's ribs hurt, due to a fall from a horse the previous Saturday. So much so that Jack Ridley, the engineer who was Yeager's alter-ego, had fashioned a small piece of broom handle so that Yeager could get the needed leverage to move the door latch into place.

As the B-29 climbed through 12,000 feet, Yeager went aft to climb into the X-1. He wore only a cloth flight suit and the trademark of a military pilot—a brown leather jacket. He recalled: *"Climbing down into the X-1 was never my favorite moment...The wind blast from the four bomber prop engines was deafening, and the wind chill was way below zero...I had to bounce on the ladder to get it going, and be lowered into the slipstream... that bitch of a wind took your breath away and chilled you to the bone."* Yeager admits to being nervous prior to each flight: *"...fear is churning around inside you whether you think of it consciously or not."* [Perhaps it was more the fear of embarrassing himself, by doing something stupid in front of his peers, than the prospect of killing himself].

The flight plan called for Yeager to move to the next plateau—Mach .97. He writes: *"That moving tail really bolstered my moral and I wanted to get to that sound barrier."* The drop speed was a little lower than normal, and Yeager had to reduce the angle of attack slightly to maintain control as he lit the four chambers. *"We climbed at .88 Mach and began to buffet, so I flipped the stabilizer switch and changed the setting two degrees... We smoothed right out."* Yeager's use of the word *we* to describe his oneness with the airplane is another trademark of the test pilot. Leveling out at 42,000 feet, with three chambers firing, he saw the Mach meter move through .96, and the aircraft was smooth and stable. Then the needle began to fluctuate and went off scale—*"We were flying supersonic."* Analysis of the data showed a maximum speed of Mach 1.07 was held for 20.5 seconds—about 700 mph at that altitude. The tracking van on the ground heard the distinct thunderclap—the characteristic sound of an aircraft traveling faster than sound. *"And that was it. I sat up there feeling kinda numb, but elated."*

Yeager took the X-1 to Mach 1.35 (890 mph) less than a month later.

These events however, were not released to the press. The story was suppressed for eight months by the Air Force because of security concerns. The Cold War was heating up, and the progress being made in breaking the sound barrier was considered sensitive information. Unlike the mass media blitz that accompanied the first astronauts, Yeager would have to wait for his "15 minutes of fame." He received the Harmon trophy, in 1954, for his work with the X-1A, but it wasn't until Wolfe's book that his notoriety spread beyond the narrow domain of the aviation enthusiast.

On January 5, 1949, the X-1, with Yeager as pilot, achieved the only ground takeoff of the X-1 program. He reached just over 23,000 feet before the limited propellant was exhausted. This was reportedly done because the international rules that governed speed records required that the aircraft be capable of taking off under its own power.

Subsequent flights established the top speed of the X-1 at 967 mph, and the maximum altitude attained (by Lt. Col. Frank Everest, Jr.) was 71,902 feet. By 1950 the Air Force X-1 number 1 had served its purpose, and it was retired to the Smithsonian Museum. The second X-1 was used by the NACA for high speed flight research, and the third was destroyed during fueling operations, after completing only one un-powered glide flight. The aircraft had a gross weight of 12,300 pounds and an empty weight of 6,800 pounds.

Following the X-1 program, several second generation X-1s, intended to fly at twice the speed of sound (Mach 2), were built. The X-1A was similar to the X-1, except for a turbo-driven fuel pump (instead of using nitrogen pressure), a new cockpit canopy, longer fuselage and increased fuel capacity. The X-1A had its first flight on July 24, 1951. The X-1B was similar to the X-1A, except for a slightly different wing, and was flown by NACA until 1958. The X-1C was cancelled in the mock-up stage, while the X-1D was destroyed in August 1951, after being jettisoned from its B-50 carrier plane, following an explosion. The second of the original X-1 aircraft was fitted with new wings, turbo-driven fuel pumps and a knife-edge windscreen and was re-designated the X-1E. It was flown exclusively by NACA from 1955 until November 1958.

With the ability of aircraft to climb into the stratosphere (above 40,000 feet), where the air pressure was a fraction of that at sea level, it became necessary to protect the pilot should the cockpit pressurization system fail or a bailout become necessary. At pressure altitudes above 60,000 feet, the blood is near its boiling point, and rapid decompression would have a catastrophic effect on the human body. The Army Air Force worked with the David Clark Company to produce the T-1, the first standardized mechanical pressure suit for use against the combined effects of depressurization and G forces. Although uncomfortable when not inflated, it became almost unbearable when it was inflated, and work continued to find better pressure suits.

The Douglas D-558-2 Skyrocket

At the same time that the Army was contracting with Bell for the X-1, the Navy was working with Douglas Aircraft Company, who received a contract to build six experimental aircraft to investigate high-speed flight using a conventional straight wing configuration (similar to the X-1) and an Allison TG-180 (J-35) jet engine—the D-558-1 Skystreak. However, the swept wing concept was advocated by Robert Jones at NACA's Langley research Center. A sweep allows the wings to remain free of the shock wave generated by the nose of the aircraft, and the compressibility effects of supersonic airflow are delayed to higher Mach numbers. In a March 1945 memo, Jones wrote: "*I have recently made a theoretical analysis which indicates that a V-shaped wing traveling point foremost would be less affected by compressibility than other planforms... In fact, if the angle of the V is kept small relative to the Mach angle, the lift and center of pressure remain the same at speeds both above and below the speed of sound.*" His ideas received considerable encouragement when the spoils of war revealed the advances made by the German aircraft industry with respect to the advantages of a swept wing for high speed flight. In August 1945 it was decided to build three of the six single-seat pure research planes, designated the D-558-2, with a much more futuristic shape. With a long pointed nose and short swept back wings and tail, the glossy white plane epitomized the shape of a Skyrocket—which was its name. Although the Skystreak and the Skyrocket aircraft shared the basic D-558 designation, there was little commonality.

The Dash-2, as it was often referred to, originally used a combination of jet and rocket power and had a 35 degree swept wing. The idea was that the jet engine would take the craft up to altitude where the rocket would be used to accelerate into the transonic region. Operating out of Muroc's High-Speed Flight Research Station (HSFRS), the three D-558-2s looked identical and had a length of 42 feet and a wing span of 25 feet. Fully loaded, the Skyrocket weighed almost 16,000 pounds, with an empty weight of 9,500 pounds. Douglas pilot John F. Martin made the first flight at Muroc on February 4, 1948. In June 1949, the number three aircraft first exceeded the speed of sound, using both the jet and rocket units. Gene May, the Douglas test pilot on that flight, noted that as the Mach meter moved past 1.0, *"The flight got glassy smooth; quite the smoothest flying I had ever known."*

The first Skyrocket was powered by a Westinghouse J-34-40 turbojet engine rated at 3,000 pounds of thrust and an LR8-RM-6 rocket engine—the Navy designation for the XLR-11 used in the X-1. This 4-chamber Reaction Motor's engine provided 6,000 pound of thrust. Initially configured only for take-off from a runway, it was soon learned that the J-34 was insufficient to power the heavily loaded Skyrocket off the ground. The technique eventually called for one or two chambers of the LR8 to fire during the take-off roll, to assist in getting airborne (JATO units were also used on occasion).

Following the initial testing, it was realized that the amount of rocket fuel available for the high speed portion of the flight program was not sufficient. When the safety issue of the long takeoff run was added, Hugh Dryden, NACA's Director of Research, recommended removing the J-34 to make room for larger propellant tanks for the rocket engine, and air-launching the D-558-2 in a manner similar to the X-1. The second Skyrocket also began its flight program with the Westinghouse J-34 turbojet, but this was removed in 1950, and it flew only as an air-launched craft with the LR-8 rocket engine. Only the third Skyrocket retained the jet engine. Two of the aircraft were modified to be carried aloft by a Navy P2B (Navy version of the B-29) and launched typically from 30,000 feet. The first air launch occurred in September of 1950 with William Bridgeman at the controls of the number three aircraft.

Test pilot Bridgeman flew the second Dash-2 aircraft to a speed of Mach 1.88 and to an altitude of 79,494 feet (an unofficial world's altitude record at the time) on August 15, 1951. Marine Lt. Col. Marion Carl flew the airplane to a new unofficial altitude record of 83,235 feet on August 21, 1953, and to a maximum speed of Mach 1.728.

As with the other X craft, data on stability and control aided in the design of the first U.S. fighters that could fly faster than sound in level flight—the so-called Century Series that began with the F-100 Super Sabre. These innovations included the movable horizontal stabilizers that were first employed on the X-1 and D-558 series. Unlike the X-1 program, the D-558 series was to have moved to a Dash-3 model that would have been a combat aircraft incorporating the features explored by the Dash-1 and the Dash-2. However, that version would never come to fruition. The three D-558-2 airplanes flew a total of 313 times, with the last flight occurring on December 20, 1956.

D-558-2 number 1 Skyrocket is on display at the Planes of Fame Museum, Chino, California. The number two Skyrocket, the first aircraft to fly Mach II, is at the National Air and Space Museum in Washington, D.C. The number three is displayed on a pedestal at Antelope Valley College, Lancaster, California.

Mach II: Deadly Competition

In 1952 Walt Williams, the head of HSFRS, had requested that NACA director Hugh Dryden approve a flight of the Skyrocket to Mach II. However, NACA had a policy of not deliberately setting speed or altitude records, although they had been setting unofficial records all along the research path. The Dash-2 had already achieved Mach 1.96 on August of 1953.

After test pilot Scott Crossfield sought and received approval from the Navy's Bureau of Aeronautics (the test program's sponsors), the NACA policy was relaxed, and an attempt was scheduled to achieve Mach II. Because the Air Force was also planning a Mach II flight in their X-1A, an air of competition had been generated. Crossfield claims that he simply *"dropped a hint"* to the Navy about how great it would be if they could beat Yeager and the Air Force to Mach II. The Navy knew that the Air Force X-1A was a much more capable plane and that it would exceed whatever speed the Skyrocket was able to reach, but the goal of being the first to Mach II, even if only for a few weeks, stirred the inter-service rivalry.

In addition to adding nozzle extensions to increase the thrust by about 6%, the fuel (alcohol) was chilled so that more could be accommodated in the tank of the Skyrocket. The propellant tank regulators were positioned so that the pilot could adjust them to provide more pressure during the flight, increasing the thrust. The aircraft was carefully waxed to reduce drag, and a flight plan was devised to fly to 72,000 feet and then push over into a slight dive. On November 20, 1953, Crossfield flew to Mach 2.005—1,291 miles per hour. This was the fastest the Skyrocket would ever fly.

Three weeks after Crossfield became the first man to fly twice the speed of sound in the Navy Douglass D-558-2, Chuck Yeager prepared to better that speed record in the X-1A. Yeager had inherited the X-1A test program after the primary Bell pilot, Skip Zeigler, was killed, when the X-2 blew up in the belly of its carrier plane.

On December 12, 1953, the flight began with the B-50 (which had replaced the B-29 as the "Mother Ship") climbing to 32,000 feet—the X-1A, more than 4,000 pounds heavier than the X-1, hanging in the cut-out bomb-bay. Yeager, in his T-1 pressure suit, waited up front with the B-50 flight crew until they passed through 13,000 feet before he entered the tight confines of the X-1A cockpit with the help of the launch crew. With the canopy bolted in place, Yeager spent some time going through the checklist to prepare the aircraft for the drop.

Following a short, 5-second countdown, the X-1A released from the B-50 at 32,000 feet at an indicated airspeed of 210 mph (about 340 mph true airspeed). Yeager lit the first three chambers and climbed to over 70,000 feet before pushing over into level flight and lighting the fourth chamber. The airspeed rose steadily, and easily slipped through the Mach II point, eventually reaching Mach 2.5 (1,650 mph).

Yeager recalled, *"Up there with only a wisp of atmosphere, steering an airplane was like driving on slick ice."* Yeager's memory of the incident in later years differs somewhat from the transcripts of the communications, but the essence is intact. *"The Mach meter showed 2.4 when the nose began to yaw left."* Yeager had been cautioned that the tail might not be able to keep the airplane aligned with the direction of flight at high Mach numbers. *"I fed right rudder, but it had no effect."*

With the fuel exhausted, Yeager encountered lateral instability that caused the plane to flip out of control. His first thought was that he had lost the tail, since his control inputs did not seem to have any affect. Although belted securely, Yeager's head hit the canopy and cracked it—he struggled to regain control. The faceplate on the helmet fogged over as the T-1 suit inflated to counter the loss of cabin pressure, and he had difficulty seeing the instruments and the horizon.

He moved the controls to a position that he hoped would result in a spin—if the tail was still attached. That would ensure the aircraft would decelerate into a slow speed configuration, and normal flight would be recoverable. At 25,000 feet, he perceived that he was in a spin and proceeded to recover from that condition to a normal flight attitude. He reported, *"Down to 25,000 feet over the Tahachapies. I don't know whether I can get back or not."* In reply to a query from the ground he simply said, *"I can't say much more. I gotta save myself."*

Soon, at lower altitudes and with the adrenalin subsiding, his head began to clear. *"Those guys* [Bell engineers] *were right* [warning about controllability at high Mach numbers]. *You won't have to run a structural demonstration on this damned thing. If I had a* [ejection] *seat you wouldn't still see me sittin in here."* Like the X-1, there was no ejection seat. The X-1A had been built to withstand 18 Gs. Yeager had proved their design limits were sufficient. He landed safely but decided that he had come too close to death, and with a several young children at home, he transferred to a fighter squadron in Europe for his next tour of duty.

The flight test environment at Edwards for the experimental rocket planes was every bit as hazardous as those the astronauts would face some ten years into the future—but there was no public acclaim nor headlines that followed each flight. Yeager's speed record of Mach 2.5 stood for the next 3 years.

The Bell X-2: Highly Temperamental

Like the X-1, the X-2 was conceived by Bell Aircraft in 1944, but with a swept wing configuration. Little was known about the characteristics of the swept wing, which appeared to delay and mitigate the effects of supersonic compressibility on the wing. However, as the Second World War ended, and results of German aeronautical research began to filter into the American aviation industry, it was obvious that the swept wing offered a dramatic increase in performance. Bell's Design 37D was proposed as a successor to the X-1 with almost twice its speed.

Although the Air Force initially rejected Design 37D, Bell continued to press for a 40-degree leading-edge sweep and actually built a modified P-63, designated the L-39, with swept wings, to explore the controllability issues. As a result of Bell's efforts and the revelation of German research during WWII, the company received an Air Force contract for two XS-2 swept-wing research aircraft in December 1945.

While the X-2 Starbuster was to explore flight at speeds and altitudes far beyond anything achievable with the first generation X aircraft, little was known of the characteristics of the swept wing. It was also recognized that an aircraft flying at Mach 3 would generate skin temperatures that would weaken traditional aluminum structures. A new material called K-monel (a copper-nickel alloy) was used in constructing the X-2, along with an advanced lightweight, heat-resistant, stainless steel alloy in those areas where significant heat build-up would occur.

The X-2 was sleeker that the X-1 series and had a more powerful rocket engine with a variable thrust rating from 2,500 to 15,000 pounds—two and one-half times the power of the X-1. The new Curtiss-Wright XLR25 two-chamber rocket engine was regeneratively cooled and throttleable: the pilot could vary the thrust being produced by the second chamber. It employed the lightest and most powerful (for its size) turbo-pump system for delivering the propellants into the combustion chamber. Moreover, because it was the most advanced man-rated rocket engine developed at that time, it encountered many significant engineering problems.

The first airframe was completed in late 1950, but the engine was nowhere near being ready for flight test. It would be another 18 month before the first airdrops of the X-2, in June 1952, when its gliding characteristics were evaluated—without the engine.

Because of the speed the X-2 was capable of, it was decided not to install an ejection seat but to use the entire nose as an escape capsule. In an emergency, the nose assembly would be separated from the aircraft, and a stabilizing parachute would be deployed. At a safe altitude and speed, the pilot would then open the canopy and bail out.

Bell test pilot Jean "Skip" Ziegler was at the controls for the first flight and he was not pleased with the control responses, evaluating them as only adequate in his flight critique. Less than adequate were the landing characteristics. To save weight, the X-2 landed on a nose wheel and a central skid that was positioned slightly aft of the center-of-gravity. The wing tips also had small skids. On touchdown, following the first flight, the nose gear collapsed, and a two-month delay followed for repairs.

On October 10, 1952, Ziegler completed a second glide flight, and two days later, Air Force test pilot Capt. Frank Everest successfully made his first glide. The X-2 was then flown back to the Bell factory in New York, under the B-50, for installation of the engine. The XLR25's numerous problems were so complex that thought was given to canceling the program. The follow-on project, the X-15, was in the design stage, and there was some thought that it might actually fly before the X-2, which was then three years behind its projected schedule. However, it was also realized that the X-15 itself would undoubtedly experience delay from its own set of problems, and so the X-2 was continued.

By early 1953 the first flight-worthy engine was delivered and installed, and a series of no-drop flight tests were conducted over Lake Ontario (adjacent to the Bell factory) in March 1953. One of these tests involved the emergency dump system for the X-2's propellants: the volatile liquid oxygen and alcohol combination. On May 12, while testing the propellant dump procedure, the X-2 exploded and fell from the EB-50 into Lake Ontario, killing Skip Ziegler and a B-50 crew member, Frank Wolko. The drop plane was damaged badly but managed to land safely. It was later determined that its main wing spar had been cracked, and the big plane was scrapped.

It was several years before the cause of the explosion was determined. Leather gaskets were used extensively in the X-2's power plant to seal the propellant plumbing joints. These were found to be highly unstable when saturated with liquid oxygen, and any significant shock could cause the gaskets to explode.

During the last half of 1954 and through the first half of 1955, several glide tests resulted in still more damage to the remaining X-2 because of its unstable landing characteristics. The main gear-skid area was increased by 300 percent, and two small skids were installed at mid wingspan, to remedy the landing-instability problem. With those improvements, it was time to move forward to the powered flight tests.

Following several aborted drops, Everest finally completed the first powered X-2 flight on November 18, 1955, igniting only the smaller 2,500-pound-thrust chamber. The maximum speed attained was Mach 0.95. Several more aborted flights preceded a second powered flight that occurred on March 24, 1956, using only the 10,000-pound-thrust chamber. The third powered flight on April 25 used both engines and attained a speed of Mach 1.40 at an altitude of 50,000 feet. However, these flights were not intended to push the X-2 to its limits, only to explore the basic handling with power. Three more flights were completed that expanded the speed envelope to Mach 2.53.

On July 23, 1956, Everest made his final X-2 flight and achieved a speed of Mach 2.87 at 68,000 feet, while gathering data on aerodynamic heating. Capt. Iven C. Kincheloe then flew the X-2 to an altitude of 126,200 feet on Sept. 7. This record stood until the advent of the X-15 program.

Milburn Apt made his first and only flight in the X-2 on Sept. 27, 1956. The X-2 performed to its maximum potential, and Apt flew a nearly faultless flight profile, allowing him to reach a speed of 2,094 mph (Mach 3.196). Apt had been cautioned about the inertial coupling control instabilities that Yeager had experienced in the X-1A, and the flight plan called for him to decelerate before attempting a turn back towards Edwards. Nevertheless, he unexpectedly initiated a sharp turn while still at a high speed. The X-2 began to tumble uncontrollably. Apt separated the escape capsule, but was unable to complete the bail-out and was killed when the cockpit impacted the ground.

With the destruction of the second aircraft, the program came to a sudden end. The X-2 was an example of the problems encountered when pushing the state-of-the art. The X-2 contributed to new construction techniques and advanced materials that were incorporated into the development of subsequent high-speed aircraft such as the XB-70 bomber and the SR-71.

The North American X-15: On the Threshold of Space

Even before the X-2 made its first powered flight, its successor had been in the planning stages. In 1954 NACA issued a requirement for a hypersonic, air-launched, manned, research vehicle with a maximum speed of Mach 6 and capable of attaining altitudes of more than fifty miles—twice the performance of the yet to be flown X-2. North American Aviation Incorporated (NAA) was awarded the contract for this new rocket powered airplane that would provide information on thermal heating, high-speed control and stability, and atmospheric re-entry. It was designated the X-15 (there had been numerous manned and unmanned X craft in the early 1950s that followed the X-1 and X-2 that had been given the intervening X designations).

A single chamber, liquid-propellant engine, the XLR99-RM-2 built by Reaction Motors, powered the X-15. Like the X-2's main chamber, the XLR99 was throttleable, and it used liquid oxygen and anhydrous ammonia as propellants to generate 57,000 lb. of thrust, with a Specific Impulse of 276 seconds. This made the X-15 more powerful than the V-2 rocket and more than 20% heavier at 33,000 pounds (15,000 empty). Because of development problems with the big XLR-99 engine, the early flights were made with two Reaction Motors XLR11-RM-5 engines rated at 8,000 lb. thrust each.

The vertical tail consisted of both an upper dorsal fin and lower ventral fin for added stability during its powered climb to altitude. The dorsal tail also contained air-brake surfaces. The ventral fin was jettisoned for landing, allowing two skids to be extended from the aft fuselage. It was later determined that the ventral fin was not needed, and it was not used after the 70th flight. A conventional nose gear completed the tricycle arrangement. Like the X-2, the plane was incapable of conventional (runway) take-off due, in part, to its unique landing gear.

The ability to climb essentially into the lower fringes of outer space, where there is no air, required that the conventional aerodynamic controls be augmented by 12 hydrogen peroxide thrusters, 4 in the wingtips and 8 in the nose. This was the first use of thrusters to control an aircraft's attitude. (They had been tested on the X-1B, however.)

With a length of 52 feet and a wingspan of just 22 feet, it was truly as much rocket as it was airplane. Constructed primarily from titanium and stainless steel, the leading edges of the wings were covered with Inconel X nickel, an alloy that can withstand temperatures up to 1,200 degrees F. Three X-15s were built as a part of Project MX-1226, with the first being completed in September 1958.

All X-15 missions were flown from Edwards Air Force base where the X-15 was dropped from a converted B-52 bomber at an altitude of 45,000 feet and a speed of 500 mph. Unlike the previous manned X-craft, the pilot had to enter the plane while it was on the ground because it hung under the right wing. This meant that the pilot would spend several hours in the confines of the cockpit for a brief ten-minute flight.

Following the drop, the pilot flew a pre-determined flight profile depending on whether high speed or high altitude was the dominant objective. An inertial guidance system, similar to that used in an unmanned missile, aided the pilot in controlling and navigating the rocket plane.

The North American Aviation X-15.

The first glide test of X-15 was on June 8, 1959. NAA engineering test pilot Scott Crossfield put the plane through a series of pitch and bank maneuvers during the all-too-brief, three-minute flight. As he slowed for landing, the X-15 began a series of severe pitch oscillations caused by inadequate rate responses of the flight control system. Crossfield's heavy breathing could be heard on the air/ground communications as he fought to keep the X-15 from stalling, and only his extraordinary skill avoided loss of the plane and the pilot.

The first powered flight with the XLR-11s took place on September 17, 1959, and it was not until the following year that the XLR-99 made its first flight on November 15, 1960. During an early ground static firing of the XLR-99 with Crossfield in the cockpit, one X-15 was blown in half when a pressure regulator and a relief valve failed and pressurized the ammonia tank beyond its structural limit. Despite the dramatic explosion and resulting fire, Crossfield was not injured.

High-speed flights were typically conducted at altitudes of less than 100,000 feet. High altitude flights employed the rocket engine to propel the X-15 into a high angle trajectory. For the X-15 pilot, the thrill began with the drop itself and the ignition of the XLR-99 that produced almost two Gs of acceleration. The brief, 90 seconds of burn time accelerated the craft to Mach 6 and four Gs. This was a busy time for the pilot, as he had to pitch-up and maintain the climb attitude until the designated pushover point. The critical parameters of speed, altitude, and air loads had to be monitored to assure the planned flight was flown correctly.

Following engine burnout, the pilot performed the engine shutdown sequence while maintaining the appropriate pitch and heading parameters using the small thrusters. The craft continued to coast upward for several minutes to its maximum altitude—exposing the pilot to weightlessness. The pilot would then reposition the X-15 for its reentry into the atmosphere.

High altitude flight returns were particularly hazardous because the vehicle had to reenter the denser layers of the atmosphere at the correct angle, or face the possible disintegration of the craft from temperatures and air loads, as was the case with flight number 191, which claimed the life of test pilot Michael Adams. Any significant deviation from the various speed, altitude, or time specifications could affect the quality of the data being recorded. The pilot experienced up to five Gs during the reentry.

When the experimental phase of the flight ended, it was time to locate the runway and position the 15,000-pound glider for landing on Rogers Dry Lake bed. The pilot could not be too low or too high on his approach—there was no power available for corrections. One or more chase planes were vectored to help the X-15 pilot during the critical final descent phase. Those pilots called out headings and altitudes and checked the aircraft from the outside to assure that any residual propellants had been dumped and that there were no anomalies that could pose a hazard.

A typical flight lasted less than 15 minutes and covered nearly 400 miles. The maximum speed achieved by the X-15 was 4,534mph (Mach 6.72), while the highest altitude was 354,200 ft. (67.08 miles). It had taken 44 years to progress from the Wright Brother's first powered flight to Yeager's Mach 1 flight in 1947. Mach 2 was exceeded 6 years later by Crossfield in the D-558-2, and Mel Apt reached Mach 3 in the X-2 only three years after that. However, the next Mach number fell quickly, in 1961, when the X-15 flew through Mach 4 in March, Mach 5 in June, and Mach 6 in November.

The first flight mishap occurred in November 1959, following a small engine fire. Pilot Scott Crossfield was unable to jettison the propellants and made an emergency landing on Rosamond Dry Lake. The X-15 was not designed to land with fuel aboard, and the heavy load of propellants broke its back on touchdown. It was repaired.

A second mishap occurred on November 9, 1962, when an engine failure forced Jack McKay to make an emergency landing at Mud Lake, Nevada. However, the flaps failed to extend, and, with a touchdown speed of 290 mph, the flare was not quite perfect, and a longitudinal oscillation caused the left main skid to collapse. The X-15-2 swerved and flipped over on its back. McKay sustained injuries but returned to flight status. The aircraft was substantially damaged, but the crash provided an opportunity to perform extensive modifications.

The resultant X-15A-2 was a test bed for development of a Mach 8, air-breathing engine – the Hypersonic Ramjet Engine (HRE). The engine was to be attached to the lower ventral of the X-15. An additional twenty-nine inches were added to the length of the fuselage between the existing tanks for the liquid hydrogen to power the HRE. Two large, external fuel tanks added alongside the fuselage under the wings increased the burn time to attain Mach 8. The propellant in these tanks was consumed first, and then they were jettisoned at Mach 2. The gross weight of the X-15 exceeded 51,000 pounds, and it was a testament to the capabilities of the B-52's ability to carry such a load under its wing.

To withstand the added heating due to increased velocity, the entire aircraft surface was coated with an ablative-type insulator. The sprayed-on ablator worked successfully but proved unrealistic because of the extensive preparation time and operational problems.

To flight-test the modified aircraft, a mock-up HRE was attached for the first, and only, maximum-speed test of the X-15A-2, in 1967. During this flight, shock wave impingement off the mock-up HRE caused severe heating damage to the lower empennage, and almost caused the loss of the aircraft. The near catastrophic situation was the result of a lack of thorough analysis as to the effect of hypersonic flow on pylons. The flight set the speed record of Mach 6.72, which stands to this day.

In November 1967, X-15-3 dropped from its B-52 at 45,000 feet, with test pilot Major Michael J. Adams, who climbed under full power for a high altitude test. Within three minutes he reached a peak altitude of 266,000 feet. During the climb, he started a planned series of slight rolling movements, which quickly became excessive, by a factor of two or three. Just as soon as these were dampened, the X-15 began a slow yaw, and within 30 seconds the plane was descending at right angles to the flight path. At 230,000 feet, the X-15 encountered rapidly increasing dynamic pressures, and at Mach 5 Adams radioed, *"I'm in a spin, Pete."* (The ground communicator was another veteran X-15 pilot, Pete Knight.)

Adams fought to recover, using both the aerodynamic control surfaces and the reaction controls. He transitioned from the spin at 118,000 feet but then went into a Mach 4.7 dive, at an angle of 45 degrees—inverted. Then a rapid pitching motion began, resulting in dynamic pressures that subjected the X-15 to 15 Gs. At 65,000 feet and a speed of Mach 3.93, the X-15 broke up. Mike Adams died, and the X-15 was destroyed.

It was decided at that point to end the X-15 program. Eight more planned missions were conducted before the final flight in November 1968. A total of 199 missions were flown, of which 109 exceeded Mach 5, and four exceeded Mach 6. The surviving examples of the X-15 are housed at the Smithsonian Air and Space Museum (Washington, D.C.) and the United States Air Force Museum (Dayton, Ohio).

The two disciplines of aeronautics and astronautics were being merged as man ventured further from the planet with wings and rocket engines, and a new term was coined—*aerospace*. The X-15 represented a new paradigm that would combine the power of the rocket with the recoverability of the aircraft to provide a reusable spacecraft—the space shuttle.

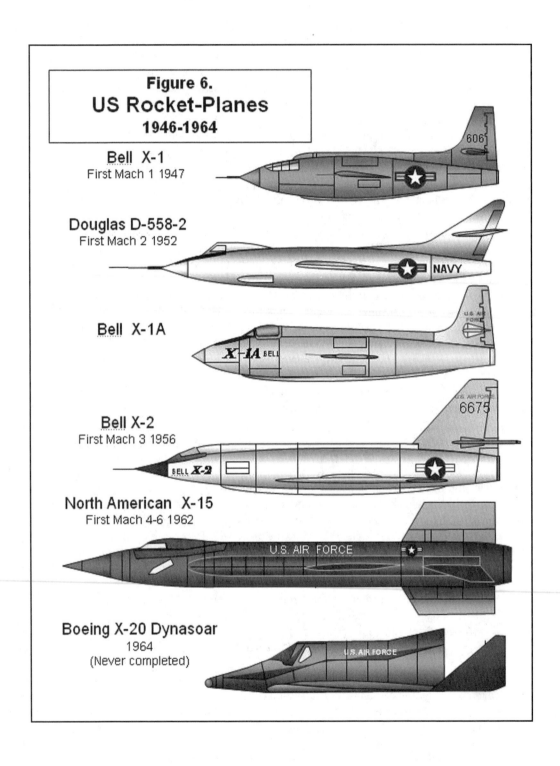

Figure 6.
US Rocket-Planes
1946-1964

Bell X-1
First Mach 1 1947

Douglas D-558-2
First Mach 2 1952

Bell X-1A

Bell X-2
First Mach 3 1956

North American X-15
First Mach 4-6 1962

Boeing X-20 Dynasoar
1964
(Never completed)

The Problem and the Promise

Down through the ages man has continually sought to perfect a weapon more lethal than those possessed by his enemy—and most of these weapons were dependent on a ballistic trajectory to deliver the deadly payload. Beginning with early man simply throwing a large rock at his antagonist, the spear, bow and arrow, gunpowder, and cannon reflected this advance towards what would someday be called the ultimate weapon. The term would be applied to a weapon that was not only more powerful than anything that came before but against which there was no defense.

By the end of World War II, the scope of German technology, as revealed by the various allied teams that pursued the spoils of war in 1945, was astounding. The impact of the V-2 on the military perception of large ballistic rockets was profound. It had been just one short year since the first of these weapons was launched against London, and only a few months since the surprise introduction of the atomic bomb. Yet a new, wide-ranging paradigm began to form that represented a drastic rethinking for the delivery of tactical and strategic destructive power. The German A-9/A-10 rocket, although just a paper design, promised the ability to send a warhead to intercontinental distances. (Five thousand miles is typically used to establish the term *intercontinental*). The combination of an atomic warhead and the intercontinental missile represented the ultimate weapon of the ages. This concept got the attention of several influential military leaders in both the United States and the Soviet Union. What had been considered as Buck Rogers science fiction just a few short years earlier was now the cutting edge of rocketry.

The American advance towards long range rockets officially began October 31, 1945, when the Army Air Force Technical Services Command issued proposals for study contracts to evaluate several methods of projecting warheads over varying distances up to 5,000 miles. But, an intercontinental ballistic missile (ICBM) was a quantum leap in technology even from the revolutionary V-2.

The power required to send several tons of explosives 5,000 miles was calculated to be at least ten times the thrust of the V-2—and perhaps more. The idea of multi-staged rockets, putting one massive rocket on top of another to achieve more efficient mass ratios, was fraught with many technical problems. The extreme temperatures of the exhaust exceeded the melting point of virtually all known metals and their alloys, while the extreme cold of minus 297 degrees below zero (F) "boiled away" the favored oxidizer (LOX). If the temperatures did not cool the enthusiasm for the ICBM, the pressures within the combustion chamber, the difficulty in achieving a "smooth" burn of the propellants, or even the acoustical shock waves that resulted from the blast emanating from the exhaust were considered insoluble to many.

Likewise, the accuracy of the guidance system over that distance would be a major challenge. Even considering the power of an atomic warhead, the missile would have to impact within a few miles of its target to be effective. This was several orders-of-magnitude improvement over the precision of existing V-2 capabilities. Because the ICBM was powered (and guided) only during the first few minutes of its half-hour flight, it had to pass within a few yards of a nominal point in space about 350 miles from the launch site. Moreover, it had to be given a velocity within a few feet per second at that point, to arrive within the desired radius of the target thousands of miles away. The difficulty here cannot be over-emphasized. Not only did the guidance system have to sense the trajectory and velocity of the missile, but also it had to be able to detect the effect of possible high-speed winds in the upper atmosphere. Imponderables on the other side of the trajectory also included encountering a jet stream on the way back down as well as the ballistic characteristics of the warhead itself.

Even the accuracy of tactical ballistic rockets (those with a range of 20 to 200 miles) of this era did not permit them to be used to destroy small targets, such as bridges or even airfields, using conventional explosives. The rocket with a conventional warhead was seen as a weapon deployed by the dozen against a target spread over several miles. The nuclear weapon changed these rules, as even a miss of several miles could wreak havoc on a much smaller target area, such as an airfield or large industrial complex. Of course, the average person usually thought in terms of cities, and the major cities as targets of the protagonists could not be ruled out.

Yet another obstacle for development of an ICBM was that its warhead arrived back into the atmosphere at 15,000 mph, and new methods of protecting it from the heat generated during reentry had to be mastered. This aspect had only been passively considered in the shorter-range rockets with velocities of 3,000 to 5,000 mph.

One noted scientists who, in 1946, considered the ICBM to be a project for several decades into the future, but not the present, was none other than Dr. Vannevar Bush—the man President Franklin Roosevelt chose to chair the National Defense Resource Committee (NDRC). This committee was an umbrella organization for coordinating most military research activities during WWII. Bush had urged the president to move forward aggressively with the development of the atomic bomb in 1941. Now, just a few short years later, he declared, *"I don't think anybody in the world knows how to do such a thing* [build an ICBM]*,"* and *"I feel confident it* (the ICBM) *will not be done for a very long period of time."*

If the technology challenges to the development of an ICBM were of a high magnitude, the promised payback was enormous. Here was a missile that arrived at such a speed that it was unstoppable. If a country could field these weapons in sufficient numbers, they could annihilate an adversary, perhaps before that country even knew it was under attack! Today, some 50 years later, the ability of any country to intercept and destroy an attack by several hundred warheads is still not a reality.

But the technologies necessary to create the ultimate weapon were also the elements necessary to propel scientific equipment, or even man, into space. As von Braun had speculated when the first successful large ballistic missile, the V-2, was flown in 1942, *"today the spaceship is born."*

Convair MX-774

Responding to a proposal for a study contract, issued by the Army Air Force Technical Command in 1945, was the Consolidated Vultee Aircraft Corporation (Convair). While normally these types of low budget contracts were avoided by the larger manufacturing concerns, the end of the war had brought aircraft production to a virtual standstill for most of the aviation industry—times were tough. The study (valued at $1.4 million) was to investigate both cruise and ballistic missile designs. Although hardware was not originally intended, Convair requested, and received approval for, some preliminary development to test proof-of-concept technologies for the long-range ballistic missile.

Convair had virtually no experience with rocketry at this point but quickly gathered what information was available, which included some V-2 documentation. The Convair MX-774 project, under the direction of Karel "Charlie" Bossart, envisioned a high mass-ratio rocket. Extreme weight-saving techniques included the use of integral fuel and oxidizer tanks that formed a part of the outer skin of the rocket and a separable warhead for reentry. The supporting structure was even more radical. By keeping the propellant tanks under pressure, even when empty, the thin walls acted like a balloon and provided the required stiffening, thus reducing the need for internal radial and longitudinal metal structures (monocoque construction). An effective analogy is the common soda-pop can, which, before it is opened, is pressurized and cannot be easily distorted. However, when the seal is broken and pressurization relieved, it can easily be crushed with two fingers. There was some criticism leveled at this structural approach. Even with these innovations, it was estimated that the rocket would weigh perhaps a half-million pounds at launch, requiring several large (yet to be developed) engines to achieve the desired thrust levels.

To prove these concepts and gain some experience in the design and construction of liquid-fuel rockets, a 31 foot *technology demonstrator* was designed that had the outward appearance of the V-2 except for a greater fineness ratio (the proportion of the width to length). High levels of propulsion efficiency were achieved by mounting the rocket motors on gimbals that swiveled for steering, thus eliminating aerodynamic fins or the need for control vanes projecting into the rocket engine exhaust as was done with the V-2. While Goddard had pioneered this concept, the MX-774 would be the first rocket of any size to incorporate this innovation.

The bulk of this project was carried out without significant input from German research—an interesting exception being the actual physical shape of the fins. The fins, although not needed, were provided as a back up to the gimbaled engines and a support for the rocket on its launch pad. Assuming the Germans had put a lot of research into the aerodynamic aspects, the Convair team emulated the shape, not realizing that the German design was actually a compromise between aerodynamics and the need to transport the rocket through railway tunnels!

As several engines would be needed to achieve the high thrust levels of the final product, the MX-774 was designed to examine the integration (clustering) of four separate rocket engines. To power the relatively small, 2890-pound rocket (one tenth the weight of a V-2), the same engine used for the X-1 rocket plane (that broke the sound barrier in 1947) was chosen. The LOX-alcohol propellant combination produced 2,000 pounds of thrust in each of the four thrust chambers for a total of 8,000 pounds. The increase from the 1,500-pound thrust unit in the X-1 was achieved by using a higher propellant flow rate that required a turbo-pump rather than a pressure feed system. The chambers themselves were not fully gimbaled, but rather, each was provided with two directions of freedom. The alignment of these directions allowed for pitch, roll, and yaw control under the direction of a simple gyroscopic guidance system. Twelve missiles were planned to evaluate the technologies involved and to gain expertise in the design, construction and firing of liquid-fuel missiles.

The first flight articles were in final assembly when the budget realities of post-war America began to require difficult choices. On July 1, 1947, the MX-774 project was canceled. Convair was permitted to use unspent allocations to flight test three missiles.

Following a series of static firings at Point Loma, California, which essentially confirmed the basic power plant integrity, the first missile was shipped to White Sands Proving Grounds in New Mexico. Erected on a modified V-2 launch pad, the missile was to follow a vertical profile that would result in it's achieving about 100 miles in altitude. The first launch occurred on July 13, 1948, and its initial climb was straight and true as the missile balanced on the four white-hot shafts of flame. Only 12.6 seconds into the flight, the engines shut down prematurely, and the missile coasted upward little more than a mile before descending to its destruction. The nose cone separation failed to occur, but the film from the on-board camera that was recording various gauges was recovered. Although the team was disappointed, they were pleased that the missile exhibited very stable flight characteristics—essentially proving the engine control design.

The second missile, launched on September 27, was somewhat of an improvement, but again the engines shut down prematurely after 48 seconds of the planned 75 second burn. An altitude of 25 miles was recorded, and again the nose cone failed to separate. The final flight was launched on December 2, but shut down at 51.7 seconds.

While none of the flights achieved the full powered duration, the basic objectives of the tests were successful. The pressure stabilized structure and control of the rocket proved the technologies that would be called on three years later when the ICBM program was again funded. Bossart was able to convince Convair management that the project should be kept alive as an in-house study.

One of the reasons for the cancellation of the original MX-774 project was the expected limitations on the accuracy of the missile. In the late 1940s, the largest fission nuclear weapon produced a yield of less than one-half megaton (about 20 times the power of the bombs used against Japan). Several ICBMs would have to be directed at a single target to assure its destruction. The essential economics worked against this strategy. However, when President Truman decided that the Soviet threat required development of a fusion or thermonuclear (hydrogen) bomb in 1951, the possibility of producing weapons with a yield of several megatons made the ICBM a more practical weapon. Following the first tests of the H-bomb in the Pacific in November 1952, a deliverable bomb began to move closer to a reality.

The "Packet" Concept

The development by the Soviet Union of the initial series of rockets (R-1, R-2 and R-5) in the late 1940's based on the German V-2 provided a firm foundation on which to build more advanced projects. Stalin savored the idea of being able to threaten the U.S. with an unstoppable weapon such as the long-range missile. This threat drove the dictator to approve projects in advance of what technology was currently capable of supporting. In an address to the Politburo in 1947, he said (with respect to the ICBM): *"Do you realize the tremendous strategic importance of machines of this sort? They could be an effective straight jacket for that noisy shopkeeper Harry Truman. We must go ahead with it comrades. The problem of creation of a transatlantic rocket is of extreme importance to us."*

Projects as massive and costly as the ICBM, were taken more cautiously in the U.S., as evidenced by the cancellation of the MX-774 because of technology uncertainty and funding. In the United States, the President had to rationalize decisions and move budgets through a Congress that represented the will of the people. Stalin had no such impediments.

However, the passion of Sergey Korolev (a founding member of GIRD) for space exploration focused beyond the ICBM. For him, these were only learning tools for the real job that lay ahead. If the KGB had viewed Korolev's talk of space flight as a distracting menace to the state, as the Gestapo had with von Braun, Korolev could have found himself back in the Gulag. Tikhonravov had prepared a proposal, in 1947, to develop a satellite launch vehicle in the near future using current Soviet technology. The report analyzed variations of multi-stage missiles using Tsiolkovsky's theories of combining stages. The conventional tandem arrangement provided for two stages that would fire successively.

There were two major drawbacks to the tandem arrangement. The first was that it required developing a second stage rocket engine that would have to go through the sequences necessary to ignite while in the vacuum of space and in a weightless condition. This was the concern that prompted the Americans to combine the V-2 with the WAC Corporal as a second stage (Bumper Project) to investigate the difficulty. Soviet designers felt that the problem might require a long-term program to solve. Another possible complication with the tandem arrangement was the extra length and structural arrangement of stages that could complicate the assembly and erection process.

To avoid these problems and others, a second and significantly different scheme was to cluster the stages in parallel. This configuration, which the Soviets called the "packet," offered several advantages. All the stages would ignite for liftoff, and the outer segments would be discarded along the trajectory as they depleted their fuel. The length of the missile would be significantly shorter, although its girth would offer substantially more drag for the first two minutes as it passed through the dense lower atmosphere.

Several variations of the packet concept, with the number of elements varying from two to five, were studied, including the use of identical boosters linked together or one with different size boosters. Calculations of performance revealed advantages of the packet scheme. When the formal report was issued, however, no mention of the packet's capabilities as a satellite launch vehicle were made; the research simply highlighted the ability to achieve very high velocities with heavy payloads. It may have been the fear of the secret police that caused the omission of the intended role.

A series of studies undertaken in the late 1940's evaluated different weapons delivery schemes with several themes specifically addressing the ICBM. The packet design conceived by Tikhonravov appeared to hold the greatest promise for early development. Despite the fact that the intermediate range R-3 was never completed, many of its design aspects had advanced enough to encourage Korolev's design team to move forward with the ICBM.

Research Project N-3, *Development requirements for a liquid rocket with a range of 5,000 to 10,000 km and a warhead of up 10 tons,* in December of 1950, was awarded to Korolev's NII-88 design team. The upper limits of the payload requirement indicated that the Soviet's newly developed atomic bomb would be the warhead.

The initial research program to develop such a long-range rocket tackled a variety of difficult problems that had to be solved to achieve the desired capabilities. Korolev had always harbored an interest in the cruise missile concept, but that would not lead to a satellite. It was also more easily defeated with defensive measures. The ICBM was considered unstoppable. Therefore, he favored the purely ballistic approach. Of course, engine power and efficiency was a prime consideration, as at least 300 tons of thrust and a Specific Impulse of over 300 seconds [as measured in a vacuum] was needed. While other more exotic propellants, such as fluorine, were considered, LOX/Kerosene, with a possible Specific Impulse of 325 seconds [vacuum], was the obvious choice.

Weapon aside, the packet concept provided a rocket that Korolev felt sure would open the heavens to man. All of the technological challenges necessary to perfect an ICBM would have to be mastered to launch and recover satellites.

Convair MX-1593 Atlas

In October 1950 the Rand Corporation completed a feasibility study begun a year earlier, which confirmed the military practicability of long-range rocket weapons (Vannevar Bush's considered opinion not withstanding). A follow-on Air Force project in 1951, labeled MX-1593, awarded to Convair, subsequently became the first American ICBM—known to the world as *Atlas* (military designation SM-65). The project was also known as Weapons System (WS)107A.

It was during this reincarnation of the ICBM project that Convair experienced the influence of some members of the former German V-2 team when Hans R. Friedrich, a physicist and guidance expert, joined Convair in 1951, as did Kraft A. Ehricke in 1954. Ehricke was to have joined Convair earlier, but his close Nazi association during the war held up the issuance of a U.S. security clearance until the later date.

By 1953 Convair had completed the initial design studies for a vehicle that would deliver the specified 15,000 lb. warhead a distance of 5,000 miles. Recognizing the problems of second stage ignition in a vacuum, the design called for a total of five engines, producing 600,000 pounds of thrust, in a unique "skirt" arrangement in which all five were ignited at launch. Four fixed booster engines surrounded a larger, main, gimbaled thrust-chamber. A single set of propellant tanks fed all engines.

There were some obvious similarities to the Soviet packet concept. However, the Convair arrangement was more efficient than the Soviet design. With the packet, any residual propellant left in any of the four booster tanks would be wasted. To avoid this, Korolev had devised a system that would attempt to empty all four-booster tanks simultaneously by varying propellant flow. The system, *Sinkhronnoye Oporozhneiye Bakov* (abbreviated SOB), worked reasonably well but was yet another complication that would eventually cause problems. With the *Atlas*, as all engines fed from the same tanks, there could be no potential waste. The *Atlas* arrangement would be referred to as a one-and-a-half stage missile.

The booster set burned for about 150 seconds, at which time the skirt with the booster engines was to be discarded. The main center engine continued to burn for an additional four minutes to achieve the required velocity. Using the thin-skinned, pressurized tank structure pioneered in the MX-774 project, it was calculated that the added weight of carrying the half-empty propellant tanks was an acceptable penalty for the simplicity of the operation. The rocket would stand almost 100 feet tall with a diameter of 12 feet.

The SM-65 was programmed as a ten-year development effort with three distinct phases that would minimize technological risk and result in operational deployment in 1963. The first phase was a single engine test vehicle designated X-11, followed by a three engine X-12, and finally the five-engine product.

With the Cold War heating up in light of announced developments with the Hydrogen bomb, the Atlas project was reviewed by the Strategic Missiles Evaluation Committee in February 1954. They recommended that its development be accelerated and that a new Air Force management group be established to oversee the project. The Ballistic Missile Division of the Air Research and Development Command was formed under Brigadier General Bernard Schriever,

H-bomb tests subsequently showed that the warhead for the *Atlas* could be made significantly smaller and lighter than expected. The first "dry" technology thermonuclear bomb that would lead to these lightweight nuclear warheads was detonated by the U.S. on March 1, 1954. This was at the same time that the new Soviet Premier, Nikita Khrushchev, gave the Soviet ICBM project the green light. Based on the anticipated weight reduction, the five-engine SM-65 design was replaced by a smaller, three-engine configuration that retained the basic one and a half stage concept. The final configuration of the *Atlas* would have a length reduced to 80 feet with a diameter of ten feet.

Of great importance to the Convair effort was research into large rocket engine technology pursued by Rocketdyne (a division of North American Aviation—NAA). After WWII, the Army Air Force shipped three V-2 engines to NAA. Some members of the German team, including Walter Riedel and Gerhard Heller, spent several months at NAA assisting in the fundamentals of design and fabrication. By June of 1951, a basic 75,000-lb. thrust unit had been increased to 96,000 lb. With confidence established, a new design, using lightweight tubular construction, was completed, and a 120,000-lb. engine was static tested in May 1952.

A new design, the Rocketdyne LR-89 with 150,000 pounds of thrust each, was selected for the two skirted boosters (later upgraded to 165,000-lb. thrust), while the sustainer engine would be the Rocketdyne LR-105 engine of 60,000 pounds of thrust. Both were fueled with RP-1 (kerosene) and LOX and used turbo-pumps to achieve high flow rates. Two small Rocketdyne LR-101 vernier engines of 1,000-lb. thrust were attached to the sides of the rockets for fine-tuning the burnout velocity and for roll control.

The LR-89 engines achieved a specific impulse of 248 seconds at sea level and 282 seconds in a vacuum. One reason rockets are more efficient in a vacuum is that there is no atmospheric drag on the exhaust, and higher exit velocities can be achieved. The LR-105 sustainer engine was optimized for operating in near vacuum conditions by having a wider nozzle. This three engine propulsion system was designated MA-1.

The availability of an engine the size of the Rocketdyne LR-89 was the direct result of another weapons program that got its start in the late 1940's. In evaluating German concepts, a Mach 3 ramjet-powered cruise missile was considered feasible, and NAA was awarded the contract to develop the WS-104 Navaho. To bring the ramjet-powered missile up to an operating altitude (60,000 feet) and speed (Mach 3), a large liquid fuel booster was developed that used two 120,000 lb. thrust engines.

NAA spawned its Rocketdyne subsidiary in 1955 that became the pre-eminent supplier of liquid-fuel rocket engines to the American space program. Thus, when Convair needed large engines for the Atlas, Rocketdyne had the expertise.

While the downsizing of Atlas meant that the design could proceed more rapidly by building to the final configuration, it was decided to proceed in a conservative manner by establishing a series of versions that added capability as testing proved each milestone. The first test articles, referred to as the "A" version, contained only the two booster engines. The configuration would verify this part of the power plant, the basic guidance and control, as well as the skirt separation process. In addition, as Rocketdyne was not yet ready to deliver the promised engines, the Atlas A would fly with two XLR-43 engines, which produced only 105,000 lbs of thrust for 135 seconds.

In an effort to pierce the Iron Curtain to confirm CIA reports of Soviet advances in rocketry, the Americans installed six powerful radar stations near Samsun, Turkey, on the coast of the Black Sea in the mid-1950s. These radars could detect missiles being test fired from sites almost 1,000 miles away as they rose above the horizon. The *Atlas* was publicly announced on December 16, 1954, and the following March, Air Force Chief of Staff Nathan F. Twining reported that the *Atlas* ICBM program was being given a high priority along with the *Snark* and *Navaho* cruise missiles because of "known" Soviet progress. It was the observations of the Turkish radars that had alarmed the U.S. military, and ballistic missile funding grew from a modest $3 million in Fiscal Year 1953 to $14 million in FY 1954 and $161 million in FY 1955.

Another advisory committee then recommended development of a second ICBM (the *Titan*) as a back-up, with a structure of more conventional design, in case the *Atlas'* tanks could not withstand high aerodynamic loads during the early part of the flight. An intermediate range ballistic missile (IRBM) with a range of 1500 miles (the Thor) was also authorized. Spending increased to $515 million in FY 1956 and $1.3 billion in FY 1957.

While the problems of accelerating a missile to achieve intercontinental ranges appeared to be resolved, the ability to protect the warhead during reentry was still a major concern. Conventional wisdom seemed to dictate a sharply pointed nose cone and the use of heat resistant structures that could absorb the heat yet retain its strength. However, there were no known materials that could withstand temperatures of up to 12,000° F.

Research directed by H. Julian Allen, head of NACA's High Speed Research Division at the Ames Research Laboratory in Sunnyvale, California, found a solution. It was understood that the kinetic energy of the incoming nose cone was divided between that generated by the compression of the shock wave at the front of the nose cone and that generated by the air friction with the skin of the cone.

Heat generated by shock wave compression was somewhat removed from the cone itself as it lay outside the boundary layer, which served as a form of insulation. Pointed objects tend to permit the boundary layer to rest against the object, allowing the heat to be readily transmitted to it along with the friction induced heat. By suggesting a blunt face to the object, the boundary layer is strengthened and somewhat distanced from the structure itself.

However, this was not the total answer, as the heat that was transmitted to the nose cone still had to be handled by the structure. The first of two alternatives was to use a heat-sink material such as copper or beryllium, which could absorb the conducted heat and retain its structural integrity. However, this choice required a relatively heavy weight—and weight was the enemy of the rocket.

A second method of handling the induced heat was to use the ablation method. Here, the outer layer of the nose cone would be coated with a material that would melt away and remove heat in the process. In effect, it was a method of a controlled burn of the outside of the nose cone to protect its content.

To examine the reentry problem of the Atlas, a small and relatively inexpensive three-stage, solid-fuel, 40-foot rocket was created: the X-17. The fin-stabilized first stage was a derivative of the *Sergeant* missile. To stabilize the rocket during the initial acceleration (before the fins became effective) the rocket would have a spin imparted by a pair of small solid-fuel rockets attached perpendicular to the longitudinal axis mid-way up the first stage. These would fire at the same time as the first stage and provide two rotations per second. The first stage fired for 23 seconds, lifting the remaining stages and the scaled down nose cone experimental package almost vertically to an altitude of about 80 miles. At that point, the rocket began an un-powered dive back into the atmosphere. Upon reaching the denser layers, the second and then third stages fired for 1.53 seconds each, driving the nose cone (with the help of gravity) to a speed of 15,000 mph.

The first two X-17 tests failed, but beginning with the third, on December 1, 1956, the program confirmed the choice of the blunt shape and use of ablative coverings. The data derived from a total of 24 flights on a variety of shapes and materials—was so precise that it was used for more than a decade as the principle measure of aerodynamic heating on structures.

The first successful static firing of the Atlas propulsion system occurred on June 22, 1956. The first flight-rated *Atlas A* arrived at Cape Canaveral in December 1956 and underwent an extensive set of tests verifying its compatibility with its ground support equipment. The stage was now set to begin flight-testing of the Atlas.

The R-7: Semerka

The initial design of the Soviet ICBM was well advanced when nuclear testing at the desert site of Semipalatinsk, in October of 1953, indicated that thermonuclear (hydrogen) bombs of significantly greater destructive power than fission weapons were feasible. To this point, the weight of the nuclear warhead had seen much fluctuation as advances were made in the design and engineering of fission weapons. The first hydrogen bombs, however, were massive in size, and initial specifications called for a warhead weight that could be as large as 20,000 lb. With the existing design, this would have limited the range of the Korolev's packet rocket, now designated the R-7, to only 3,500 miles. Subsequent design reviews in early 1954 focused on weight reduction of the rocket.

Combustion instabilities continued to plague the engineering efforts to produce large engines of more than 100,000 pounds of thrust. However, the R-7 design could not wait for Soviet technology to catch up with the requirement and schedule. The single thrust chamber

The Soviet R-7. The world's first ICBM.

(engine) originally envisioned for each module was replaced with a set of four chambers, but a single turbo-pump was retained for the unit.

Because control vanes within the exhaust reduced its velocity (and thus engine performance), the task of gimbaling (swiveling) the engines to control the rocket became a prime focus. Because it is easier to control the motion of smaller rockets, it was decided to leave the large thrust chambers fixed, and incorporate smaller, vernier rockets around the periphery of the base. Thus, each of the outer modules would have two 10,000-lb. thrust, vernier engines that would provide for pitch, roll, and yaw in concert with the verniers of the other modules. The core module would contain four of these small engines which would later become the basis for upper stages in advanced Soviet space exploration.

By February 1954 the final design was submitted, and in May the government authorized development of the R-7 (official designation 8K71) intercontinental ballistic missile. Following some additional changes, the design was frozen in March 1955. The vehicle in the final design used a core stage that was significantly longer than the booster modules that surrounded it. Its diameter was also increased at the point where the top of the other conically shaped modules attached so that it had an unusual 'hammerhead' configuration.

The ignition sequence called for the four booster modules to fire initially. Only if all 16 primary thrust chambers and all of their 8 verniers ignited would the four engines in the core (and its four verniers) be ignited. A unique launch pad design suspended the rocket above the flame pit allowing technicians to service the rocket and shielding it from high winds. This unique *Tyulpan* launch concept involved no hold-down clamps. When sufficient thrust built to overcome the weight of the rocket (which implied that the core stage had ignited), it lifted off, and the suspension arms rotated away on counter-weights.

Initially, the warhead was enclosed in the German "sharp point" design: a 16 degree cone almost 25 feet long. This would change, as experience with the blunt shaped ablation technique for protecting the payload would ultimately be chosen.

It was estimated that the rocket could boost a 12,000-lb. warhead to almost five miles per second, resulting in a 5,000 mile range, with a maximum miss distance of 5 miles. The rocket had a gross lift-off mass of 280 tons, and an empty mass of 27 tons. The first stage burned out at 4,700 mph and the second stage at 14,300 mph. Thrust at lift-off was over 900,000 pounds. The R-7 incorporated ingenious solutions in ground handling of the large rocket. The rocket would be assembled horizontally in a large building several miles from the launch pad. It would then be rolled out to the launch pad on a railroad track, raised to the vertical position, and quickly fueled.

There were several possible guidance systems to consider. The all inertial (like the V-2) used a preprogrammed path that gyroscopes and accelerometers used to keep the missile on course. A second was a radio system that established a set of electronic beams that the missile would track (much like the Instrument Landing System of an airplane). A third method was the use of radar, which would monitor the missile's track and send corrective signals to maintain the desired course. The first R-7s would employ radio beam technology because it would provide greater accuracy more quickly.

The selection of the radio beams also required that a new and larger firing range be chosen to accommodate three radio control stations, each 100 miles away on either side of the launch site, and the third 200 miles behind the launch site. The Soviet Union was so vast that it was possible to test a missile with intercontinental range within its borders. In January 1955, Tyura-Tam (more commonly known today as Baykonur) was selected as the ICBM test range, with a warhead impact area on the Kamchatka peninsula that jutted into the Pacific on the eastern coast of the Soviet Union. Tyura-Tam is just east of the Aral Sea at 45 degrees North Latitude and 63 degrees East Longitude.

Construction of the rocket began in early 1956 with the objective of making the first test flight by the end of the year. A project of this magnitude, working under the constraints of the Soviet industry, ensured that schedule slippages would occur. But the resourceful nature of the many individuals involved and their high degree of motivation allowed the rapid solution to the myriad of problems that plagued the project.

The radio guidance system was flight tested on the R-5R with launches in May and June of 1956 to prove the system. Rocket engine tests were begun in July 1956 to determine the optimal arrangement of engines and their components and to minimize thermal and vibration effects.

Static firings of the booster stages were completed in January 1957, while core stage tests took place in December 1956 and again in January 1957. Finally, two static firings were conducted using the complete rocket with four booster stages and the core stage. The R-7 was now ready to begin flight testing.

Testing the ICBM

The first R-7 flight-worthy vehicle was delivered to Baykonur in March 1957 and was launched on May 15, 1957. A fuel leak at liftoff caused a fire in the engine compartment of one of the booster modules and destroyed the rocket after 98 seconds of flight. Although disappointed, the Korolev team was able to trace the malfunction, and the second rocket was launched on July 12, 1957. This time failure came much earlier as a short circuit in the control system power supply resulted in a rapid roll that caused all four outer modules to tear away from the core after only 33 seconds of flight. The team was now highly demoralized.

The first launch of the Atlas took place on June 11, 1957, from Cape Canaveral, less than a month after the first Soviet ICBM test. The 180,000-lb. missile, significantly underweight, because it lacked the center sustainer engine and a full fuel load, rose for about 10 seconds. Then the brilliant white exhaust flame turned to yellow smoke as the engines shutdown prematurely. A problem in the fuel system had doomed the missile. With no inherent stability, once the engines stop firing, the missile turned several summersaults in the air as it proceeded to fall from its maximum climb of about 10,000 feet. The Range Safety Officer sent a radio signal to rip open the fuel tanks with an explosive charge to allow the volatile propellants to express their destructiveness high above the ground. Although the bulk of the test objectives had not been met, the structural integrity of the thin skinned missile was proved by its unexpected gyrations. Few would doubt Charlie Bossart's folly after viewing the films.

An American Atlas ICBM takes flight.

The test range at Cape Canaveral is only a few miles from the populated areas along the east coast of central Florida, and launches can be plainly viewed by anyone. Thus the relative success or failure of a launch was not only observable, but the press began to report launches with spectacular photos and film. This gave a definite advantage to the Soviets. They could easily see the relative progress of the Americans, while Soviet launches were veiled in the tightest secrecy.

With the knowledge of the cause of the second R-7 failure, the third Russian missile was launched on August 21, 1957. This time all systems performed as expected, and the missile warhead flew the entire route, achieving an altitude of 600 miles before reentering the atmosphere over the Kamchatka Peninsula and disintegrating. This last aspect had been anticipated as the nose cone material was not yet perfected. But the missile had flown a complete configuration test on only its third attempt, although it was far from being an operational ICBM.

The dilemma for the Russians now was if and how the success should be reported to the world. A saber can't be rattled unless the opposition knows of its existence. In a brief announcement on August 26, 1957, the TASS News Agency of the Soviets reported *"a super long-distance intercontinental multistage ballistic rocket was successfully tested a few days ago."* Much of the world was unimpressed as there were no pictures and no impartial observers to confirm this latest Soviet boast.

Korolev's team launched another R-7 on September 7th that was also successful. Those who were intimately connected with the development of the R-7 now adopted a more personal and affectionate name for their creation; it was called *Semerka*- Old Number Seven. This moniker would remain with the rocket throughout its more than 50 year lifetime.

Meanwhile, a second flight of the Atlas in September failed in an almost identical manner to the first. After much investigation, the problem appeared to be excessive heat at the base of the rocket. Some of the hot exhaust gasses were being drawn up into the skirt area and overheating the hydraulic lines that controlled the gimbaling of the engines. This was resolved, and the third flight of a limited range of 500 miles was finally accomplished on December 17th. But this was not a fully configured missile. The Atlas, employing as it did many high-tech features, suffered a series of highly publicized failures before the 10th flight (number 4-B) achieved a successful test of all primary flight systems on August 2, 1958—one year behind the R-7.

If the Soviet announcement of August 26, 1957 was to be believed (and many elected not to), they were far ahead of the American ICBM effort. The stage was set for launch of the first artificial earth satellite—Sputnik I.

Contrasting the Contenders

The Soviet's R-7 weighed a total of 600,000 pounds at lift-off with a thrust of more than 900,000 pounds. The Atlas was virtually half that weight at 267,000 pounds with a thrust of 360,000 pounds. It stood 82 feet tall (varying with the warhead configuration) and was ten feet in diameter. The R-7 also varied in absolute height depending on the payload but was typically 95 feet tall when configured as an ICBM, with a diameter of just over 34 feet.

Test flights of the first lot of 12 prototype R-7 missiles were completed on January 30, 1958. By that time the Soviet Union had used the R-7 to demonstrate the first full-range ICBM and to orbit the first two artificial satellites of the earth. The ability to use the missile as a satellite and lunar launch platform at such an early stage in its development was a tribute to the perseverance of the entire Korolev team. The next two lots of missiles were test flown through December 1959. These demonstrated a flight configuration capable of carrying a nuclear warhead. The first R-7 missiles were placed on nuclear alert (operational) on October 31, 1959, and then only about six were available for the next several years.

Of the first eight flights of the Atlas A, only four were successful. Following the last flight of an A model on June 3, 1958, testing of the *Atlas B* with a three engine MA-1 propulsion assembly began. The MA-1 consisted of two XLR-43 engines rated at 105,000 lb. for 100 seconds and a single XLR-43 rated at 120,000 lb. for 220 seconds. This combination was successfully flown in August of 1958, with a full range of over 6,000 miles being achieved on November 28, 1958. However, significant events in the Soviet Union had already diminished the accomplishment.

The Atlas C was the last test version, and only six were fired in this weight reduction program (an even thinner skin) to improve the payload carrying capability and to test a new guidance system that coupled General Electric radar to a Burroughs computer. The same MA-1 propulsion unit generated 330,000 pounds of thrust. The next to last C was the first Atlas nose cone recovered.

The Atlas D was the first operational version that began its career with three failures. However, it then went on to demonstrate 90% reliability with a 9,000-mile range and a one-mile accuracy (CEP) factor. During a flight to its maximum range, it would rise to a high point of 760 miles and reenter the atmosphere at 16,000 mph. It became the rocket that would carry the manned Mercury spacecraft as well as the Atlas Agena that boosted lunar and planetary missions. It used the MA-2 propulsion unit up-rated to 368,000 lb. and later production variations had the MA-3 unit that produced 392,000 pounds of thrust. The booster burned for 140 seconds and the sustainer for an additional 130 seconds.

In July 1959 the operational variant SM-65D (the military designation of the *Atlas D*) successfully flew. The first squadron went on Combat Alert on October 31, 1959, at Vandenberg AFB. The three missiles were in unprotected open launch pads, with the W-49, a 1.4-megaton yield warhead. By a strange set of circumstances, this was the same day that the Soviet R-7 was declared operational. The later Atlas E and F models carried the larger, 4-megaton W38 warhead. The Atlas E contained the first *all-inertial* guidance unit and was deployed from 1961 through 1965. After being retired, they were stored at Norton AFB in California to await assignment as the first stage of a wide variety of satellite launch vehicles.

The *Atlas,* as an ICBM, had several significant operational shortcomings. It had to be fueled immediately before launch, leading to a slow reaction time (about 15 minutes) after launch order. The radio-command/inertial guidance system of the *Atlas D* was susceptible to jamming and restricted the launch frequency of an *Atlas D* squadron to one missile every 5 minutes (there was only one set of radio guidance antennas, and thus only one missile could be guided at a time). This was rectified with the all-inertial E and F models.

The technique for assembly of the two rockets was significantly different, as the R-7 is mated with all of its booster and payload segments in a horizontal orientation on a railcar. This allowed the checkout to be accomplished within a climate controlled building. The severe weather in much of the Soviet Union may have been a consideration in using this technique, along with the ability to use a single pad to erect another fully assembled and validated rocket ready for flight within hours of a previous launch. This process minimized the amount of activity performed at the launch pad itself.

But, the test pads used by the Americans also provided the ability to static test the completed missile after assembling it. The Soviets had no such capability, and the R-7 had to be static tested at another location prior to being transported to the launch site. The Soviet system was more like an operational site than a test facility.

The Americans elected to vertically stack and prepare the rocket on the launch pad (or later in protected underground silos) which had been the case for most all their large rockets since the earliest days at White Sands. This required more launch pad facilities and a more massive, 150-foot-high launch pad gantry crane with enclosures to allow work in inclement weather. The test launch pad itself was a 3-story structure with intricate erection and servicing facilities rising from the barren Florida beaches. Tons of water cooled the flame bucket and pad during a launch to keep the exhaust from destroying the steel and concrete facilities.

The Soviets elected to use a dry launch pad, in part because of the scarcity of water in the region. A large umbilical tower stands next to the rocket to feed electrical power and exchange data until, as its name sake implies, it disconnects its life supporting cables and gives birth to flight.

Both the *R-7* and the *Atlas* would serve only a short time as ICBMs before being replaced by smaller and more cost-effective missiles. In the case of the Atlas, it would give way to the solid-fuel Minuteman, and the larger, storable-propellant Titan. However, the Atlas and R-7 would serve their countries for more than 50 years as satellite launch vehicles. Most of the 130 Atlas missiles manufactured as ICBMs were refurbished for this purpose. An additional 350 missiles were manufactured specifically as satellites launch vehicles with a variety of configurations, for a life that would finally end in 2005 (the current launcher identified as Atlas IV and V are no longer of the original lineage).

Although the designation R-7 is used to describe the generic rocket, each of the many configurations that were to come had a more precise nomenclature. The Sputnik I launcher was designated 8K71, while the operational version of the ICBM was the 8K74.

The U.S. Department of Defense has its own numbering system for identifying various Soviet launch vehicles and assigned the SL series (SL-1, SL-2 etc.) for satellite launchers. A congressional researcher, Dr. Charles Sheldon, used yet another series that began with A (A-1, A-2 etc.). This publication will simply identify all rockets that were derived from the original ICBM as R-7s to avoid any confusion on the reader's part. The R-7 is the only Russian vehicle that has been used for manned flights and it is still in production with more than 1000 having been launched.

The actual demonstrated orbital payload of the Atlas and the R-7 (without adding upper stages) was essentially the same. The largest satellite orbited by an Atlas was the 2981-pound Mercury spacecraft. The largest satellite orbited by the R-7 was the 2919-pound Sputnik III. The impact of technology was considerable.

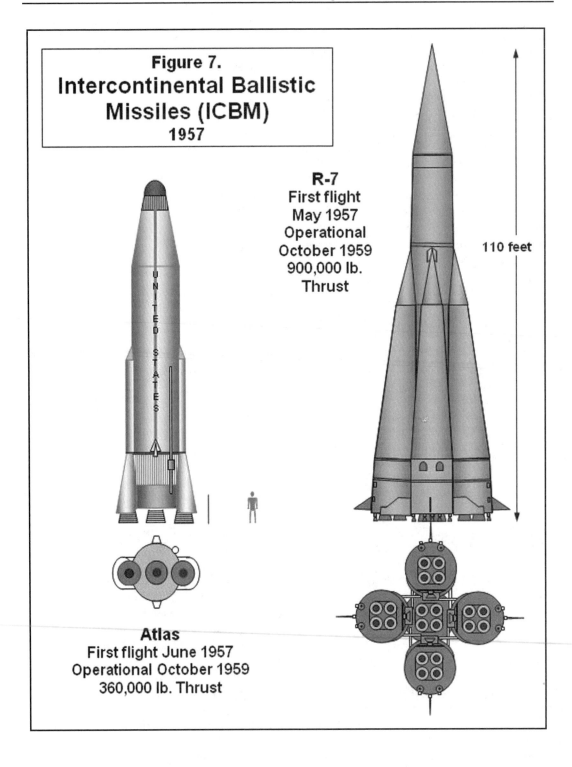

Figure 7.
Intercontinental Ballistic Missiles (ICBM)
1957

R-7
First flight
May 1957
Operational
October 1959
900,000 lb.
Thrust

110 feet

Atlas
First flight June 1957
Operational October 1959
360,000 lb. Thrust

First Thoughts

Manned space flight had been put forth by many visionaries over time, but the idea for a manned artificial earth satellite was first described by Hermann Oberth of Germany in his 1923 publication *Die Rakete zu den Planetenräumen* (The Rocket Into Interplanetary Space). Because the capability of radioing data automatically (a process called telemetry) was not developed until shortly before WWII, the concept of an unmanned satellite was not defined in a scientific paper until the 1940's. The possible uses of low earth orbits (LEO: within a few hundred miles of the earth's surface) for unmanned satellites as envisioned during this period included weather observation and communication platforms, with the author Arthur C. Clarke being in the forefront of speculation. He produced several articles during the 1940's that culminated with his book *The Exploration of Space* that was published in 1952. A later work would become the popular movie *2001: A Space Odyssey*.

The United States military also recognized the significance of artificial satellites for reconnaissance. The U.S. Army Air Force issued a request (classified secret) to the principal aviation companies for the design of an *earth orbiting satellite* in February 1946. The one million dollar contract, issued to the Douglas Aircraft Company in July of that year, was transferred to the newly created RAND (Research ANd Design) Corporation. It concluded that a satellite launch vehicle capable of placing 500 pounds in low earth orbit was possible with existing technology. This weight was considered the minimal useful size for a military application.

The RAND report, *Preliminary Design of an Experimental World-Circling Spaceship* stated that *"The achievement of a satellite craft by the United States would inflame the imagination of mankind, and would probably produce repercussions in the world comparable to the explosion of the atomic bomb."* This last observation was particularly perceptive but was greeted with more than a bit of cynicism by many in the scientific and military community. For the two years that followed, there existed a loose alliance between the Army and the Navy, and several serious moves to proceed with the project.

However, when Rear Admiral Leslie Stevens requested formal R&D funding for the project, it was declared that *"no military or scientific utility commensurate with the costs"* existed. However, it was not difficult for visionaries of this period to see that the technology of the German V-2, then being launched in the southern desert of New Mexico, was advancing the day when man would move into outer space.

The International Astronomical Federation (IAF) held its first meeting at the Sorbonne in Paris, in 1950, to discuss possible satellite projects. The leading groups included the British Interplanetary Society (BIS) and the German Interplanetary Society. The United States was conspicuously absent. However, a group of American scientists, led by Dr. James Van Allen, met in Silver Spring, Maryland, that year to discuss the possibility of an international scientific program to study the upper atmosphere and outer space via sounding rockets, balloons and ground observations. This would ultimately become the foundation for the International Geophysical Year (IGY), which would prove to be the impetus for a satellite program.

A paper presented at the Second International Astronautical Congress in 1951 by Gatland, Kunesch, and Dixon, members of the BIS, was titled *Minimum Satellite Vehicles*. It stated that a relatively small satellite of 20 pounds could yield useful information about the environment immediately beyond the Earth's atmosphere by orbiting at an altitude of about 100 miles. Frederick C. Durant, who represented the American Rocket Society, presented a paper, *Exploration of Space—A Job Calling for International Scientific Cooperation,* prepared by Wernher von Braun (leader of the team that designed the V-2 of WWII), who was unable to attend.

The First Symposium on Space Flight (sponsored by Colliers Magazine) was held at the Hayden Planetarium in New York in October of 1951. It was one of the first opportunities that von Braun had to move out into the public arena, and his presence was met with some protestation. However, the antagonism gradually waned during the course of the meeting as von Braun's charismatic personality smoothed away much of the war-time animosities. The meeting resulted in the exchange of many ideas on not only what the exploration of space might mean to mankind but what might be some of the most appropriate steps for getting there.

At the Fourth IAC in 1953, Dr. S. Fred Singer, of the University of Maryland, presented the "MOUSE," a *Minimum Orbital Unmanned Satellite of Earth*, which was a comprehensive analysis on the subject. A satellite with miniaturized components could provide data on cosmic rays, atmospheric density, and other items of interest to science. Using newly invented transistor electronics, whose thirst for battery power was significantly less than that of large and heavy vacuum tubes, a satellite of perhaps 10 to 15 pounds could be constructed that could be orbited with a V-2 sized launch vehicle. With MOUSE, the scientific community was making a concerted effort to interest governments (specifically the United States) in investing in a satellite program.

The international scientific community had periodically formed cooperative efforts to study various aspects of the Earth and its environment. Prodded by such proposals as Singer's MOUSE , America's National Science Foundation (NSF) in 1954 became involved with planning an International Geophysical Year, a cooperative effort among 50 nations. Set to begin in July of 1957, it would be an 18-month period of scientific exploration of the Earth and the upper atmosphere.

Orbital Mechanics

Orbital mechanics is the science (physics) that defines how a satellite can remain in space. The dictionary defines a satellite as *"A device that orbits a planet: an object put into orbit around Earth or any other planet in order to relay communications signals or transmit scientific data."* A second level definition says, *"A moon orbiting anther body: a celestial body that orbits a larger one."*

To understand orbital mechanics, it is necessary to understand the force of gravity and the acceleration that it imparts to a falling body. Sir Isaac Newton observed and recorded the basic laws of gravity and motion and explained how objects move on Earth as well as through space.

The force of gravity near the Earth's surface results in a form of straight-line acceleration of 32 feet per second (fps), per second. Every second that an object falls, its speed increases by 32 fps. A rock dropped from a rooftop, for example, would start with zero velocity. At the end of one second, it would have a velocity of 32 feet per second, or 21 miles per hour. After two seconds, it would be moving at 2 x 32 or 64 fps.

With respect to the attraction that gravity has on an object, a rock that weighs 10 lb. is pulled toward the earth with a force of 10 lb. at the Earth's surface. In principle, the 10 lb. rock and a feather would both fall with the same acceleration if there were no other forces acting on them. In reality, however, air friction exerts a greater retarding force (drag) on the falling feather than on the rock, and the feather will fall more slowly. The rock will eventually show the effects of drag and will fall at what is called its *terminal velocity*—the point at which the acceleration force is balanced by the drag of air friction produced by the object at a specific speed. The human body has a terminal velocity of about 130 mph. If a skydiver shapes his body into a more streamlined position, he can fall at a faster rate of up to 200 mph.

The gravity of the Earth continues forever (theoretically), but gets weaker the further an object is from the surface. The radius of the Earth is about 4,000 miles. At a distance of 4,000 miles above the surface (twice the radius from the center of the Earth), gravity is only a quarter as strong as it was at the surface (the inverse square of the distance). The rock that weighed 10 lb. at the surface would weigh only about 2.5 lb. at a distance of 4,000 miles above the Earth. Because of the reduced weight force, the rate of acceleration of the rock at that altitude would be only one quarter of the acceleration rate at the surface of the earth.

A second aspect that must be considered is *acceleration* (the rate of change of velocity over time). An accelerating object may be speeding up, slowing down, or changing direction. Acceleration is a vector quantity—it has both direction and magnitude. Newton's Second Law states that the acceleration of an object results from the application of a force. Objects do not speed up, slow down, or change direction unless a force acts on them. The acceleration (a) of an object with mass (m) produced by a given force (F) can be calculated using the equation $F = ma$. A larger force produces a greater acceleration; a larger mass results in a smaller acceleration given the same force.

Placing a satellite in orbit requires accelerating it to an orbital velocity corresponding with the selected altitude. For example, at an altitude of 100 miles above the earth (the lowest altitude that permits a stable orbit), a speed of approximately 17,500 mph (5 miles per second) is required, and the satellite will take about 90 minutes to complete each revolution. As the altitude increases, the required orbital speed decreases. At 500 miles, the satellite will need 16,600 mph and will take 100 minutes for each orbit. At 10,000 miles above the Earth, it travels at 9,400 mph and requires 10 hours to orbit. The moon, at 230,000 miles, is moving at 2,400 mph and requires about 29 days for its orbital journey.

Once a rocket achieves orbital velocity appropriate to the desired altitude, it shuts down its engine and is essentially in a free fall. Because of the curvature the Earth and the high speed of the satellite, the satellite falls around the Earth. The centrifugal force of the satellite balances the gravitational force of the Earth, and the occupant of the satellite is weightless.

Because of the lack of precision in achieving the desired velocity parallel with the earth's surface, most satellites do not achieve perfectly circular orbits. That is, their altitude varies from a high point on one side of the orbit to a low point on the other side of the orbit—essentially an elliptical shape. When this occurs, the low point is referred to as the perigee and the high point as the apogee.

Because there are still some air molecules at these extreme altitudes (separated by several feet), there is some drag on the satellite. Each time the satellite hits one of these molecules, energy is lost— slowing the satellite down and causing it to drop into a lower orbit. In moving to a lower orbit, the speed of the satellite actually increases. Because of descending into a lower orbit, the satellite encounters even more air molecules (higher density), and the increasing drag continues to lower the orbit of the satellite. Thus, a satellite with an initial orbit of 100 miles can expect a lifetime of only a few hours, depending on its size (the amount of drag).

When the altitude of a satellite is lowered to a critical value (slightly less than 100 miles), it will decelerate more rapidly and return to Earth. Because of the substantial amount of kinetic energy required to place a satellite in orbit, this energy is dissipated during the reentry through the aerodynamic force of friction. This force results in the generation of extremely high temperatures. The process is called "re-entry," and typically the satellite will burn up unless it is protected from the heat.

The angle formed between the Earth's equator and the inclination of the satellite's orbit is the orbital plane. Once established at launch, changing the orbital plane even a few degrees requires high levels of energy. Change to the perigee or apogee requires relatively less energy for a change of several hundred miles.

Because the Earth is turning on its axis, the speed of its rotation causes one complete revolution each 24 hours. As the circumference of the Earth is about 24,000 miles, the speed of rotation at the equator is approximately 1000 mph (24,000 miles divided by 24 hours). This speed drops off as the latitude increases—effectively becoming zero at the poles. Thus if a satellite were launched eastward at the equator, the rocket would only need to add 16,500 mph to achieve a 100 mile high orbit. Launch sites further north or south from the equator will have less speed gained from the Earth's rotation. At Cape Canaveral, with latitude 23 degrees, about 800 mph is gained. The Russian launch site at 65 degrees latitude has a gain of only 250 mph. Of course, a satellite can be launched into a higher inclination from the latitude of its launch site and accept the loss of rotational speed as in a polar launch, which has an inclination of 90 degrees and receives no gain from the Earth's rotation.

There are two very special orbits that are available based on altitude. A geostationary orbit at 22,000 miles above the Earth allows a satellite to remain over the same point on the Earth if the launch is into an equatorial plane. If launched at any inclination greater than the zero degrees, the satellite will scribe a figure eight over a specific point which is a geosynchronous orbit. The center of the eight will be on the equator with the top and bottom excursions of the eight representing the orbital inclination.

The second type of special orbit is a 'Mid-Earth' orbit, which is about 12,500 miles above the Earth. Here the satellite will require twelve hours to complete each orbit, causing it to appear over the same point on the Earth every two days.

Travel from the Earth to the moon requires placing the spacecraft in a highly elliptical orbit so that the spacecraft's apogee will intercept the orbit of the Moon at the desired point.

Every satellite passes directly over a path or track that can be traced on a map of the Earth (Figure 1). In most instances, a Mercator map is used which distorts the size relationship of the Polar Regions. Thus, Greenland looks like it is much larger than the United States for example, but is actually about one-half the size. However, because position is more important than size relationships, the Mercator works well for this application. Note in Figure 1 the ground track of a satellite as launched from Cape Canaveral into a 100 mile circular orbit inclined 25 degrees to the equator. Its first orbital track is depicted as a solid line. As it completes the first orbit by crossing 80 degrees west Longitude, it begins its second orbit shown with a dashed line. Note that the ground track between the two orbits is displaced by 1,500 miles at the equator—illustrating the rotational affect of the earth over the 90 minute period of the orbit. Also observe that a 25 degree inclination will result in the satellite never passing over any part of the earth with a higher latitude.

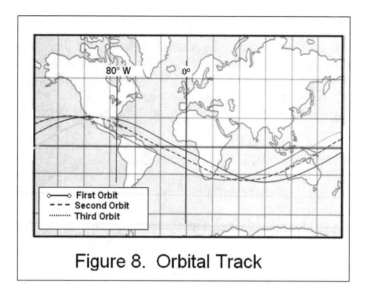

Figure 8. Orbital Track

There is another set of five special orbits called the Lagrangian or L-points. These are the result of the gravitational force exerted by two independant bodies on a third (and typically much smaller) mass. These L-points allow the third object to remain in the same relative proximity of the two larger bodies. In particular, a space probe orbiting the Sun between the Earth and Venus should have, according to Kepler's laws of orbital mechanics, an orbital period less than the Earth's 365 days. But if that object is positioned relatively close to the Earth (within a few hundred thousand miles), then the Earth's gravitational field will counter some of the Sun's pull, and the probe will orbit with the same period as the Earth. This position, shown in Figure 2, is known as the L-1 point and will have constant sunlight.

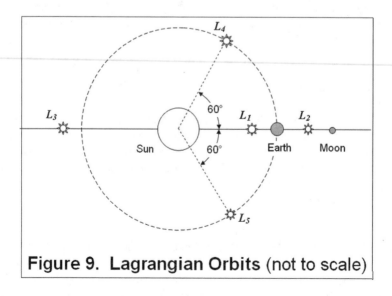

Figure 9. Lagrangian Orbits (not to scale)

The second point, L-2, is on the outside of the Earth's orbit. Here the cumulative affect of both the Sun and the Earth's gravity fields will cause a satellite in this posiiton to again have the same orbital period as the Earth, despite being farther from the sun. This position will have the satellite in the constant shade (occulatation) of the Earth and would be ideal for a space telescope.

The L-3 point lies on the opposite side of the Sun from the Earth and has the same period as the Earth. Thus, it is never visible from the Earth and is the theme of several science fiction stories of a "counter-Earth planet".

The L-4 point is defined as the third point of an equilateral triangle where the first two points are the Sun and the Earth. This third point is defined by a line extending out 60 degrees from the central body (the Sun), and intersecting the side formed by a line extended from the secondary body (the Earth). The L-5 point is determined in a like manner on the other side of the Sun-Earth base line.

Von Braun's Vision

For Wernher von Braun the years between 1946 and 1950, although filled with a variety of assignments in support of Project Hermes, did not provide an opportunity to move beyond the V-2 in developing a new and larger rocket. He returned to Germany briefly in March of 1947 to marry Maria von Quistorp. The more relaxed pace at his quarters in Ft. Bliss, Texas, was a significant departure from the hectic wartime schedule he had pursued in Germany, and it allowed him to reflect on the direction for mankind in space.

During this period, von Braun embarked on a study of sending a large expedition to Mars, using the existing technology of the V-2. He expressed the scenario in the form of a novel, but provided extensive notes in an appendix to support all of the technical assertions. *Mars Project* was an insightful look at what could be achieved if (and when) man decided to travel beyond the gravitational bonds of the earth. The expedition that he envisioned would consist of ten ships, each weighing 4,000 tons and operated by 70 men. Making the trip in formation, the flotilla would carry all the provisions to be self-sufficient for more than two and one-half years.

Following publication of *Mars Project*, von Braun joined with author Cornelius Ryan to write a series of articles on manned space flight for Collier's Magazine, a popular monthly publication. They appeared in serial form in 1952. To illustrate these articles, artist Chesley Bonestell was called upon to bring to life von Braun's vision, coupled with Bonestell's own creative imagination of what these interplanetary vehicles might look like in the realistic environment of space, the moon and Mars. The results were startlingly realistic paintings that depicted places that man had yet to visit, and would fire the imagination of many.

What may have made the Collier's articles so enthralling to the nonprofessional was that von Braun was not simply addressing the hardware of fire and steel, but man himself. While he recognized that some aspects of space exploration could be achieved with unmanned probes, he felt strongly that a human crew must be present to observe, correlate, and report. However, the simple presence of man requires extensive life support systems to allow him to breath, eat, and dispose of waste products. Additionally, his frail body might be adversely affected by radiation, heat and cold, weightlessness, and acceleration during powered portions of the flight.

As a stepping-stone to the moon and Mars, von Braun proposed a manned space station in the shape of a large wheel, orbiting about the earth at an altitude of 1,000 miles. Because of the then unknown affect of weightlessness on the human body, the wheel (some 300 feet in diameter) would rotate slowly to generate its own gravity (one-forth that of Earth) through centrifugal force. In the March 22, 1952, issue of Collier's, von Braun estimated that it would require ten years and $4 billion to achieve the space station. He had the preliminary design of a massive, 265-foot, three stage, 7,000-ton rocket needed to carry 36 tons of cargo. It would be powered by 51 improved V-2 engines that would each generate 500,000 pounds of thrust.

In the days before solar cells and nuclear isotopic generators, electrical power for the space station was provided by a set of polished mirrors on top of the wheel, which directed sunlight to turn liquid Mercury into vapor that would drive turbines and generators to produce the electricity.

In von Braun's vision, one of the primary reasons for the space station being in a polar orbit was that every point on earth could be seen (and photographed) at least once every 24 hours—a point that he hoped would not be lost on his military readers.

Walt Disney, the man who made Mickey Mouse a household pet and who was an aviation enthusiast, had read the Colliers articles with great intensity. In 1954 he was eager to round out a new weekly TV series called *The Wonderful World of Disney* with futuristic segments that would highlight "Man in Space." In collaboration with von Braun, Disney's animators brought to life his imaginary creations. Yet another younger generation was being acquainted with the prospects for the future.

All three of these endeavors (the Mars novel, the Colliers articles, and Disney) were viewed by von Braun as critical to getting his ideas of going into space before the public (and especially the unencumbered minds of the young) so that government funding would be easier to obtain. The first of three shows, 'Man In Space' was aired in March of 1955 and President Eisenhower is reported to have requested a copy.

The famous "Roswell Incident" that claimed an alien spacecraft had crashed in an isolated part of the New Mexico desert, not far from the White Sands Proving Grounds in 1947, had re-ignited the wide spread speculation that we are not alone in the universe. Several movies in the early 1950's provided additional motivation to this end, including *The Thing*. The spacecraft envisioned by the serious futurists, such as von Braun, were also a part of the entertainment media's contribution to spurring the imagination. Both *Destination Moon* and *The Conquest of Space* used realistic spacecraft and reasonably good special effects that presented some of the perils of space flight with a degree of technical accuracy. The public were slowly being drawn to the possibilities of space travel.

Early Soviet Planning

Before the Soviet hierarchy discovered that the ideas of several of its visionary scientist, such as Mikhail K. Tikhonravov, were in fact bordering on reality, they did not actively try to hide their work nor their presence as they eventually would. An article, in the October 4, 1950, edition of the New York Times by Tikhonravov, said U.S.S.R. science made feasible space flight and creation of an artificial earth satellite. He also reported that U.S.S.R. rocket advances were equal to or exceeded those in the West. *"More groundless boasts from the Soviets,"* was a common response from those in the West.

In mid-1953, press coverage about the possibility of satellites, especially von Braun's Collier's series, prompted Korolev to ask Tikhonravov to prepare a presentation for the Academy of Artillery Sciences that would support a request for an official Soviet space program. The information collected included a large number of clippings from the American and European scientific and popular press, and tentative American plans for artificial satellites.

Accompanying the presentation was a detailed description of a proposed program complete with calculations that showed that Soviet technology was capable of achieving these goals. And, by using the *R-7*, then in the planning stages, a Soviet satellite could be more than ten times the weight of the satellites being described by the Americans. With Stalin's death in March of 1953 and the subsequent execution of Beriya, Korolev felt more open to move these ideas forward without fear of reprisals. However, he also realized that success of the ICBM program was the key to the satellite proposal.

The presentation was impressive, using references to Tsiolkovsky's 1903 mathematical proofs that a device accelerated to a specified velocity could achieve orbit. Nevertheless, Academy president Anatoli Blagonravov was divided in his thoughts. The idea of the Earth satellite was still science fiction as far as most of the attendees were concerned, and he was reluctant to embrace the concepts for fear of ridicule. Although lacking enthusiastic support, approval was given in September 1953 for a satellite research study.

Korolev sought support for his satellite proposal from the Soviet Union's Academy of Sciences; as his R-7 funding was based on military needs, he needed broader justification. He was aided by mathematician Mystislav Keldysh, who, beginning with his appointment in 1946 as director of NII-1, held a series of progressively more important positions in the Soviet scientific hierarchy. By May of 1954, Dimitri Ustinov, Minister of Armaments, approved the satellite plan only a week after the ICBM project was given the go-ahead. Korolev was as determined as von Braun was to move mankind into space.

The Soviet's expectations were quite high with their first proposals. The description of what was termed the simplest satellite included an evaluation of various possible orbits, its ability to be viewed in the twilight sky, attitude control systems, power sources, and onboard instrumentation. Even the possibility of recovering exposed film of the Earth's surface and a 660-pound television system for transmitting images of the earth were considered.

With a total weight capability of 3,000 pounds using the R-7, it was anticipated that an animal container might be installed on later "simple satellites." Follow-on aspects included manned satellites, a space station, and flights to the moon. The report was the equivalent of von Braun submitting his visionary articles to Congress rather than Colliers. The proposal concluded with a note that creation of an artificial satellite could be of great importance to the defense of the Soviet Union (the military tie-in for high priority use of the R-7) and that the first satellite would also have vast political significance (to get the backing of the Communist Party). The memo, entitled *A Report on an Artificial Satellite of the Earth,* resulted in only modest funding and lukewarm priority for the satellite program, which continued at a slow pace through the mid-1950s.

With such an aggressive plan, it appears somewhat surprising that the Soviets allowed the deadline for participation in the IGY to expire in May 1954 without submitting any formal satellite proposal. In Rome in October 1954, Soviet scientists passively witnessed the IGY committee approve a US-sponsored plan to orbit satellites during that eighteen month period. It may have been simply that the delegation had no real knowledge of the Korolev plan.

This situation was apparently corrected on the delegation's return to the Soviet Union when the Soviet Academy of Sciences established a commission to coordinate information on the satellite program in April 1955. Academician Leonid I. Sedov, a well-known gas dynamics expert, was appointed chairman. In the coming years, Sedov would be looked to by the west as the spokesman for the Soviet space program. However, in reality he was given a minimum of information and lacked any real knowledge or expertise in the program itself. Sedov reportedly ordered a copy of Disney's 'Man In Space' TV episode.

The Redstone/Jupiter C

While von Braun was creating visions for the American people in the early 1950s, he was also finally creating a new, large rocket for the Army at the revitalized Redstone Arsenal in Huntsville, Alabama. In August 1949 General Toftoy took a trip to Huntsville to review the possibilities of relocating the growing Army ballistic missile team. He obtained the Army's old Redstone Arsenal, and in 1950 preparations were made to move the von Braun team to its new home with the U.S. Army Ordnance Missile Command.

The small rural town in northern Alabama somewhat reluctantly accepted the influx of the German team when they transferred from Fort Bliss in 1950. The Federation of American Scientists was not quite as friendly. They proclaimed the German presence was *"an affront to the people of all countries who so recently fought beside us."* Although the Germans had made a good impression on the people of El Paso, the town closest to Fort Bliss, the residents of Huntsville remained divided for several years.

The Air Force was separated from the Army to be a co-equal branch of the military in 1947 in a reorganization of the *War Department,* which itself was renamed the *Department of Defense.* Among the many changes taking place was a new adversarial relationship for funding. Inter-service rivalry between the Army and Navy had been a part of the military arm-twisting in Congress for a century. Now three branches had to vie for the tight defense dollar. Gold Braid bickering in the Pentagon over the assignment of roles and missions of the respective services saw the responsibility of the guided missile restricted to *tactical* ranges for the Army. This meant that von Braun's team, the most experienced rocket technologists in the world, would be constrained to developing weapons of not much greater range than the V-2.

An outgrowth of the Hermes C-1 program of the late 1940's was a ballistic missile of 500-mile range. Limited funding and higher priorities resulted in lagging progress on the C-1 during that period. The advent of the Korean War in 1950 caused a sharp upward reversal in these trends and the C-1 project transferred from General Electric to the Army's new Guided Missile Center at Redstone Arsenal. The missile received its formal designation of SSM-G-14. With the anticipation of a large thermonuclear warhead, the payload requirement significantly increased, reducing the planned range from 500 mile to 250 miles.

Design work of the 69 foot, 60,000-pound *Redstone* (as officially named) was completed in 1952. It employed a single North American Rocketdyne NAA75-110 liquid-fueled rocket engine of 75,000 pounds thrust. Although it was a much more capable missile than the V-2, it still used liquid oxygen and employed vanes in the rocket exhaust for control. Firing the *Redstone* in combat conditions required 20 large vehicles and 8 hours to complete.

On the surface, the *Redstone* didn't appear to present much in the way of advanced technology as compared to the Soviet's R-5. However, the *Redstone* would be the first American combat missile with an all-inertial navigation system, consisting of a stabilized guidance platform with air bearing gyros, air bearing accelerometer, and air bearing leveling pendulum. Air bearings are almost frictionless and allow for less precession or drift in the gyro and result in very high levels of precision. This precision allowed the *Redstone* to deliver its separable, four-megaton, W-39 thermonuclear warhead to within 1,000 feet of its intended target.

In reviewing the basic design philosophy of the Soviet and German teams, it was obvious that both had a preference to using existing, reliable technologies as evidenced in the *R-7* and the *Redstone*. The more typical American approach, as illustrated in the *Atlas*, was to develop and employ the highest levels of technology to arrive at the least expensive product. This method entails higher research and development costs but significantly lower per unit costs. It also reflected the "commodity" mentality ingrained into American society.

The first successful flight of the *Redstone* occurred in August 1953, and by 1955 the Chrysler Corporation began production with its first missile launched in July 1956. The first fully operational *Redstone* unit was deployed to West Germany in June 1958.

Prompted by the 1953 IAC Conference Proceedings, representatives of the Army (including von Braun) and Navy met in December 1954 to consider establishing a satellite program. Adding to the urgency was intelligence that indicated the Russians were considering such an endeavor. The *Redstone*, being the largest US missile successfully flown, represented a capability that could be extended by upper stages to reach the required orbital velocity. Using clusters of small Loki solid-fuel rockets to form second and third stages, and a single Loki as the fourth stage, an inert slug, approximately two feet in diameter and weighing five pounds, could be injected into orbit. The Loki itself was only three inches in diameter and 60 inch long and produced about 1,000 pounds of thrust each—thus the need to combine so many (37 in total) to achieve the required power.

Although there were more capable upper stages that could have been considered, the use of the relatively inexpensive Loki arrangement was simple and would accomplish the minimal task in the shortest possible time. This proposal was known as *Project Slug*. A low-cost, un-instrumented satellite such as *Slug* could confirm launching techniques, basic orbital behavior, and allow for the perfection of tracking methods. More sophisticated satellites could subsequently be placed in orbit, with the capability to use telemetry to gather data on conditions outside the sensible atmosphere.

At about this same time, the Jet Propulsion Laboratory produced a scaled down version of their *Sergeant* solid-fuel rocket, which provided higher Specific Impulse than the Loki. Using a cluster of "scaled" *Sergeants* to replace the Loki, a payload of 10 to 15 pounds was possible and allowed for limited scientific experiments to be placed in orbit. Dr. Singer's MOUSE was one-step closer to reality. This plan also took on a more positive name—*Project Orbiter*. Because the technology was available, a concerted effort could have produced a satellite in 1955 for less than a million dollars, and a more conservatively funded project could have launched in the summer of 1957.

As *Project Orbiter* began to gather momentum in early 1955, another missile program was introduced that would have both positive and negative impact on it. The Army had convinced Congress that an intermediate range ballistic missile (IRBM) of 1,500 miles range would be a good follow-on to the *Redstone*. Basing missiles farther from possible initial attack by Russian dominated Warsaw Pact countries, an IRBM could be launched from less exposed locations in England, Italy or Turkey. These missiles could reach not only tactical targets in a European war but also targets well into the eastern part of the Soviet Union itself. The *Jupiter* IRBM (as it was named) would be powered by single Rocketdyne LR-89 engine. Although from the start *Jupiter* was challenged by the Air Force as intruding into its strategic domain, it was allowed to continue while the Air Force developed a virtually identical missile, the *Thor* that used the same LR-89 engine.

When the Ordinance Corps requested funding for *Project Orbiter*, higher echelons in the Army refused, using the commitment to *Jupiter* as taking higher priority. However, von Braun, encouraged by General John B. Medaris, the new head of the Army Ballistic Missile Agency (ABMA)—the new name for the facility at *Redstone* Arsenal—was not to be denied.

To develop and test various *Jupiter* systems, General Medaris authorized von Braun to modify a set of *Redstone* missiles with lengthened fuel tanks and up-rated engines. The first of these was designated *Jupiter A* and outwardly looked identical to the *Redstone*. A second group of twelve, designated *Jupiter C*, would prove the technology for protecting the warhead during heat of atmospheric re-entry. To achieve the high velocities needed to replicate the 10,000 mph speed of the IRBM re-entry into the atmosphere, von Braun configured the *Jupiter C* to the same specifications as he had envisioned for the launch vehicle for *Project Orbiter*.

Eleven "scaled" solid-fuel *Sergeants* rockets sat atop the *Redstone* (as its second stage), arranged around the perimeter of a cylindrical aluminum "bucket" that was 30 inches in diameter and four feet long. Each of these rockets was only six inches in diameter and three feet long and produced 1,800 pounds of thrust. Nestled within these were an additional three *Sergeants* as a third stage. The nose cone to be tested was then set on top of the bucket instead of a fourth stage! The entire bucket was spun-up by an electric motor prior to launch, and this rotation was maintained by on-board batteries during the first stage firing.

Following burnout of the *Redstone* first stage, the forward segment that contained the guidance system, topped by the spinning tub, separated. The unit continued to coast to about 150 miles altitude. Aligned parallel to the Earth's surface, using small gas jets operating under the command of the guidance unit, the second stage would then fire. The spinning solid-fuel rocket assembly would accelerate the nose cone, leaving the guidance unit behind to follow the *Redstone* booster's plunge back to earth. Each solid-fuel stage burned for 6.5 seconds before separating, to allow the remaining segment to ignite and proceed. The reason for the spin imparted to the solid-fuel stages was to provide a crude, but effective means of stabilizing the rocket's direction. Since there is virtually no atmosphere at 150 miles at which the ignition takes place, fins would not have been effective. If a fourth stage were used in place of the nose cone, orbital velocity could be achieved.

In yet another effort to initiate a satellite program, the American Rocket Society was working with the National Science Foundation (NSF). The NSF was involved with planning the IGY, the 18-month period of scientific exploration of the Earth and the upper atmosphere that was set to begin in July of 1957. The NSF considered three satellite proposals: the Army's 15-pound *Project Orbiter* using the *Jupiter C (Redstone)*, the Naval Research Laboratory's 30-pound Project *Vanguard* using the *Viking* sounding rocket as a basis, and an Air Force proposal to use the *Atlas*. The last-mentioned, which could orbit several thousand pounds, was not expected to be available until 1960.

When Werner von Braun started to lobby for a satellite program, he discovered that the Eisenhower administration was not only tight with a dollar, but they shared little of his enthusiasm for the project. Eisenhower was also very sensitive to the use of military hardware for

A Jupiter-C launch

this peaceful scientific endeavor, especially since an Earth satellite would pass over the airspace of other countries. Satellites presented new interpretations of airspace sovereignty because they were technically not in the *airspace* but above it in *outer space*. Not to be overlooked in this decision was the notion that several members of the selection committee could not accept *"an American space project riding into orbit on a rocket designed by engineers hitherto employed by Nazi Germany."*

On July 29, 1955, White House Press Secretary James Hagerty officially announced that the United States would launch an Earth satellite as a part of the IGY that was scheduled for July 1957 through December 1958. The Administration had been very generous in its funding of the IGY endeavor, and after much soul searching, Congress and the administration agreed to fund a satellite project as a part of it. To von Braun's dismay, the Navy's *Vanguard* proposal was accepted—his *Jupiter-C* would not be used. There were four dissenting votes by the committee that had made the selection. One of them was the "Lone Eagle," Charles A. Lindbergh.

Several historians claim that Eisenhower actually preferred that the Soviets launch first to establish the *right of over-flight*. The primary reason for Eisenhower's sensitivity was that the U.S. Air Force was well along with secret plans for a series of military reconnaissance satellites (Project WS-117L) which would be scanning the Soviet Union from orbit by 1960, and he wanted the Soviets to set the precedent of freedom of space much like freedom of the seas.

While this assertion has never been fully proven, Ike's focus on military surveillance of the Soviets is apparent with his support for both the U-2 and A-11 (SR-71) spy planes. While the U.S. may have had the upper hand with more capable bombers and more sophisticated and lighter warheads, the vastness of the Russian country made targeting and navigation to a target, a very difficult problem for U.S. Military planners. Ike was determined not to allow the U.S. to be subject to a nuclear "Pearl Harbor."

The impenetrable Soviet Union had to be mapped and essential military targets located. The U.S. had already pushed the limits with "illegal" flights into Russia by reconnaissance aircraft such as the RB-29 and RB-47 in the early 1950's as well as the U-2 in the later part of that decade. Several of these flights had been shot down, and American lives had been lost. America's claim that these aircraft were "off-course weather surveillance flights" was growing thin on the diplomatic side as well. For Ike, the ability to over-fly the U.S.S.R. legally with satellites was of paramount importance.

The Soviet Satellite Program

One day after the announced plan for a small earth satellite to be orbited by the United States, Leonid Sedov convened a press conference at the Soviet Embassy in Copenhagen, Denmark, where he was attending the Sixth Congress of the International Astronautical Federation.

Sedov stated, *"In my opinion it will be possible to launch an earth satellite within the next two years and that the realization of the Soviet project can be expected in the near future."* Referring to the press reports of the United States' announcement, he added, *"From a technical point of view, it is possible to create a satellite of larger dimensions than that reported in the newspapers which we had the opportunity of scanning today. The realization of the Soviet project can be expected in the comparatively near future. I won't take it upon myself to name the date more precisely."*

He was unwilling to elaborate to the press about Soviet plans because he had no real knowledge of them. His pronouncement was typical of Soviet propaganda of the time. He was simply parroting what he had been instructed to say. Nevertheless, back in the Soviet Union, work was preceding at an increasing pace that was accelerated by the American announcement. Yet official government approval had yet to be granted.

It was not until January 30, 1956, that the Council of Ministers finally issued a decree authorizing development of the satellite. One of the caveats was apparently the need for a military photoreconnaissance satellite based on the capabilities of Korolev's proposed satellite. Korolev also understood that only when the *R-7* had completed its initial flight-testing would a satellite follow. He was attuned to the political reality and noted, *"the creation of a satellite will have an enormous political significance as evidence of the high development level of our country's technology."*

Korolev arranged for Mikhail Tikhonravov to join his team, transferring from Special Design Bureau #385. Korolev then met with Tikhonravov and Keldysh, to discuss proposals for satellite instruments that scientists had submitted to the commission during the past year. The study of the ionosphere, cosmic rays, magnetic fields, and luminescent in the upper atmosphere were high on the list.

Even with the January decree, it was difficult for Korolev to get the other institutes to meet schedules for the wide variety of components needed to produce their satellite. It seemed that he, or one of his high-level assistants, always had to sell the project to members of the other design bureaus as well as other government agencies. They also had to deal with the objections of the military generals, many of whom felt that the satellite project would hinder the development of the ICBM.

Project Vanguard

The resulting IGY satellite decision in America gave the task of launching "the first" artificial Earth satellite to a Navy team. They would use the rather modest, non-military *Viking* sounding rocket, built by Martin Aircraft, as a basis for developing a new and extremely efficient three stage vehicle defined by Milt Rosen, Director of the Viking Program. The name of the program, the rocket itself, and the satellite it was to launch was *Vanguard*. The dictionary defines the word as "in the forefront" and was suggested by Rosen's wife.

Vanguard would have to endure all the teething problems of a major high-tech endeavor on a shoestring budget. *Vanguard* had to overcome the staging problems that the *Atlas* and the Soviet *R-7* had avoided. In addition, it had to place into orbit a satellite capable of doing useful scientific work but weighing less than 30 pounds.

To accomplish its task, *Vanguard's* 44-inch-diameter first stage would be powered by the X-405, a 27,000 lb. thrust engine—half the power of the 15-year-old V-2. The engine was developed by General Electric as a part of the Hermes program. The second stage was based on the Aerojet Corporation's *Aerobee* sounding rocket. A solid-fuel third stage, initially built by Grand Central Rocket Company, completed the 22,000-pound, 72-foot-tall rocket. With a mass ratio of .86, the rocket was truly state-of-the-art although virtually the minimal size for the task to be accomplished. On the surface, *Vanguard* was an impressive and elegant program, whose initial schedule would orbit a satellite within 30 months (by the end of the IGY in 1958). The world was told that the first Earth satellite would be American, it would be called *Vanguard*, and it would weigh 30 pounds.

Two additional *Viking* rockets were built by the Martin Company, following the launch of *Viking XII* in 1955, to support *Project Vanguard*. Initially referred to as *Viking XIII* and *XIV*, these would be re-designated

A Vanguard launch vehicle at Cape Canaveral

TV-0 and TV-1—Test Vehicles for the *Vanguard* program. TV-0 launched from Cape Canaveral in December 1956 with a basic radio package that simulated the instrumentation of the satellite itself. Down range tracking facilities verified the usability and fidelity of the signals as the rocket arched over into the initial trajectory that *Vanguard* would follow, reaching an altitude of 125 miles, before intentionally falling into the Atlantic Ocean.

TV-1 was more ambitious as it contained the unguided solid-fuel third stage and the spin-up table used to stabilize its flight. Its launch, in the summer of 1957, proved these systems, and *Project Vanguard* was poised for the flight of TV-2, scheduled for October of 1957. It was the first vehicle to resemble the final satellite launch configuration, but only the first stage would be tested. Events at Baykonur, some eight-thousand miles away, would shortly change the complexion of *Project Vanguard*.

Second Thoughts

The Army continued to make known that its *Jupiter C* was available as a backup for *Vanguard*. Presentations were made to the Ad Hoc Study Group on Special Capabilities in April 1956, but the group concluded that *Vanguard* was not encountering any serious difficulty, and backup was not necessary. The Army was directed not to make plans for either the *Jupiter* or *Redstone* for satellite launches.

When Dr. Ernst Stuhlinger presented a paper to the Army Science Symposium, in June of 1956, on the ABMA potential to launch an earth satellite, he inadvertently placed General Medaris in an awkward position. Major General Andrew P. O'Meara of the Army's R&D directorate questioned General Medaris on how ABMA had funded its satellite program. General Medaris quickly replied that the *Jupiter C* was a re-entry test vehicle for the *Jupiter* program that could provide a dual role as a satellite launcher.

Almost a year after the decision was made to go with the *Vanguard* program, a memo was drafted in June 1956 to the assistant Secretary of Defense for Research and Development. It reported on the status of the *Vanguard* project and raised the possibilities of allowing the Army to launch a satellite in late 1956 or early 1957 using the *Jupiter C* (the modified *Redstone).*

A response to the "second thoughts" memo, dated July 5, 1956, discouraged consideration of the *Jupiter C* proposal, in part because adequate tracking and data handling facilities were yet to be completed. It also noted that an early launch, before the start of the IGY, would not satisfy the U.S. IGY obligation.

These recently declassified memos from the Eisenhower administration indicate that there was little or no awareness on the part of key Eisenhower advisors as to the impact that the first satellite would have on world opinion. There was, however, mention of the morale problem that an Army launch would impose on the Navy team!

Yet the psychological and political impact of a first launch by the Soviets had already been assessed. Former Presidential Science advisor James R. Killian writes, *"In 1955, Nelson Rockefeller presented a study to the Operation Coordinating Board predicting psychological shock if the Soviets were first* [to launch a satellite]. *"*

Knowing how adamant the von Braun team was to launch a satellite, the Commanding General of ABMA (von Braun's boss), John B. Medaris, was advised by the administration that the Army was not to launch a satellite "by accident" during its nosecone tests with the *Jupiter C*. This was not really possible under the conditions of the test program, but the people in Washington did not want their lack of technical knowledge to be used to von Braun's advantage.

The first flight of the *Jupiter C* occurred in September 1956 when the missile flew more than 3,300 miles, reaching an altitude of over 600 miles. Following successful completion of the re-entry test program objectives in the summer of 1957, after only three *Jupiter C* flights, the remaining nine vehicles were carefully stored away. The flight hardware necessary to accomplish the orbiting of a satellite was essentially on hand by 1956. All that was needed was the authorization.

There were some who recognized that both the *Vanguard* project and *Project Orbiter* were entirely too modest. Dr. I.I. Rabi, a distinguished physicist and then chairman of the Office of Defense Mobilization Scientific Advisory Committee, alerted the Eisenhower administration in 1956 that the *Vanguard* satellite was *"too small and that we should be making bolder plans for much larger satellites... in view of the **competition*** [emphasis added by the author] *we might face".* His advice was to prove prophetic. Here was perhaps the last opportunity for America to defuse a possible space race. However, no one who could do anything was listening. The CIA reported in the summer of 1957 that the Soviets had the capability. Nevertheless, complacency was dominant in America as October 4th drew near.

Object D

Nikita Khrushchev had finally succeeded in subduing most of the opposition to his leadership of the Soviet Union by the time of his February 1956 visit to review progress with the *R-7* and other rocket programs. Unlike Stalin, who had initiated the rocket program and kept his finger on its very pulse through his NKVD hit-man Beriya, this was to be Khrushchev's first top-secret briefing on the direction being taken in the ballistic missile program.

On viewing a mock-up of the *R-7,* he was overwhelmed by its size. He stood in silence as Korolev explained the United States satellite program and derided the U.S. launch vehicle *Vanguard*. Korolev emphasized the capabilities of the *R-7* and how it would position the Soviet Union as a leader in space exploration. Korolev also noted that the major costs of *R-7* development were being borne by the military; the added costs of its use as a launch vehicle would be comparatively small. Khrushchev, still somewhat awestruck, simply nodded his approval: *"if the main task* [of developing the ICBM] *doesn't suffer, do it."* With that, the Soviet satellite program had been elevated to the highest political levels and had survived.

The Soviets used a simple method of designating the payloads for the R-7; the progressive letters of the Cyrillic alphabet. Thus the warheads were identified as A, B, C, and etc.. But the Russian alphabet is not a one-for-one replacement of English characters. Thus the "D" is the fifth in the series, and it became the designation of the Soviet satellite.

By the time of Khrushchev's February 1956 visit, most of the technical requirements for building the satellite had been established, and design and fabrication began in March. While the *R-7* was hardly a viable ICBM because of its size, cost, and need for elaborate launch preparations, it would provide the Russians with a relatively reliable vehicle that

Soviet Premier Nikita Khruschev

could be used to launch large earth satellites. The velocity needed to throw a warhead 6,000 miles is about 15,000 miles per hour. To achieve low earth orbit (100 miles), a velocity of 17,500 miles per hour is required. The changes to the *R-7* would be minimal but did require up-rated main engines and a new payload faring.

The primary scientific goals for the satellite were formidable but built on experiments devised for the vertical sounding rockets. The basic shape would be a cone weighing about 3,000 pounds. Because Object D would be out of direct contact with the radio receivers on the ground for long periods of each orbit, some form of storing the scientific information had to be engineered—a tape recorder. The first American satellite would simply radio its findings continuously, using a virtual worldwide network to receive and record the data for later transcription and analysis.

Electrical power for Object D would be supplied by a combination of solar and chemical batteries, with an emphasis on maintaining a reasonable operating temperature by the use of fans. A test model that was expected to be completed by October of 1956 was not completed until late November. Schedules were beginning to slip.

Sputnik: Fellow Traveler

Chief Designer Sergei Korolev was driven by the desire to beat the Americans into space. He had been able to gauge the intellect and passion of Wernher von Braun through the Colliers articles and recognized that he faced a more than formidable opponent. When he heard the reports of the first *Jupiter C* launch in September of 1956, he expected that von Braun would make an effort to secretly launch a satellite (with or without official approval of the American government). Although he was disdainful of the capabilities of both the *Jupiter C* and *Vanguard*, he did not want his creation, no matter how superior it was to the Americans in size or complexity, to be *second* to orbit.

Korolev was very critical of the support organizations that consistently were unable to meet their schedules for components of Object D: *"...it would be good if the Presidium were to turn the serious attention of all its institutions to the necessity of doing this work on time...we all want our satellite to fly earlier than the Americans'."* Attempts to motivate key engineers and managers included the phrase *"we must beat the Americans."*

Korolev suffered additional anxiety when the first static tests of the RD-101 engine unit revealed a specific impulse that was about 2% below what was expected.

Of course, all of the satellite plans were dependent on the *R-7* completing its first successful tests. In retrospect, Korolev had set an extremely optimistic schedule. Most large rockets developed to that date had required dozens of launches to achieve a reasonable reliability. The *R-7* represented an order-of-magnitude in complexity beyond anything that had left a launch pad.

With the development of Object D falling behind schedule and the *R-7* not meeting its anticipated performance levels, Korolev decided that a smaller satellite might be the answer to all of his known problems and perhaps some that he had yet to encounter. In late November of 1956, he requested that he be permitted to launch two smaller and less sophisticated satellites during the early summer of 1957, before the official start of the IGY on July 1st, with the intent of heading off the American effort.

In support of his revised plan, Korolev reported: *"It is known that the USA is preparing in this nearest months a new attempt to launch an artificial earth satellite and is willing to pay any price to achieve this priority."* Where Korolev got this information is of some interest. From all outward appearances, *Project Vanguard* was slowly progressing according to its publicly announced schedule towards its first satellite launch no earlier than December of 1957 (more than a year in the future) and probably later. A satellite launch by the *Jupiter C* had been blocked by the Eisenhower administration, but, as noted, there were still some in the government who recognized that a "race" was in the offing and were still promoting *Project Orbiter*.

Therefore, a smaller Soviet satellite, now designated Simple Satellite Number One (Prostreishiy Sputnik) or *PS-1,* came into Korolev's plan. This plain, polished 184-pound sphere would contain only two radio transmitters, batteries (sufficient for ten days), and temperature measuring instruments. Within a month the design was approved, but it would be launched only after one or two successful *R-7* flights. Object D was rescheduled for April 1958. Two PS-1 flight article spheres were assembled, one for the launch and another as a back up, which would also serve for ground testing. The second one would later be launched too, but with more dramatic results than its precursor.

Calculating the appropriate trajectory was a significant problem for the Soviets because they lacked basic electronic computing facilities, which were then in their infancy. Most of the calculations were done by hand.

In keeping with the nature of a closed society and tight military secrecy, the Russians were not about to announce when, or what, they intended to launch. Instead, they made a series of generalized statements that were again taken as groundless boasts by much of the world. The first official announcement of the Soviets intent to place a satellite in orbit during the IGY was made in September of 1956.

In May 1957 the Soviets published the frequencies that they would use for the satellite transmitters. The two radio transmitters operated on frequencies of 20.005 and 40.002 megacycles with the signals on both frequencies lasting 0.2 to 0.6 seconds and providing the famous beep-beep sound that would forever separate the old from the new. The lower frequencies were chosen (as opposed to the higher 108 megacycles used with *Vanguard*) because they were available on most short wave sets in use around the world—hundreds of thousands of persons would be able to pick up the signal.

In September 1957 the Soviets indicated that they would launch a satellite "soon" and that it would be considerably heavier than the American *Vanguard*. Much of world opinion reflected on the statement as Russian propaganda. Aware of the effect that a failure would have on their credibility, the Soviets held secret the time, place, and vehicle that they would use.

The PS-1 satellite was originally to be launched on the 100th anniversary of Tsiolkovsky's birthday on September 17, 1957, but a variety of problems intervened, and the launch was rescheduled for October 6. But then Korolev heard there was a conference in the U.S. with a presentation entitled *Satellite Over the Planet* on that date. He believed the conference was timed by the United States to announce the actual launch of a U.S. satellite by the *Jupiter C*. He moved the new launch date up by two days to October 4.

The *R-7* was transported to the launch pad from the assembly building, about a mile away, on the early morning of October 3. It was reported (decades later) that Korolev walked in silence ahead of the giant rocket with his staff. His thoughts of that moment were never recorded. His untimely death in 1966 occurred well before Glasnost opened the Soviet enigma, which would have allowed much of his feelings and actions to be revealed to the world. As the final preparations continued, he had cautioned his engineers that they should halt the launch if they had any doubts on the operation of any of the systems.

On the night of October 4, 1957, floodlights illuminated the massive rocket as the technicians made the final checks of all the systems. At 2228 hours (Moscow time), the engines ignited, and the 600,000-pound booster lifted off the pad, propelled by a thrust of over 900,000 pounds. The packet performed flawlessly as the rocket accelerated into the night sky. Shutdown and separation of the boosters occurred on schedule. A turbine pump failure due to a malfunctioning fuel sensor in the SOB system resulted in core engine cut off one second early. However, the necessary minimum velocity had been achieved, and at T+324 seconds the first man-made satellite entered orbit around the earth.

As the satellite passed over the Kamchatka tracking station, its signals were received. But Korolev cautioned that they should not celebrate until the satellite completed its first revolution. When it did, the ballistics experts determined that the satellite was in orbit with a perigee of 142 miles and an apogee of 592 miles (50 miles lower than expected due to the early engine shutdown). The inclination of the orbit was 65.6 degrees, while the orbital period was 96 minutes. Korolev was quoted as saying, *"I've been waiting all my life for this day!"* The first satellite was called Sputnik, it was Russian, and it weighed an astounding 184 pounds.

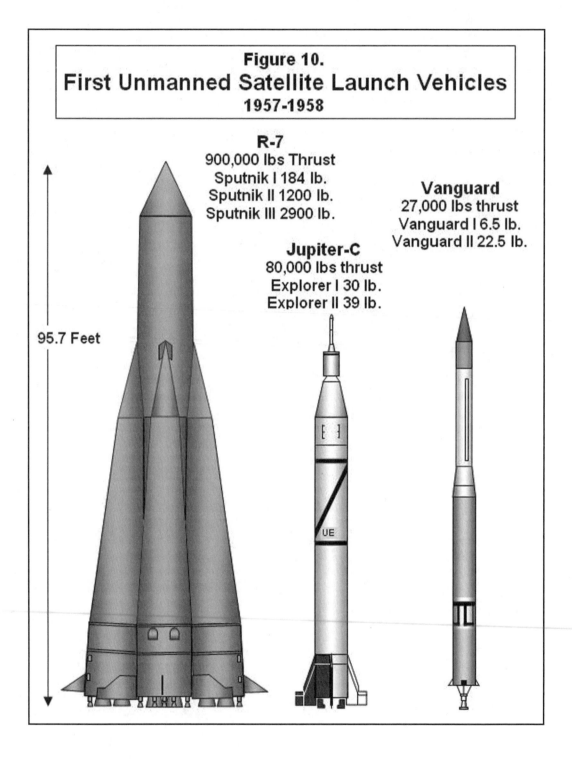

Figure 10.
First Unmanned Satellite Launch Vehicles
1957-1958

R-7
900,000 lbs Thrust
Sputnik I 184 lb.
Sputnik II 1200 lb.
Sputnik III 2900 lb.

Vanguard
27,000 lbs thrust
Vanguard I 6.5 lb.
Vanguard II 22.5 lb.

Jupiter-C
80,000 lbs thrust
Explorer I 30 lb.
Explorer II 39 lb.

95.7 Feet

UE

Caught Flatfooted

As the countdown reached zero, the base of the huge rocket was bathed in the shimmering glow from a raging sea of fire that few had ever witnessed. Twenty rocket engines clustered in the tail of the 100-foot beast quickly built to full thrust. Then huge steel arms that supported it on the launch pad lay back like the fingers of a giant hand releasing a mechanical bird from captivity. Almost a million pounds of thrust propelled this creation upward as the flame from a massive ball of incandescent gasses licked into the black night sky. Twelve smaller rocket engines around the periphery of its base moved in precise response to the guidance system to balance the 660,000-pound marvel. While those on the ground, who stood gazing in wonder at this creation come to life, knew they were witnessing a special event in the annals of mankind, no one could foresee the ultimate effect on history of the launching of the world's first artificial earth satellite—Sputnik!

And so it was, on a Friday evening, that the Russian news agency TASS startled the world with its announcement: *"On October 4th, 1957, the first artificial Earth satellite was successfully launched by the USSR. The Soviet Union proposes to launch more artificial satellites in the course of the International Geophysical Year. The Sputniks will be larger and a broad range of scientific experiments will be carried on them"*.

Initial reaction around the world reflected uncertainty. Sensing that the announcement held some significance but unsure exactly what that was, radio and television news interrupted their regular programming to convey the terse 30-second TASS news item. Then they went looking for an "expert" to tell them what was happening. By the next morning, the front page of the October 5th newspapers had large headlines of the size reserved for major news events.

```
U.S. Tracks Red Moon 558 Miles Out in Space
Orbit Spans U.S. 7 Times A Day
U.S. Caught "Flatfooted"
```

Some of the experts were themselves a bit confused. Rocket expert Willy Ley, who had participated in the early German rocket experiments back in the 1920s (before immigrating to the United States in the 1930s), declared the Russians had scored *"a great propaganda victory."* But then went on to say that he was *"unimpressed."* He was confident that the *Vanguard* program would launch within the next three months. Ley also expressed doubt about the reported weight, believing that a mistake had been made in converting from metric to English units. As the reported size of the sphere was only 23 inches, *"Surely the satellite did not weigh 184 pounds! Why, that would be almost ten times the weight of the Vanguard."*

The director of *Project Vanguard*, Dr. John P. Hagen, said the Russian Moon could not be compared to *Vanguard "which is a much more 'precise' scientific instrument"*. He was of course speaking out of complete ignorance, as he had no knowledge of what information this first Sputnik was transmitting. He was correct in that the Sputnik was simply transmitting the inside temperature and pressure readings of the satellite and was not making any active scientific measurements. Had he known what was waiting in the wings, he might have chosen his words more carefully.

In the nation's capital, news correspondents seemed to sense that there were big changes in the offing even before any pronouncements were made by congressional representatives. In an article headlined "Moonstruck D.C Talking in Billions," reporter Ted Lewis predicted that, based on initial queries with lawmakers, *"...Congress, if not the administration, will demand faster development of the Pentagon's various missile programs."*

President Eisenhower's press secretary, James Hagerty, declared the Russian announcement was *"no surprise"* and emphasized that the US was *"not in a race with the Soviets."* Both of these statements were naive at best.

To the average American and for many in the US Congress, the significance of Sputnik was initially rather vague. Most had little appreciation of what it took to get a satellite into orbit. However, within 24 hours newspapers were feeding their readers new space age jargon, technical details, and observations by knowledgeable experts. Much of the information had already been published over the past year as *Project Vanguard* had progressed towards its goal. Now the information took on added significance.

Sputnik I was traveling at a speed of almost 18,000 miles per hour in an elliptical orbit that ranged from 150 miles at its perigee (closest point to the earth) to 558 miles at its apogee (farthest point). It circled the earth every 96 minutes (the orbital period) and was inclined at 65 degrees to the equator. The inclination resulted in a ground track that covering most of the inhabited areas of the world over the course of a day. Its path over the earth would shift with each orbit as the Earth rotated under it.

The Soviet satellite was visible only during the twilight periods as a sixth magnitude object (almost invisible to the naked eye). Because a satellite is reflecting sunlight to make it visible, it has to be illuminated by the sun while the observer must be in relative darkness. Thus, only the two hours before sunrise or the same period after sunset allow for observing a satellite. Of course, the altitude of the object is also a factor since the higher it is, the longer the period it will be in sunlight.

However, the huge spent core stage of the Sputnik launch vehicle, that was some 85 feet long, was easily the brightest object in the sky when it passed over. Many newspapers printed the schedule, released by the Soviets, of its expected appearances, and it was more than unsettling to see all the major U.S. cities listed. Sputnik I would slowly "decay" to a lower orbit over the weeks and months to come, and eventually it would encounter the more dense layers of the upper atmosphere and burn-up from the friction with the air molecules.

Even the name 'Sputnik' was different. As the Soviets had not given the satellite a formal name it had simply been referred to by its Russian descriptor for a 'Moon' or 'fellow traveler' or 'companion'.

As the launch of Sputnik I was being analyzed by scientists, politicians, and the man-on-the-street, many who could not accept the Soviets as equals, were downplaying Sputnik as a fraud, a *"neat scientific trick"*, or the results of *"captured Germans"* from World War II. What was ultimately to surface as truth was that the Soviets had accomplished a significant milestone in the annals of science and technology.

The Russians were masters of inflated language to make their accomplishments, particularly the Sputniks, appear to be more than what they were (when in fact they were indeed spectacular). Several 'experts' came forward to proclaim that the Soviets had faked the launch and several impressive books quickly appeared in an effort to discredit their claims. But, when the rhetoric abated, it was clear that the Soviets had orbited the first earth satellite.

What embarrassed the Americans was simply that they had been beaten to the record books for a very important entry. Moreover, they would be beaten to many "firsts" in space flight over the next 8 years.

What angered the Americans was that those who were in control politically (essentially the Eisenhower administration) allowed the Soviets to achieve this success when, as the story quickly unfolded, the United States could have placed a satellite in orbit more than a year earlier than the Soviets.

What shocked the Americans was the weight of these earth satellites—the first being nine times heavier than the long anticipated American entry. It was convincing evidence that the Russians (as the Soviet Union was often referred to in that era) did have the power, the guidance, and the ability to launch intercontinental ballistic missiles. This was seen as a threat to the very existence of the free world.

The complacent belief in American technical superiority, which had waned during the early stages of World War II and again with the MiG in Korea, had become ascendant in the 1950's. The reasoning was quite simplistic: *"we won the war, so we must have had the better weapons."*

Dr. I.M. Levitt, Director of Franklyn Institute's Fells Planetarium, ominously declared, *"... it* [Sputnik] *would remain in the sky for a thousand or a million years and there is nothing man can do to bring it down"*. Sputnik I actually "decayed" from orbit and re-entered the atmosphere to meet a fiery end 92 days after launch. However, the scary rhetoric of these pseudo-experts continued to raise the hair on the back of the neck of many Americans.

While there were cooler heads, their comments were often less than illuminating. The outgoing Secretary of Defense, Charles Wilson, called Sputnik *"a nice scientific trick."* Presidential advisor

Sherman Adams spoke of the competition between the two powers as an *"outer-space basketball game."*

The President himself indicated that Sputnik did not raise his anxiety "one iota." Ike had carefully crafted military R&D funding along with sufficient military combat capabilities during his preceding term to avoid creating a large military industrial complex funded by an inflated federal budget. Now, with the Soviet Sputnik, his entire military and economic strategy would start to unravel early in his second term.

The President, once the indomitable spirit of the American people, lost much of his influence and trust, not only with the people of the United States but as leader of the free world. As the opposing political party, the Democrats were quick to endorse the perception of the American public that there was a "gap" between the missile capabilities of the U.S. and those of the Soviets. This gap, they claimed, was caused by the inept handling of American priorities by the Republican administration.

Lyndon Johnson, then Senate Majority Leader, demanded a Congressional investigation of the impact of Sputnik only a few days after its launch. On the Senate floor a few months later, he made recommendations in his Preparedness Subcommittee to "start work at once on the development of a rocket motor with a million pound thrust." He also wanted to "put more effort in the development of manned missiles (satellites)," and "accelerate and expand research and development programs, provide funding on a long-term basis, and improve control and administration within the Department of Defense or through the establishment of an independent agency."

This alleged "Missile Gap" was to cost the Republicans significant losses in the 1958 congressional races. As history was later to reveal, U.S. technology in virtually every critical defense area was ahead of the Soviets at that time and six long-range strategic missiles were being developed with the highest national priority. Nevertheless, technological image was paramount in 1957.

Eisenhower (via U.S. intelligence efforts) had reasonable proof of America's lead—evidence which he had shared in confidence with the Democratic leadership. Despite these assurances, the missile gap became the key issue of the 1960 Presidential campaign that was won by a thin margin of victory by John Kennedy, a young liberal Democrat from Massachusetts. Sputnik moved the Cold War into a new phase where technology, once the solid base of America's dominance, became a focal point of ideological supremacy. While the Soviet R-7 ICBM on which Sputnik was launched was a triumph of Soviet technology, it was a not an effective weapon, and less than a dozen R-7s would ever be operational as ICBMs. Yet the presidential candidacy of John F. Kennedy in 1960 was closely tied to the misconception that the Republicans, under Eisenhower, failed to keep the American defense posture ahead of the Russians.

Eisenhower's failure was his lack of sensitivity to the impact of the first Earth satellite and the effect that satellite weight would have on the perception of technology in world opinion. His aversion to the German rocket scientists was tied to his first-hand knowledge of the Nazi concentration camps and the manner by which these "guest" scientists had been allowed to enter and pursue their careers in the United States.

Affect on the Soviets

Khrushchev was in Kiev to discuss economic issues when the meeting was interrupted at 11:00 PM by a telephone call notifying him of the success of the Sputnik launching. It was reported that Khrushchev's initial reaction to the launch was not as ebullient as might be expected, a strong indication that perhaps he did not immediately appreciate its significance. More than likely, he was waiting to see world reaction. He phoned Korolev to congratulate him. As the evening wore on, some attendees, including his son, recall that he finally announced the launching to those assembled and proceeded to describe a new era in which *"we can demonstrate*[to the Americans] *the advantage of socialism in actual practice."*

Even the first edition of the October 5th issue of Pravda, the official Soviet newspaper, had only a few short paragraphs about the launch. With little information coming from their own official sources, Pravda (and TASS) simply parroted what the world was saying about the accomplishment. However, in subsequent issues, after world reaction expressed awe and amazement, it was in large bold headlines. But something very subtle was occurring that even the Soviet leaders had not quite considered. The

Western press, and most of the scientific authorities (even some within the American government) were lauding the accomplishment. The editors of Pravda and the readership of the Soviet people were surprised by the generous praise. How could the leaders of the United States allow such a response?

Literally overnight, Khrushchev achieved credibility with world leaders, allowing the influence of Communism to move with unprecedented assurance into many third world countries—particularly Cuba. Some Political analysts have long felt that had it not been for the prestige of Sputnik, Fidel Castro might not have been lured into the Soviet sphere of influence; the Bay of Pigs debacle and the Cuban Missile Crisis of 1962 all might have been avoided.

Khrushchev was further bolstered by the subsequent success of the dog-carrying Sputnik II a month later on November 3rd, and the headlines around the world, which proclaimed that the Russians were leaving the West behind in technology. In a speech on November 7, 1957, he stated that the Soviets would soon overtake the United States in heavy industry and production of consumer goods.

Khrushchev, while not very technically inclined, was politically astute and now sensed that the launching of the first satellite would give credibility to Russia's emergence as a superpower. Credibility was not only important with respect to the Western democracies but within the Soviet Politburo. Anything that Khrushchev could do to enhance the prestige of the Soviet Union abroad would also strengthen his power base at home where he still faced some opposition.

What was not known by the West at the time was that the Sputnik aura allowed Khrushchev to consolidate his power base within the Presidium, in part by removing the hero of The Great Patriotic War, Marshal Georgi K. Zukov, and many of the other old Stalinist generals from power.

Historian James Oberg summarizes Khrushchev's perceptions of the Earth satellite program: *"...it would signal to dissident political forces within the Soviet Union that Khrushchev was really leading the country to a glorious future; second, it would overawe the traditionalist 'artillery generals' in the Red Army and allow reorganization of the armed forces, including a reduction of obsolete ground forces (saving money and pulling the rug out from under possible opposition leaders); last it would demonstrate in an unequivocal manner the existence of the long-range missile system, which was intended to discourage a potential attack from the United States".*

At the launch site there was much celebrating and many speeches the following day. As reaction from the Western press began to filter in, Korolev was told, *"the whole world was abuzz"* with the launch. He was reported as saying, *"well comrades, you can't imagine—the whole world is talking about our satellite. It seems that we have caused quite a stir."*

As the media riot, as some had termed it, over Sputnik began to mount in the West, the Soviet leadership began to realize the trump card they held. On October 9 a report detailing the construction and design of the satellite was issued—but no details about the rocket itself. Because of the closed nature of Soviet Society and the "oneness" of the Soviet space program and its military capability, secrecy has caused the achievements of many contributors to the Soviet space program to be lost to history. Those responsible for the success were not named, and even Tikhonravov, who had been published under his own name throughout the 1950's, suddenly disappeared from official histories. Their names would not be mentioned again until they were deceased. With the launch of Sputnik, all contributors were simply referred to as "the chief designer." This in itself caused a more heightened sense of mystery.

With the coming of Glasnost and the demise of the Soviet Union more than three decades later, new insights into this history have appeared. Many individuals who were involved in, or had knowledge of, the early space spectaculars have come forward with important pieces of information to help fit this giant puzzle together.

With respect to the veil of secrecy that surrounded the early Soviet space program, specific plans for future space missions were expressed only in vague generalities, and failures were never mentioned. Although Khrushchev liked to rattle his ICBM sabers, military operations were never implicated in the space program. Details about the spacecraft, specific mission objectives, and even the launching sites, were never divulged. Most experts in the West assumed that the Soviet rocket was similar in concept to the three-stage *Vanguard* (just ten times bigger), and so that became the defacto description of the Soviet rocket. Even the designation *R-7* was never mentioned.

Only the chairman of the Commission of Interplanetary Communications, Sedov, was allowed to travel and speak with authority about a space program of which he had little knowledge.

The leap into space became a focal point to the Soviet citizenry. It established pride and confidence in the Soviet people. It assured them that their government was making progress to improve their standard of living. For nearly a decade, it temporarily shifted their focus from exceptionally poor economic conditions to the promise of the heavens above.

Secrecy was an important ally to the Soviets during this period. It cloaked their limited technology and hid their failures. Had the R-7's technology been revealed, the anxiety that gripped America for several years following Sputnik could have been diffused.

Sputnik II

While most of the engineers and technicians involved with the Sputnik launch enjoyed a brief holiday as they celebrated their success, Korolev and several of his key people were called to the Kremlin to accept the accolades from the Premier himself. Almost from the start, Korolev was asked if something could be launched to celebrate the 40th anniversary (on November 3rd) of the Russian Revolution of 1917. That date was less than four weeks away.

Korolev may have anticipated the request; he indicated that there was a chance that an even larger and more spectacular Sputnik might be prepared in time for that date. What he had in mind was not Object D, which was still lagging, but something more than just scientific equipment: something that would not only stir the imagination but give the Americans another shock. Korolev proposed sending an animal as a passenger aboard the next Sputnik.

There has been much speculation as to whether the construction of such a sophisticated satellite could actually have been accomplished in the short time available. Given that much of the hardware was in existence, the scenario is possible although somewhat improbable. The rocket itself was available since a backup was prepared in case of a failure with the first Sputnik launch. The spherical communications container, essentially the backup satellite for Sputnik I, was available. There were animal containers built for vertical biomedical launchings over the previous two years. Whatever the conditions and requirements, Korolev accepted the challenge and passed it on to his co-workers. Much of the connecting hardware was fabricated in the shops at the launch site and there was little in the way of formal drawings.

The passenger selected was a small female mongrel Husky named Laika, who had been retrieved from the streets of Moscow as a stray. She weighed about 8 pounds, and the padded, pressurized cabin provided room for her to lie down or stand. The environmental system provided oxygen, food and water which were dispensed as a gelatin. Laika was fitted with a bag to collect waste, and electrodes to monitor her vital signs. The total mass of the new satellite was 1,118 pounds, six times heavier than Sputnik I! The launch occurred on the designated day, November 3, 1957.

The Soviets reported that the initial telemetry indicated Laika tolerated the weightless condition and ate and slept normally, and there were strong hints that Laika would be returned unharmed. However, there was never any intention to return the dog back to earth, and when it became clear that she was not to be brought back, the official Soviet statement was that she had been painlessly euthanized after several days of electronic observation of her condition.

However, Laika's demise was not as humane as the Soviets reported at the time. After reaching orbit, the thermal controls did not function as planned, and the interior temperatures quickly reached 120 degrees F. For almost the entire period of her flight, Laika suffered from these high temperatures and succumbed to heat exhaustion on the fourth day of the mission, on November 7. The dog's body burned up along with the satellite as it returned to the Earth's atmosphere after 162 days in orbit.

Laika's story remained a state secret for more than 35 years. The animal activists groups who protested her initial announced fate would have been incensed had they known the whole truth. However, there are few in the world's scientific community who would not have sacrificed Laika for the advances that might have been achieved with a test flight such as Sputnik II.

When it became obvious decades later that the flight was made with virtually no testing of the apparatus because of the schedule to produce another "neat scientific trick" on queue, many questioned the ethics. But those who were responsible felt more than justified in taking Laika's life for the knowledge that she was able to impart during her brief flight as the first living animal to orbit the Earth.

Laika's flight appeared to prove that man could probably survive in space, and the excitement of this new era reached even higher levels. It was also apparent that a set of progressive milestones represented the ability of a nation to exhibit its scientific and technological prowess. These "firsts" became the focus of the space race. The orbiting of Sputnik I represented the first achievement and signaled the beginning of the race.

While the public had little idea of the technology required to perform these feats, the news media and the aerospace journals conveyed what constituted significant milestones. They were fed by recognized experts (or, in some cases, pseudo experts) who declared what was significant. The ability to achieve a milestone "first" provided significant insights into the level of technology of the United States and Russia. Likewise, the failure to accomplish a milestone could have significance, especially to the intelligence community who carefully monitored launch vehicles and the satellites themselves. Because of this aspect, the Soviets were particularly sensitive to what they would reveal for each of their activities.

Now Sputnik II, because of its size, seemed a precursor to a manned flight and again the headlines were alarming:

```
Soviet Launches Second Satellite - Dog Aboard Heavier Moon
Unmanned Flights to the Moon and Back Are Now Possible
Concern Increases in Washington
Confidence in U.S. Is Held Impaired
British Concede Soviet Is Ahead
Soviets say New Power Sources Were Used In Second Satellite
Reds Quit All Arms Talks
```

The last two headlines were an obvious attempt by the Soviets to "stampede" the American public and perhaps the military as well. A short animated film was released by the Soviets that showed a huge winged rocket being launched from a mile-long, curved launch rail as had been proposed with the Sanger-Brendt antipodal bomber of the 1930's. That depiction left such a lasting memory that there are still many today who believe that technique was the "advanced method" of launch almost 50 years later. No film, pictures or drawings were released of the actual Sputnik launch vehicles until a decade later.

Soviet rhetoric embellished their accomplishments to enhance the propaganda value. Following Sputnik II, statements by various Soviet authorities indicated that new sources of power were used. Speculation in the West that perhaps nuclear powered rockets had been developed was without foundation. Even the use of Fluorine or Liquid Hydrogen fuels would be beyond the reach of Soviet technology for more than 25 years. Soviet rocket engine development was several years behind the U.S. when Sputnik launched, and it continued to fall farther behind as the U.S. developed ultra high performance liquid-Hydrogen engines and the 1.5 million-pound thrust F-1 that powered the Saturn V Moon rocket. Nevertheless, none of this was evident in the fall of 1957.

Psychological Shock Sets In

If Sputnik I was an affront to the national pride of the United States, the launch of Sputnik II was overwhelming. Now the press and the experts openly talked of missions to the moon and to Mars, and of manned spaceflight. Suddenly, within a period of 30 days, science fiction had become science fact. The phenomenon known as *future shock* was born.

Dr. Joseph Kaplan, chairman of the *US National Committee for the IGY,* stated: *"I am astounded by what they have managed to do in the short period of time they had. It seems to me that this is a remarkable achievement. They did it—and they did it first."*

Hungarian born Edward Teller, father of the American H-bomb, stated with despair in his voice in a television interview that if the Russians *"pass us in technology there is very little doubt who will determine the future of the world."*

The prestigious New York Times editorialized on November 10, 1957, that *"it must be hoped that the National Security Council... will not only be receptive to new ideas, but will take immediate steps to remedy deficiencies and put the US again in a race that is not so much a race for arms or even prestige, but a race for survival."*

The assumption that there were deficiencies in the U.S. program was quick to surface because the Russians were first, and Sputnik I weighed nine times that of the yet to be launched *Vanguard*. The use of the word *race* came quickly to the forefront in the lexicon of the press within the first 24 hours after the launch of Sputnik. The Cold War *arms race* had evolved into the *space race*. The final word of the Times editorial, *survival*, seemed to sum up the attitude of many Americans. The expression *"we will bury you"* that Khrushchev used in a speech earlier in the year was resurrected by the alarmists.

Americans recognized that, when the *Vanguard* satellite was launched, it would only weigh about 22 pounds. The experts being interviewed were aghast that Sputnik I had weighed 184 pounds. Less than a month later, Sputnik II weighed in at over 1,100 pounds and carried a live dog. The core stage of the R-7 was also in orbit, with a length of over 80 feet and a weight of about 14,000 pounds (although this weight was not disclosed for many years). Without knowledge of a comparison of the various launch vehicles and the levels of technology involved with each, the only conclusion that could be drawn by world opinion was that the United States was way behind the Russians.

In an attempt to re-assure the American public that the President was in-tune with science and technology, MIT president Dr. James Killian was appointed as the Presidential Science Advisor on November 7, 1957. Unfortunately, the press and much of the American public saw this as an indication that the president needed outside assistance to understand science and technology. Headlines declared Killian as the new "Missile Czar," borrowing from Russian culture the image of a supreme dictator who would bring order from the alleged chaos of the American missile program. The press had once again misled the American public. Killian was a policy advisor, not a technology whip, and the American missile program, while often-exhibiting uncertainty, was not in chaos.

Elements of the Democratic Party quickly assessed the situation, and one front-runner emerged to press home the attack on the Republican Administration. Encouraged by the "media riot" and selected party members, Senator Lyndon B. Johnson opened a Senate *Inquiry into Satellite and Missile Programs* on November 25, 1957.

There was no lack of criticism of Eisenhower's handling of any specific program as both military and industrial leaders came forward in an attempt to widen the federal coffers. While Johnson would not to be successful in obtaining the 1960 Democratic Presidential Nomination, he was successful in undermining the confidence of the American people in the Republicans' abilities to lead the country into the new space age.

American Recriminations

Almost before the exhaust plume of Sputnik could dissipate, there was an abundance of self-analysis as to "what was wrong with America?" and "why had America failed?" One of the first sacrificial lambs was the educational system. Had America gone soft on science? Was America more interested in the materialistic aspects of life? A plethora of committees from Congress down to the local school boards began a critical review of many aspects of the American educational system and its approach to science.

American education in the 1950's reflected significant changes brought about, in part, by World War II. Many soldiers had seen an aspect of life that, except for their military service, would have gone unnoticed. There was a thirst for education and for jobs in a new and growing sector of the economy that related to high technology. President Franklin D. Roosevelt had signed the Servicemen's Readjustment Act of 1944, better known as the "GI Bill of Rights" (the term GI was coined by former President Franklin Roosevelt's wife, Eleanor, when she referred to the troops and their equipment being "Government Issued"). This Act of Congress provided for educational benefits for discharged veterans who had completed their service. Millions of GIs took advantage of the opportunity, and by the mid-fifties, a new generation of Americans armed with a college degree had entered the workforce in unprecedented numbers. The problem, therefore, was not necessarily in the educational system but in the focus and funding of the scientific programs themselves.

On the Soviet side, education was actually severely limited for the common man. The drudgery of daily life for most Russians was a 12-hour day on a collective farm or at a brutal factory job. Those who had the capability for learning were motivated by the desire to stay warm and eat well—in addition to using their intellectual talents. Of course there was also the large "stick" that KGB chief Beriya carried, should anyone fall short in their endeavors.

During the decade that followed Sputnik, several innovative math and science programs were introduced into America's educational systems. The *new math* in particular brought out as many critics as proponents. A new emphasis on math and science presaged a movement to federally funded programs and a secular approach to education. Over the next 40 years, declining SAT scores and growing chaos in the classrooms would leave America wondering whether it had made the right adjustments.

Nevertheless, all was not negative with the impact of Sputnik. America's youth (as well as many around the world) were excited by this new frontier. Aviation enthusiasts became backyard experimentalists and began building small homemade rockets. Most, lacking the facilities of a machine shop and knowledge of mixing chemicals to formulate rocket fuels, were content to use spent CO_2 capsules (used to inflate life rafts) loaded with the heads off a few dozen matches to power their creations. The "epidemic" of model rocketry grew so fast that accidents were inevitable. Some states and communities, hoping to keep the kids from blowing off any more hands or enduring more fatal consequences, felt compelled to enact legislation. However, the enthusiasm and motivation to learn was a wonderful byproduct of Sputnik. The contemporary biographical movie *October Sky*, the experiences of young Homer Hickam, is an accurate portrayal of this period of American history.

Edwin Land (of Polaroid fame) was a member of the ODM Science Advisory committee that met on October 15, 1957, with President Eisenhower. He stated that it was the President who could kindle the essential enthusiasm for science among the America's young people. Eisenhower, although a fatherly figure, was unable to relate to the younger generation. However, Land's remarks were prophetic of the effect that a youthful President Kennedy would have in stirring a new generation of Americans into action four years later with his Peace Corps and Moon Race challenges.

An article entitled "Battle for Science Lead" in a November 1957 issue of the New York Times, was purported to be "An analysis of Difference Between U.S. and Soviet Strategy on Research." It effectively summed up how many viewed the situation when it stated: *"A superior Soviet post-war strategy for guiding scientific research appears to go far in explaining the Soviet lead in the conquest of space. Far sighted decisions taken shortly after WWII are now resulting in a much wider area of gain than simply the two earth satellites."*

Noting the Soviets possessed the *"most powerful jet engines"* and *"largest atom smasher,"* the author criticized the American system of *"fits and starts"* and slow decision making. It concludes with the statement: *"Soviet leaders have been able to build a system of incentives and a system of education that guarantees... that very large numbers of the most able people ... will get intensive scientific training and education."* These misleading overstatements were not based on a broad spectrum of Soviet technology but solely on isolated slices of Soviet science. Uninformed but widely circulated opinions such as these continued to plague the U.S. throughout the cold war.

Thirty years later the Soviet system would self-destruct economically and politically. The *"most powerful jet engines"* did not guarantee reliability or cost effectiveness. The technology of the *"largest atom smasher"* did not prevent the Chernobyl nuclear reactor meltdown.

While few Americans are totally content with a political system that tends to change its mind every four years, this political and economic process has shown a remarkable resiliency to recognize when it has been bested and to respond to that challenge. Today the technology challenge is still being played out with the Japanese in electronics and automobiles, and with the Europeans in the commercial aviation market.

When Eisenhower left office in 1961, his farewell address to the nation included a warning about the undo influence of the "military-industrial complex" in the United States. He had seen quite clearly how some key players in the corporate board rooms and the military hierarchy fanned the fires of the missile gap to ensure large and lucrative weapons contracts so that America could catch up with the Russians.

Notwithstanding the credible knowledge of American missile supremacy, the incoming Kennedy administration was engulfed by the wild fires of the space race and the arms race. It embarked on a greatly expanded military missile production program that ultimately resulted in thousands of strategic and tactical nuclear warheads of questionable need being fielded by both sides.

Vanguard TV-3

If the emotions of the Americans had not yet known the depths of despair, the attempted launch of the first *Vanguard* satellite on December 6, 1957, achieved the ultimate humiliation. *Project Vanguard* had realized three successes with its first three Test Vehicles, but none had involved the complete three-stage rocket. The most recent launch, TV-2 on the 23rd of October (two weeks after Sputnik I) represented the first time the entire configuration had flown. With the two Russian Sputniks wheeling around the Earth, the *Vanguard* team was feeling pressure. Although it is doubtful that they *rushed* the TV-3 launch before it was ready, there is no doubt that they worked longer hours to prepare the rocket at an earlier date than it might otherwise have flown. Likewise, the expectations of the World had been fanned by the press in an attempt to play-up America's pending accomplishment.

According to the development plan, the second stage was being tested for the first time. The third stage, which had flown successfully on TV-1, was live, and if everything worked as designed, orbital speed might be achieved. Because the probability of achieving an orbit on the first try was considered remote, a small 3.5 pound, 6-inch diameter, solar powered satellite was used in place of the more expensive 22-pound scientific package. Even this aspect had not been considered in the initial schedule and the grapefruit-sized satellite had been included only the previous July. The past year had seen many problems with the delivery of the various stages as the contractors struggled to engineer the technologies into a usable rocket.

The plan called for six "test vehicles" before the full sized satellite would be aboard. The project officials attempted to downplay the possibility of success on the very first try (something the Soviets had accomplished). In keeping with the unclassified civilian nature of *Project Vanguard*, the test schedule was publicly announced, and there were hundreds of reporters from dozens of nations around the globe present at Cape Canaveral when the countdown reached zero. With the whole world watching, TV-3 lifted only a few feet before a break in a fuel line caused it to loose thrust and fall back into the launch pad. A series of violent explosions marked the end of the rocket and America's hopes for getting into the record books with a satellite in the same year as the Russians. The little satellite was thrown clear of the blazing inferno and continued transmitting its signal until someone recovered the crying infant and turned its transmitter off. The satellite may be viewed in the National Smithsonian Air & Space Museum, bent antennas and all.

However, failures at the Cape were a common occurrence in those days. During 1957 the U.S. had been launching the Thor and Jupiter IRBMs, the Atlas ICBM, and the Navaho cruise missile (which was launched vertically). A half dozen other smaller missile projects were also underway, so there was lots of activity with a launch occurring about every week or two. Many of these tests exploded in flame within sight of the launch complex and the surrounding towns along the Atlantic coast. The technology was still immature with both random failures and systemic failures. A systemic failure is one caused by a poorly designed or built part, whereas a random failure occurs when the statistical accumulation of reliability of a part or series of parts results in failure. *Vanguard* TV-3 was termed a random failure.

The Soviets, as well as some of America's "friends" used the flaming debacle to further humiliate the American effort. Pravda reproduced the front page of the London Daily Herald which had two photos. The first was the tiny 6-inch sphere being enclosed in the nose shroud before the launch. The second showed the explosion. Above the bold Herald headline, which read, *"Oh, What A Flopnik!"* was Pravda's notation *"Reklama and Deistvitelnost,"* that translates to *"Publicity and Reality."*

The Army Moves into Space

The group of distinguished attendees at a cocktail party and dinner in Huntsville Alabama, on the evening of October 4, 1957, was impressive. The recently appointed Secretary of Defense Neil McElroy and Secretary of the Army Wilbur Brucker were the honored guests of the party being hosted by the Commanding General of the Army Ballistic Missile Agency (ABMA) John B. Medaris and his top rocketeer, Wernher von Braun. The occasion had just gotten underway when ABMA's public relations representative, Gordon Harris, rather abruptly rushed in to announce that *"...the Russians have just put up a successful satellite."*

Following a period of deafening quiet, von Braun began to speak. If the others needed more time to digest the significance of the event, von Braun was about to help direct their thoughts. After briefly berating the decision that left his *Jupiter C* and *Project Orbiter* on the shelf gathering dust for the past two years, he abruptly told the new Secretary of Defense that von Braun's team could orbit a satellite in 60 days. General Medaris was a bit more conservative and suggested that it would take 90 days to prepare for the launch. Of course, von Braun knew that there could be no unilateral decision by McElroy but continued to "button hole" him at every opportunity for the rest of the evening.

Nevertheless, Eisenhower was not going to be stampeded by 184 pounds of polished batteries traveling at 18,000 miles an hour. The decision to allow the von Braun team to launch was not forthcoming in the weeks that followed Sputnik I. But Medaris approved some preliminary steps that could be taken without ruffling too many feathers. The *Jupiter C* with the designation UE was removed from storage, and initial refurbishment was begun. It wasn't until Laika, a stray dog from the streets of Moscow, raced across the skies on November 3, and the discomfort in Washington, D.C., was elevated to the level of pandemonium, that the decision to use the Army's *Jupiter-C* was revisited. While the nation's leaders were still confident of America's secure defensive superiority, the Democrats were making noise; and America's allies, who looked to the military strength of the United States as a shield against Communist aggression, needed reassurance.

While history would prove Eisenhower's basic premise correct—that the American defense posture was more than adequate—the spectacular achievements made by the Soviets in space exploration demanded a more positive response. On November 8th the Defense Department released the following statement: *"The Secretary of Defense today directed the Department of the Army to proceed with launching an Earth satellite using a modified Jupiter C...."* The Army was allocated $3.5 million to launch two satellites. The *Vanguard* program would consume $22 million before it was completed.

There were a multitude of challenges to be addressed if the *Jupiter C* (now designated *Juno I*, but that moniker was rarely used) was to keep its rendezvous with destiny. The missile itself, Number 29, would need to be moved to the Cape and thoroughly checked out. (Each of the discrete letters in the name "Huntsville" was given its equivalent numeric value beginning with H=1. Thus, *Jupiter C* serial numbered 29 had the identification letters UE painted on its sides). Its elongated propellant tanks would provide power for 155 seconds instead of the 121 in the *Redstone* configuration. It also had the hydrogen peroxide tank enlarged to provide for the increased 34 seconds of turbo pump power. A new hydrocarbon fuel called Hydyne would raise the thrust from 75,000 pounds to 83,000 pounds resulting in an increase in Specific Impulse of 15%.

The many modifications that had been made to the *Redstone* to turn it into the *Jupiter C* had been overseen by Willy Mrazek of ABMA's Structures and Mechanics Laboratory. The Guidance and Control was under the direction of Dr. Walter Haeussermann, while the Aeroballistics Laboratory was headed by Ernst Geissler. All were veterans of Peenemünde. Even the Launching and Handling Equipment Laboratory was in the care of Hans Heuter, who had been with von Braun since the earliest days of the VfR in 1932 when a burning rag tossed over the end of the rocket's exhaust nozzle provided the ignition sequence. As Eisenhower had lamented several years earlier, the first American satellite would be launched by former Nazis now being underwritten by the American taxpayer.

Caltech's Jet Propulsion Laboratory was selected to coordinate the overall construction of the satellite package, which would be a 30-inch cylindrical extension of the final fourth stage. The six-inch diameter would limit the size of the instruments, while the overall weight had to be constrained to less than 20 pounds. The full configuration of the yet-to-be-launched *Vanguard* satellite was a highly polished, 21-inch perfect sphere to enable consistent estimates of the drag produced by remnants of the atmosphere at orbital altitudes. The ABMA/JPL team was not concerned with that aspect at this critical point in history.

It was decided that the primary instrument would be a Geiger counter developed by Dr. James Van Allen of Iowa State University. Van Allen had been a pioneer in the study of cosmic rays with his instruments being the first to fly on the V-2s a decade earlier. He was one of the primary advocates of the International Geophysical Year and had been an early promoter for the launching of satellites. Dr. Ernst Stuhlinger, one of von Braun's staff, had contacted Van Allen before ABMA had received its approval to launch and had encouraged him to redesign his radiation detection instrumentation to fit within the physical confines of the *Jupiter C's* satellite.

The second experiment would be the recording of micro meteor impacts, and the final set of instruments would transmit the temperatures at three points on the outside of the satellite and one on the inside. The instrument package would have to be more rugged than those designed for the *Vanguard* as they would have to endure the 40G acceleration of the solid-fuel upper stages rather than the more modest 15Gs of the *Vanguard* third stage. The instrument section was painted with alternate strips of white and dark green to provide passive temperature control. The realtive size of these stripes was determined by the expected sunlight of the anticipated orbit. Nickel-cadmium chemical batteries provided the electrical power and were 40 percent of the payload weight. These would operate the high power transmitter for 31 days and the low-power transmitter for 105 days.

Although ABMA had been given launch authorization, the *Vanguard* team had cleaned up the debris and repaired its launch pad from the failed attempt the proceeding month and was ready with TV-3BU, a back-up rocket. The Army was directed to defer launch dates to the *Vanguard* team.

A launch countdown for TV-3BU (on the 22nd of January) ended just four minutes and thirty seconds before the scheduled lift-off when the weather and a poor telemetry signal cause a postponement for two days. The attempt on the 24th was scrubbed just nine minutes before firing as the weather continued to disrupt the schedule. The following day the countdown reached T-22 seconds when an umbilical plug feeding helium to pressurize the second stage propellants failed to eject in sequence. This was fixed and the countdown was recycled and again halted at T-14 seconds due to a frozen LOX valve. The attempt continued the following day, the 26th, but a flaw in the second stage RFNA system nearly caused a major explosion. Two technicians were injured by the toxic propellants and the second stage engine needed to be replaced. These were typical of the types of problems that the early rocketeers faced in trying to launch these new, high-tech marvels. The "firing range" for the next week was now the Army's.

Explorer I

The launching of *Jupiter C* "UE" was originally scheduled for January 29, but the high altitude winds of the jet stream were directly over the Cape producing velocities of over 200 miles per hour between 30 and 40 thousand feet. For two days, the winds kept the launch postponed. On the night of the 31st, the forecast called for winds of 157 mph, still quite high but within the tolerance of the *Jupiter-C.*

As the countdown passed through T-13 minutes, the bucket at the top of the missile that contained the solid-fuel upper stages began to spin-up to 550 rpm. As the countdown reached zero, the ignition sequence began that resulted in a lift-off at T+15 seconds (at 10:48 PM EST). The brilliant white flame pierced the black sky. Even if this was not the first satellite, the excitement among the missileers and others in attendance could not have been higher.

At T+70 seconds, as the rocket continued to accelerate upward, the rotation of the bucket increased under program control to 650 rpm. This was again increased to 750 rpm at T+115 seconds to avoid possible destructive resonances between the bucket and the booster. The rocket had by now completed its pitch to a trajectory angle of 40 degrees.

A pressure gauge in each of the propellant tanks was set to detect a sudden decrease as an indication that the tank was sending its last gulp of fuel downward towards the engine and to issue the engine cut-off command. The event occurred at T+157 seconds; a two second bonus in energy had been realized. Five seconds later the explosive bolts that held the guidance compartment to the booster fired, freeing it and the satellite attached in the spinning bucket above.

Now the most critical part of the launch sequence was in progress as this upper segment of the rocket continued to coast upward to its optimal attitude of 225 miles. The guidance system aligned the bucket parallel with the Earth far below using small spurts of compressed air from the attitude control system with an accuracy of one-tenth of a degree. As the upper stages reached what had been determined to be the most precise point for initiating their ignition, Dr. Ernst Stuhlinger (yet another Paperclip acquisition) pressed the button from his location in the tracking hanger at the Cape.

The eleven solid-fuel rockets came to life for 6.5 seconds in the second stage, followed immediately by the three in the third stage and then the final stage. The spinning that had been imparted to the rotating bucket was all that was guiding the burning trail across the night sky. When the exhaust flame subsided, the instrument package was now traveling at almost 18,000 mph, just 7 minutes and 30 seconds after leaving the launch pad. Stuhlinger could see from the Doppler shift of the 108 megacycle signals he was receiving from the satellite's two 60 milliwatt transmitters that all four stages had ignited. But there were still some, perhaps unknown, elements that could thwart the effort, and the team would have to wait for the completion of the first orbit to confirm success. The signals from the satellite quickly faded as it passed over the horizon out of radio communication range.

General Medaris, von Braun and Dr. William Pickering, Director of the JPL, were in Washington, D.C., with Army Secretary Brucker during the launch. President Eisenhower was in Georgia playing bridge with some friends on this Friday night. Although von Braun was elated that the launch had gone well, he, too, was quite anxious for the first signals to be heard from the tracking station in California that would indicate that an orbit had been achieved.

If everything had gone as planned, the satellite should have an orbital period of 106 minutes with an inclination of 33 degrees to the equator. However, as the designated time came and went, the tracking station in California reported no signals. The minutes crawled by as von Braun and the contingent in the Pentagon waited with growing impatience. Dr. Pickering held an open telephone line to the engineers in California. Finally, eight minutes after the expected time, the voice on the other end of the phone reported and Dr. Pickering exclaimed, *"They hear her Wernher, they hear her!"*

The satellite had been given more energy than needed, and the orbit was thus higher than planned (perigee of 224 miles, apogee 1,575 miles). This resulted in a longer orbital period of 114 minutes. Total weight in orbit (including the spent casing of the fourth stage) was 30 pounds. Due to its small size, the satellite (now named Explorer) could not readily be seen with the naked eye. The successful launch occurred 86 days from the date the Army was given approval. The President was advised of the success, and he interrupted his card game to make the announcement to the world.

Vanguard's Belated Success

The hard-luck TV-3BU, which was unable to progress through a complete countdown during the last week of January, finally reached that point on February 5, 1958. In the blackness of the 2:30 AM launch, the rocket proceeded upward, and the emotions of the exhausted launch team were momentarily lifted. Suddenly, at about 22,000 feet, a spurious signal from the guidance unit caused the first stage engine to gimbal hard over, and the resulting aerodynamic loads caused the missile to break in two.

TV-4 was immediately assembled on the launch pad over the next few weeks while an analysis was made of the TV-3BU failure from some of the wreckage that had been recovered from the shallow waters off the Cape. A countdown was begun on March 8th, but again several technical difficulties and a heavy fog postponed the flight.

More delays ensued until March 17th. This time (perhaps with the "luck of the Irish" as it was St. Patrick's Day) the missile flew a perfect trajectory into the cloudless blue sky at 7:15 AM. The first stage depleted its fuel after imparting a velocity of almost 3,800 mph in just 145 seconds. It provided only 15 % of the speed necessary to achieve orbit but consumed 65% of the total energy. The 31 foot long, 32-inch-diameter, 7,500-pound- thrust second stage (a descendent of the *Aerobee* rocket) now fired. The missile continued to accelerate, and the nose cone that had protected the satellite during its acceleration through the atmosphere was jettisoned to reduce the weight being carried as it climbed to over 140 miles and a speed of 9,000 mph before its fuel was exhausted at 258 seconds after take-off.

Now the spent second stage with the third stage nestled into its nose coasted to an altitude of over 400 miles while the attitude control jets (using the helium that had pressurized the fuel tanks) maintained a precise parallel alignment with the Earth below. On reaching the high point in its trajectory, a set of small solid propellant rockets fixed to the sides of second stage (mounted perpendicular to the direction of travel) fired for a few seconds to impart a longitudinal spin to the stage. Moments later (585 seconds after launch) the solid-fuel third stage ignited and propelled the tiny satellite to its orbital speed, some 1,000 miles down range. At 9:56 AM, more than two hours after the launch, a MiniTrack Station (the name of the worldwide network of optical and radio reception stations established for *Project Vanguard*) in San Diego picked up the signals confirming the success.

Because of the extreme lightweight of the test satellite, a very high perigee of 405 miles and apogee of 2,463 miles was achieved resulting in an orbital period of 2 hours and 15 minutes. While the satellite itself provided only a temperature sensor, its exceptionally high orbit meant that it would have very little perturbations of its orbit due to the rarity of molecules of the atmosphere, and its life would be measured in hundreds of years as opposed to months. Because one of its two radios was powered by solar cells, it continued in operation longer than any other satellite. This factor allowed its radio transmissions to be used to accurately deduce the size and shape of the Earth over the next few months and resulted in the revelation that the Earth was actually slightly pear shaped by some 50 feet!

For the 135 men on the *Vanguard* team, success was a welcomed relief from the pressures they had endured since Sputnik I. The *Vanguard* spokesman, Deputy Director J. Paul Walsh, commented quietly to a reporter following the press conference: *"I want to get across one point. Any one of a thousand things could have gone wrong and made monkeys out of us. That is what must be understood about a test program. We could have gotten within inches, and then had something go wrong. It's a little like making a shoestring catch in baseball. If you make it, you look like a hero. If you miss you're a bum."*

Vanguard was a state-of-the-art rocket for its time as von Braun conceded when he said that it represented a large technical advance over the *Jupiter C*, the latter having two-and-one-half times the thrust in its first stage but only able to orbit the same weight as *Vanguard*. While failures would continue to plague the rocket, it orbited two more scientific packages. *Vanguard II* was launched in February 1959 and the last, the 50 lb. *Vanguard III*, in September 1959 into a 319 by 2,329-mile orbit. With more capable rockets being introduced, the last *Vanguard* was not fired but relegated to the Smithsonian where it can be viewed today along with one of the *Jupiter C*s that was likewise retired before all had been used. Only 3 of 11 attempts for *Vanguard* and 3 of 6 attempts for the *Jupiter C* were successful.

With the success of *Vanguard I*, and *Explorer III* that followed on March 26th, the United States now had three satellites orbiting while both the Soviets Sputniks had ceased transmitting, and the first had reentered the atmosphere and burned up (the second would follow on April 14th). The press played the "numbers game" to give some allusion to the United States being more competitive, but the total weight of the instrumentation of all three was still less than 55 pounds.

Sputnik III

The Soviets had been busy finishing Object D and preparing for launch which occurred on April 27, 1958, almost six months from the success of Sputnik II. Khrushchev was applying pressure for the launch to occur immediately before elections scheduled in Italy. He wanted the Italian communist party to have the full political weight of Sputnik III. Unfortunately, a failure at T+88 seconds in one of the boosters caused the rocket to disintegrate at about 50,000 feet. The cause was apparently longitudinal resonance that would continue to plague the R-7 over the next eight months. The remains of the satellite were recovered. This was the first failure the Soviets had experienced in their space program, and, of course, it was not announced to the world. A back-up *R-7* was immediately rolled out to the launch pad, and a second attempt (with a duplicate satellite) on May 15, 1958, was successful. The 1.5-ton satellite was more than twice the weight of Sputnik II. There was no doubt about the Soviet's ability to send a nuclear warhead into the heart of the United States or anywhere else in the world.

Object D, which now was called Sputnik III by the world, contained an impressive collection of scientific instruments, which the Soviets reported as functioning normally. It had the ability to detect micrometeorites and survey the density of the upper atmosphere. It could record the intensity of cosmic rays and solar radiation as well as the presence and effect of high-energy particles. The Soviets had the opportunity to score some major victories in science over the United States except for one small problem.

As had Explorer III launched a month earlier, Sputnik III carried a tape recorder so that the data from its instruments could be stored while it was out of communication range with Soviet tracking stations. However, during the final testing prior to launch, a problem was detected with the recording unit. Some members of the satellite preparation team suggested that the launch be scrubbed until the problem could be more definitively pinpointed and resolved. Perhaps spurred on by the stigma of not wanting to be the one to cause another delay in a program that was already a year over due and which had just suffered a launch failure two weeks earlier, the engineer in charge of the unit expressed his confidence. He believed the problem was simply electrical interference with the many other test signals that were being propagated at the same time and called for the launch to proceed.

Once in space and beyond reach, the tape recorder did not work. While each experiment aboard the spacecraft performed to specification, the vast amount of data being sensed by the instrumentation could never be analyzed by the scientists who eagerly awaited it in the data centers below. Without a more complete picture as provided by the recorder, scientists could not determine if the data they did receive was characteristic of a local region of space or of global significance.

The Van Allen Radiation Belt

As information from Explorer I instruments began to be received, the data analysis group at Iowa State University, headed by Van Allen, were almost immediately startled by the radiation levels they were seeing. In the days before personal computers with their handy spreadsheet applications that can readily present data in a variety of graphic representations, the available computers of the day were quite limited. Thus, the information being received from the on-board instruments had to be laboriously laid out on graph paper to get a picture of what was being detected.

Sometimes the instruments would report the expected cosmic-ray count (about 30 counts per second) based on vertical sounding rocket experience. But at other times the Geiger counters would show nothing (zero counts). The researchers noted that all of the zero counts were from altitudes near the satellite's apogee (in excess of 1,200 miles) and near the equator over South America. It was also noted that the zero counts occurred immediately after the count reached high levels. Reports from near the perigee of 300 miles would show the expected level of cosmic rays.

Because the satellite did not have any way of storing its data while it was out of range of ground stations, much of the radioed information was lost. A revised instrumentation package designed for the next Explorer launch attempt included a tape recorder that would store the information during each orbit and send it to a ground station on command.

The next *Jupiter C* launch with Explorer II occurrred on March 5, 1958, but it failed to achieve orbit when the fourth stage inexplicably failed to ignite. However, Explorer III, launched on March 26th, quickly provided a much more comprehensive view of the radiation phenomina. In analyzing the data from Explorer III, it was determined that the Explorer I Geiger counter had been overwhelmed by a strong radiation field. The zero indications were actually the Geiger-Mueller tubes being over-saturated. Van Allen later commented about the tape recorders that they had *"functioned beautifully in response to ground command and fulfilled our plan of providing complete orbital coverage of radiation intensity data."*

What emerged from the mapping of the radiation was a toroidal (donut) shaped field (or belt) of charged particles around Earth that was trapped by Earth's magnetic field. When the belts become "overloaded," particles striking the upper atmosphere are energized to fluoresence, which causes the aurora borealus or "northern lights." A more complete mapping of these belts was achived by Explorer IV and Pioneer III. The "Explorer" name would henceforth be used for general scientific research satellites and the designation "Pioneer" would reflect deep space probes.

While Sputnik I might have yielded some clues to what became known as the Van Allen Radiation Belt if it had carried radiation detection instrumentation, its apogee of 558 miles was not high enough to penetrate the upper level regions of the intense radiation. Sputnik II did have a higher apogee and the required instruments which operated properly and yielded data for seven days. However, Soviet scientists failed to effectively analyze the data and thus missed the first big discovery that satellites provided. Much to the chagrin of the Soviets, the huge Sputnik III also missed its chance to discover the radiation belts because of the failed tape recorder.

Perhaps the most disappointed scientist was reported to be Sergei Vernov, a distinguished physicist whose detectors on board Sputnik III had sensed periods of high levels of radiation, but these regions could not be adequately mapped.

The name Van Allen Radiation Belt was applied for the first time at an IGY conference in the summer of 1958 that was held to discuss some of the findings of the first satellites. The discovery of an outer belt, and confirmation of the origin of the radiation, came two months later when Explorer IV continued to map the belts in the late summer of 1958.

When the data that from Sputnik III was correlated with that of the Explorers, it was clear that the Soviets could have determined the existence of the belts had they been able to effectively interpret their data earlier. In an effort to save some scientific credibility for the Soviet people, Vernov produced an illustration of the radiation belt in Pravda in the spring of 1959 and implied that it was presented at the IGY conference. However, when a curious Van Allen reviewed that Soviet paper, the data was conspicuously missing. Because of Soviet secrecy, almost no scientific result was conveyed to the world from the first Sputniks.

Desperation Reaches for the Moon

The Department of Defense had recognized its shortcomings in the area of coordinating many of its high-tech activities and had formed the *Advanced Research Projects Agency* (ARPA) early in 1958. It was through this organization that the von Braun team launched Explorer IV on July 26, 1958. This satellite was propelled into an orbit with a higher inclination (51 degrees) in an effort to provide more information on the mysterious radiation that surrounded the Earth. This new track required the trajectory head northeast from the Cape rather than southeast, as that would have required the launch vehicle to over fly the more populated islands in the Bahamas. The satellite weighed in at 38 pounds with two Geiger-Mueller tubes and two scintillation detectors designed to distinguish energy levels of the radiation.

Realizing that the Russians would undoubtedly be looking to establish another "first" by reaching out to the Moon, another of ARPA's initial assignments was to see what might be done to produce a launch vehicle capable of the lunar mission. A variety of existing launch vehicle combinations was evaluated and it was decided that the *Thor-Able* was a likely candidate.

At the same time that the Army's *Jupiter* IRBM received approval for development in 1955, the Air Force sought an intermediate range missile to serve until the *Atlas* would be available as America's first line of "unstoppable" defense. By basing the shorter-range missile in countries friendly to the United States, such as England, Turkey and Italy, the IRBM would provide access to Soviet targets. The *Thor*, built by the Douglas Aircraft Company, was that stopgap measure for the Air Force. The same LR-79 engine as was in the Atlas booster and the Jupiter powered the 65-foot liquid fueled missile.

Following a series of tests with the usual number of spectacular failures, the *Thor* was approaching operational certification in the summer of 1958. As the Atlas was still in its early test phases, its team was looking for a vehicle that could accelerate full-scale re-entry test models to generate temperatures of up to 20,000 degrees Fahrenheit to validate the *X-17* data. The *Thor* was mated with the second stage of the *Vanguard* to produce the *Thor-Able* rocket. Three tests in the first half of 1958 validated the *Atlas* reentry warhead enclosure and proved the viability of the *Thor-Able* combination.

Working with Douglas Aircraft, Aerojet General, and the Space Technology Laboratories (STL), it was determined that the two upper stages of the *Vanguard* mated with the *Thor* could loft an 80-pound spacecraft to escape velocity—25,000 mph.

The excitement of this project was overwhelming to those involved. In the period of little more than 8 months since Sputnik I, dreams that had anticipated decades to reach fruition were now brought to reality in the period of a few months. The big question was, "Would the Soviets again beat the US to this critical event in the history of space travel?" All associated with preparing the Thor-Able rocket held their breath as each lunar launch opportunity during the summer of '58 came and went without a launch announcement from the Soviet Bear. ARPA could hardly believe its good fortune.

Each month presents a lunar launch period when the position of the Moon provides the optimum opportunity to reach it with a minimum of energy. The rocket must be launched during this brief time period, called a "launch window," or be delayed until the next day, and only a few days each month allow for this window.

The trajectory for a Moon shot is essentially a highly elliptical satellite orbit timed to intercept the orbit of the Moon some 240,000 miles away. Since it takes about three days for the spacecraft to make the journey, the spot aimed for is well ahead of where the Moon is when the launch occurs. Thus, with the launch pad sitting on a spinning earth that is rotating almost 1,000 miles per hour, and the Moon orbiting around the earth at 2,400 miles per hour, and with no mid-course guidance available, these early attempts were rather speculative as to whether the spacecraft would get close enough to be captured by the gravitational field of the Moon.

This first ARPA/STL spacecraft was quite sophisticated, having several scientific instruments and a photocell to determine its proximity to the Moon. A fourth stage was also included that would allow the spacecraft to be slowed enough to be captured by the Moon and, with a tremendous amount of luck, orbit it.

On Sunday morning, August 17, 1958, man's first attempt to reach the Moon lifted off from Cape Canaveral. The launch was picture perfect as the tall, slim *Thor-Able* rocket rose swiftly on a shaft of white-hot exhaust. Reaching into the stratosphere, the signature condensation trail of "frozen lightening" began to form and follow the accelerating vehicle. Then, suddenly, at 77 seconds into the flight, at about 50,000 feet, something went wrong and the vehicle exploded. Man's first attempt to reach the Moon had failed.

Three *Thor-Able* vehicles had been allocated to the project, and the second was delayed while the failure was analyzed. The September launch window was missed, but a truly miraculous effort saw the second *Thor-Able* launched on October 11th at 4:42 AM EST. All stages of the 88-foot, 52-ton rocket fired successfully, but the trajectory was imperfect and the required velocity of 24,000 mph, was short by several hundred miles per hour. The 30-inch-diameter 78-pound payload, christened Pioneer I by the scientists, soared 78,000 miles into space, only one-third of the way to the Moon. A failure to be sure, but the boost to American morale was significant.

Although nationwide television was still in its infancy, live coverage by the three networks (ABC, CBS, and NBC) was provided for these launches. The third launch on November 8, was scheduled for the very early morning and experienced a series of delays. As the prospects for launch dwindled, only NBC and a young reporter named Roy Neal, elected to continue the coverage. When all of the problems were finally sorted out, the launch occurred shortly after 2 AM, and all appeared to go well. ARPA announced that all three stages had fired—the lunar probe was on its way—and applied the name Pioneer II to the spacecraft. However, a later review of the telemetry data from the rocket indicated that the third stage had failed to fire, and the spacecraft re-entered the earth's atmosphere over Africa and burned-up in a fiery plunge.

The Army had also gotten into the Moon shot program and had received approval to use their *Jupiter* IRBM mated to the solid-fuel upper stages of the *Jupiter C* to complete a configuration called Juno II. The rocket was another example of desperation engineering. However, the "minimal approach" had gotten America into space with Explorer, which had made important discoveries.

The December 6, 1958, launch looked perfect, but the *Jupiter* booster had shut down 3.5 seconds too early. The 13-pound Pioneer III, instrumented to survey the radiation between the earth and the Moon (cis-lunar space), soared to 63,000 miles before plunging back to earth after a flight of 38 hours. It did make one important discovery: the Van Allen Radiation Belt, originally revealed by Explorer I, had a second component farther out in space between 10,000 and 40,000 miles above the Earth.

Although the first lunar program of 1958 was essentially a failure, the excitement of watching the launch "live" enthused the American people. A launch could be replayed on the evening news with the newspapers providing in-depth coverage, but the ability to be a part of the event by watching it as it happened was beginning to play a major role in the effect the space program was having on Americans and their willingness to pay for these expensive "failures."

NASA Established

It became obvious to President Eisenhower that the greatly expanding effort in space exploration to compete with the Russians was fragmented. The Navy had its assignment with the *Vanguard* while the Army was moving forward with the *Jupiter C*. Both programs were considered within the unclassified civilian domain. The Air Force had been given new impetus to proceed with its satellite reconnaissance project (no countries were complaining about satellites passing over their territory as Ike had feared), and it had provided the first three lunar attempts.

With several other space exploration projects in the works, it was important that a single agency oversee the activities and funding to assure that duplication of effort was minimized while new important projects were not neglected. There was a precedent for such an organization—the National Advisory Council for Aeronautics (NACA). NACA had been in existence since World War I and had made significant contributions to aviation progress. It operated three major research laboratories (Ames, Lewis, and Langley) with a staff of about 8,000. NACA was strongly supported by both the military and the civilian aeronautical industry. The military received the funding to build experimental aircraft (such as the X-1) and provided oversight of the construction while NACA provided the instrumentation of the aircraft and analysis of the data. After initial testing, these aircraft were often turned over to NACA for further evaluation.

NACA provided an advisory and analysis function that had resulted in the construction and use of several important facilities such as wind tunnels. With the magnitude of the space projects on the horizon, a much more capable and higher level organization was needed—a civilian space agency.

Along with his Science Advisor, James Killian, President Eisenhower directed the current head of NACA, Dr. Hugh Dryden, to create a new organization and prepare the appropriate legislation that would bring it into existence. Public Law 85-568 was signed on July 29, 1958, and created the National Aeronautics and Space Administration (NASA).

A Big SCORE for America

With the launch of the 3,000-pound Sputnik III on May 15, 1958, the Soviets continued to widen the perceived technology gap. A few weeks after the launch, Roy Johnson, Director of ARPA, was visiting the Convair facility where *Atlas* production was in full swing, despite that fact that it had yet to achieve its full 5,000-mile range in an incremental testing program. Johnson expressed frustration that our pitifully small satellites, which, while gathering valuable scientific information, were not garnering world acclaim because of their small size. *Atlas* Program Director, Jim Dempsey, mentioned that the *Atlas* was capable of putting itself into orbit. Johnson's interest was piqued, and following some preliminary discussions, he set in motion a project to orbit an entire *Atlas*. As a personal commitment to seal the bargain, he wrote his name in a broad marking pen on one of the vehicles, number 10B, lying inert among the "yet to be born" giants on the production line.

Johnson returned to Washington and posed the possibilities to Eisenhower. Ike, taking a page from the Russian secrecy script, gave his approval with two provisions. First, *Atlas* had to demonstrate its ICBM potential by flying a full 5,000-mile mission, and second, the project must remain a secret until a successful launch or he would call it off.

To this point the largest American payload in orbit was only 32 pounds, and that included the weight of the spent fourth stage of Explorer III. If this project were successful, the entire upper portion of the Atlas would be in orbit. A hunk of thin steel tankage and its rocket engine some 85 feet long and weighing 8,750 pounds would at last give the impression that American rocketry was competitive. The down side of the project was that the actual payload would be limited to about 150 pounds. This was due, in part, to the weight of the test equipment flown on each Atlas. In addition, there was the desire to ensure that this first orbital effort would be successful, so the conservative payload would be slightly less than the weight of Sputnik I.

A communications experiment was assigned to the mission that involved transmitting the human voice between a satellite and ground stations. The U.S. Army Signal Corps was called on to develop the electronics. Atlas number 10B began a six-month journey to fame.

While *Thor-Able, Juno II,* and their Pioneer spacecraft were taking center stage during the fall of 1958, *Atlas* 10B completed its assembly in Convair's San Diego plant and made the long trip to Florida by trailer. The secrecy was intense. Even the launch director, Curt Johnson, was not privy to the final destination of his "bird." Several activities had to be accomplished in secret to put the *Atlas* in orbit, not the least of which was disabling the cut-off signal that would shut down the engines to ensure that the *Atlas* did not over-fly its intended impact point in the South Atlantic.

However, before 10B could make its attempt, Convair had to demonstrate that the *Atlas* could achieve its primary role as an ICBM and fly the full intercontinental distance per Eisenhower's edict. *Atlas* 12B accomplished this on November 28th and cleared the way for 10B sitting on launch pad eleven. Several of the 10B launch team had strong indicators that this missile was headed for a special destination, but they kept their hunches to within the small group, understanding that there was a reason for the security.

On the evening of December 18, 1958, just 55 years and one day since the Wright brothers had first flown a powered aircraft, *Atlas* 10B erupted to life in the cool Florida air and headed for space. As the sequence of flight events transpired, and the predicted impact point on the plotter in the blockhouse moved further down the Atlantic, those in-the-know were expectant, while the outsiders waited in wonder. Finally, the plotter could no longer predict an impact point! The secret was revealed to the blockhouse crew when the guidance station operator reported that the recording pen had run off the graph. The missile had exhausted its fuel, and was traveling at 17,300 miles per hour—10B was in orbit!

Two hours after the launch, when the first orbit was completed and success assured, President Eisenhower announced to a late evening meeting of diplomats at the White House that an Air Force *Atlas* had been placed in orbit and it weighed 8,800 pounds. Because the Russians had never revealed the weight of the core stage of the *R-7* that entered orbit with its Sputniks, the weight of 10B came as surprise to the world. Many simply looked at the reported weights of the Sputniks and that of 10B and jumped to the conclusion that the Americans had surpassed the Russians; an illusion that the Americans, and the Eisenhower administration, were more than willing to feed.

It was interesting that Ike, long a believer in keeping the military presence in space on a low key, had finally been convinced that this space race required an all-out effort and had accepted the military capabilities: the Thor and Jupiter IRBMs for the Moon shots and the Atlas for this heavy weight contender.

The following day, with more surprises up his sleeve, Ike's voice was heard from space as the onboard electronics of the Signal Communications by Orbiting Relay Equipment (SCORE) began to broadcast his Christmas message to the world:

"This is the President of the United States speaking. Through the marvels of scientific advance, my voice is coming to you from a satellite circling in outer space. My message is a simple one. Through this unique means I convey to you and to all mankind America's wish for peace on earth and good will toward men everywhere."

Eisenhower ended 1958 on high ground. A year that had begun in desperation, that had seen the launch of the first small American earth satellites and the first failed attempts to reach the Moon, had concluded with a note of triumph, primarily due to the surprise of the secret launch and the political "spin" on the illusive aspect of weight in orbit. Nevertheless, the Russians, although only recording one launch in 1958 (Sputnik III), had not been sitting on their hands as the first days of 1959 would reveal.

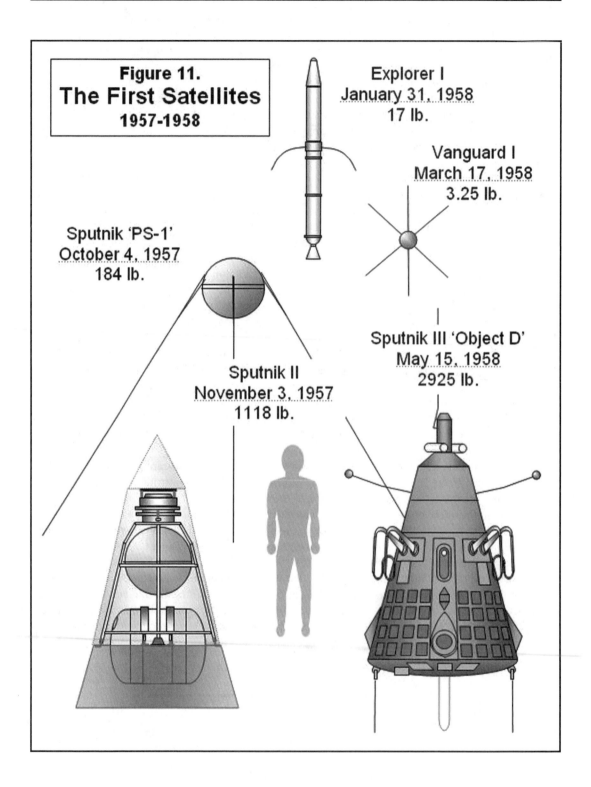

Figure 11.
The First Satellites
1957-1958

Explorer I
January 31, 1958
17 lb.

Vanguard I
March 17, 1958
3.25 lb.

Sputnik 'PS-1'
October 4, 1957
184 lb.

Sputnik III 'Object D'
May 15, 1958
2925 lb.

Sputnik II
November 3, 1957
1118 lb.

Luna 1: First Successful Lunar Probe

No one could deny that the Soviets had given the Americans their greatest opportunity to recoup some of their lost prestige in the latter half of 1958. Yet, with the exception of Project Score, the U.S. had been unable to capitalize on months of apparent Russian inactivity since Sputnik III in May. Four American attempts at the Moon had come up short.

As each lunar launch cycle rolled around, the eyes of the world kept waiting for the announcement that all knew must eventually come. Finally, on January 2, 1959, the Soviet news agency TASS announced that a successful probe was on its way to the Moon. Once again, the announced weight of the probe, almost 800 pounds, left the Americans and the world in awe! The American Pioneer I that made it only a third of the way to the Moon weighed about 80 pounds. The Army's *Juno II* payload, Pioneer III, was a mere 13.5 pounds. What the world was not told was that this first announced Russian Moon shot was not their first attempt.

To the world Lunik I, which missed its target by a mere by 3,728 miles, was a spectacular success. It was, as expected, another first for the Chief Designer: the first man made object to go into solar orbit. (Virtually all lunar probes—unless specifically designed to return to the earth, land on the Moon, or orbit it—eventually pass beyond terrestrial and lunar gravity boundaries and end up as satellites of the Sun.)

Of course, Korolev had had his eye on the Moon for a long time. As early as August 1956, a report generated by Korolev's NII-88 revealed the possibility of sending a probe to the Moon using a modified version of the R-7. At that time lunar probes were considered science fiction by the Soviet hierarchy. However, the success of the early Sputniks paved the way to approval for Korolev's team to begin building an upper stage for the *R-7* that would enable it to accelerate a probe to escape velocity. Three scientific payloads that had been prepared for Earth orbit following Sputnik III never made it into space. Korolev had determined that that aspect of the space race was over and it was "no contest." Now it was on to the next arena of the competition to determine the master of rocket technology. For the next two years, the team gave most of its effort to the lunar probes and the *R-7s* that would propel them there.

The new 10,000 lb. thrust upper stage engine was a derivative of the vernier engines of the R-7, eight of which controlled the pitch, roll, and yaw. The RD-105, as it was designated, was integrated into a nine-foot long LOX/kerosene tank section that was eight feet in diameter and provided a burn time of 440 seconds. The new stage extended the length of the *R-7* to slightly over 100 feet and added 17,000 pounds to its gross weight. It was actually referred to as a second stage because the *R-7* boosters are defined as stage zero and the core as stage one.

Because one goal for the Soviet's initial lunar probe was to prove to the world that the Soviets had reached the Moon, some thought was given to exploding something on the lunar surface that would be visible from the Earth. Goddard had suggested this in his writings 40 years earlier. On examining the size of the explosion needed to be viewable to the Earth with the naked eye, it was determined that only a nuclear blast would be sufficient. This was ruled out because of the possible negative political implications that might result.. It was finally decided that simply having a powerful transmitter on board that would suddenly go silent while sending a stream of data, would be sufficient.

Racing against the Americans, Korolev actually had his first Moon rocket ready for launch on August 18, 1958, but with the failure of the American *Thor-Able* on the 17th, it was rescheduled to the following launch period in September to allow more time for testing. A series of malfunctions during pre-launch preparations were strong indicators to the Korolev team that their rocket needed more development. The trajectory of the Soviet probe would have allowed it to reach the Moon before the Americans even if launched the following day.

This rocket was launched during the next lunar window on Sep 23rd, but the *R-7* disintegrated 93 seconds into the flight, apparently due to longitudinal resonance of the booster. Another attempt on October 11th again ended in failure at 104 seconds, apparently due to the same cause. The November lunar opportunity was skipped while more investigations attempted to resolve the problem.

A third attempt on December 4th saw the boosters perform flawlessly; the "fix" instituted was apparently successful. However, the core stage engines shut down prematurely at 245 seconds due to a loss of lubrication to the hydrogen peroxide pump. Korolev was doubtless thankful for the secrecy with which his embarrassing failures could be kept from the world. Thus, it was the fourth attempt by the Soviets on January 2, 1959, that the world heard about.

The primary instrument package, officially named "Luna Ye-1," was a polished aluminum-alloy sphere, 4 feet in diameter. Its pressurized interior, designed to maintain a temperature of 68° F, contained the instrumentation and battery power for about 60 hours. Luna 1 actually weighed 423 lbs., the additional 350 lbs. is another part of the story.

As originally envisioned, the *R-7* and its new upper stage would loft an estimated 400 pounds to the Moon, the weight to which the Luna probe was designed. However, several improvements to the Block E upper stage (as it was designated) along with some improvements in performance to the *R-7* itself allowed almost double the weight to be carried. As a result, another experiment, proposed by Soviet astronomer Iosef Shklovsky, was included in the spent upper stage rather than making any changes to the completed probe. This consisted of vaporizing some 2.2 lbs. of sodium at a point almost halfway to the Moon (110,000 miles) that was illuminated by the sun to produce an artificial comet. The 400-mile fluorescent orange trail of gas was visible over the Indian Ocean for five minutes with the brightness of a sixth-magnitude star and marked the location of the probe.

Among the experiments included in Luna 1 was a magnetometer mounted on a one-meter long boom. This was to be one of the more significant Soviet discoveries: the Moon does not have a magnetic field as does the Earth. It also observed the solar wind, a strong flow of ionized plasma emanating from the Sun. A micrometeorite detector and cosmic radiation sensors were also included.

Luna 1, also known as Mechta (Dream) and Lunik, passed within 3,700 miles of Moon on January 4th after 34 hours of flight, but was planned to hit the Moon (although this was never officially announced). Its package of various metallic emblems with the Soviet coat of arms was to be "distributed" over several miles of the lunar surface by the impact. However, it ended up in orbit around the Sun, between the orbits of Earth and Mars.

Luna's batteries expired after 62 hours on January 5, 1959, at a distance of about 300,000 miles. While it did not achieve all of its unannounced goals, it did establish another set of firsts for the Soviets. The first man made object to achieve escape velocity, the first close encounter with the Moon, and the first to enter an orbit around the Sun. Without knowledge of the exasperating failures that preceded Luna, the world was once more at the feet of Soviet science and technology.

With the achievement of Luna 1, von Braun and NASA pressed to make the final *Juno II* effort a success. The tiny Pioneer IV, almost identical to its unsuccessful predecessor, was launched at 1:45 AM on March 3, 1959. To ensure that the probe did reach escape velocity, the *Jupiter* engine of the first stage was planned for a longer burn. However, the slightly higher velocity coupled with some errors in the guidance package sent the little spacecraft well off course—missing the Moon by 37,300 miles two days after launch. The batteries that powered the transmitters died when Pioneer IV was 407,000 miles from the Earth.

Luna 2: Lunar Impact

The next Soviet attempt on June 18th failed when the guidance system malfunctioned 153 seconds into the flight and was destroyed by range safety. To this point the Soviets had succeeded only once in five attempts, but as the failures were not reported, the world stood in wonder of the unerring Soviet technology.

Luna 2, launched September 12, 1959, became the first probe to impact lunar surface on September 13th, delivering its cargo of Soviet pennants close to the Archimedes crater. It was speculated that perhaps the Soviets might use the arrival to claim the Moon. A resolution in the United Nations had strongly suggested that the Moon and the Planets should be considered international territory open to all nations in the same manner as Antarctica.

Luna 2 was a virtual duplicate of Luna 1 with some subtle enhancements. As did its predecessor, the Luna 2 spacecraft released a bright orange cloud of sodium gas at the halfway mark to aide in spacecraft tracking and as an experiment on the behavior of gas in space. On September 14th, after 33.5

hours of flight, its radio signals abruptly ceased, indicating impact. The instrumentation confirmed the Moon had no discernible magnetic field, and found no radiation belts. Thirty minutes later, the second stage of the launch vehicle also crashed into the Moon.

It was, of course, no coincidence that Khrushchev had timed his state visit to the United States to coincide with the launch. He just happened to have replicas of the emblems carried on Luna 2 to present to President Eisenhower. Khrushchev was again crowing loudly, *"America sleeps under a Soviet Moon."* Ike was gracious in his acceptance of the gift and wished the Soviets success in their space exploration.

Luna 3; Photographing the Far Side

Luna 3, launched the following month, was much more complex in both its mission and instrumentation and was, by far, the most ambitious undertaking of the Soviets. Not only was accurate guidance and launch timing necessary, but also the sophistication of the experimental package and its mission was unique—obtain photos of the Moon's far side. What would make these images so intriguing is that the orbital period of the Moon being 29 days, coupled with its rotational period of 28 days, means that the Moon always presents the same face towards the Earth (or approximately so). Thus, since creation, man has only been able to see about 58% of the Moon's surface.

Lifting off on the second anniversary of Sputnik I (October 4,1959) the 613 lb. spacecraft flew a figure-eight trajectory (a barycentric orbit) which brought it within 3,900 miles of the far side of the moon, which was sunlit at the time. An orientation system stabilized Luna 3 and pointed its cameras at the far side of the Moon that had never been seen by man. On October 7, 1959, Luna 3 obtained 29 photographs over a period of 40 minutes.

As the spacecraft returned toward the Earth, the film was developed automatically on-board and then scanned by a TV camera and transmitted to ground stations on October 18, 1959. The quality of the photos was not very good, and Luna 3 was commanded to retransmit the photos several times as it continued to loop in its orbit around the Earth. The images were finally enhanced by hand to provide acceptable quality, and the first pictures of the lunar far side were revealed to the world. The indistinct images showed a somewhat different terrain from the near side and were clear enough to allow the Soviets the privilege of naming the prominent features. Among these were two dark regions which were named Mare Moscovrae (Sea of Moscow) and Mare Desiderii (Sea of Dreams)—a propaganda coupe that would forever be etched in the annals of mankind—Luna III had accomplished its mission.

Two more lunar attempts made by the Americans in the latter half of 1959 failed. The first occurred during a static run of the booster in September when it exploded, while a second attempt on November 26th resulted in the payload shroud disintegrating while the vehicle was passing through maximum dynamic pressure (Max Q) within two minutes after launch, and it was destroyed.

Two additional attempts in September and December of 1960 also ended in failure when the booster exploded in flight. These four attempts represented the only *Atlas* launch vehicles configured to use the Aerojet *Vanguard* second and third stages. designated *Atlas-Able* they carried a significant payload of almost 400 pounds, but reliability was lacking. It was also decided that the *Thor-Able* and *Juno II* were just not capable of sending payloads to the Moon that would result in a high level of scientific return, and the *Atlas-Able* did not have the "growth" necessary to handle missions that were more advanced. As the Soviets had essentially accomplished the initial set of lunar "firsts," there would be no propaganda value for the United States in repeating their success. Further exploration would have to wait until a more capable launch vehicle was available.

Two more failures concluded the Soviet's first round of lunar exploration when the second stage engine failed to provide the required velocity on April 15, 1960. The probe reached an altitude of about 120,000 miles before meeting its fiery end in a plunge back into the atmosphere. Four days later, booster segment B on the next attempt reached only 75 percent of thrust at ignition and less than a second after liftoff, broke away from the core and exploded.

Korolev at this point decided that he had indeed won the second round of the space race for the Soviet Union and directed the efforts of his team to manned space flight and the exploration of Earth's nearest two planets. However, trouble lay ahead for the man who had brought the Soviets to the pinnacle of technological and political power.

Not a Good ICBM

The Soviet Ministry of Defense had planned for 50 ICBM launch pads with multiple rockets available at each pad so that, after the first had been fired, a second and third could be erected and launched within hours. However, as the costs began to be counted, it was obvious that the size and support infrastructure required for the R-7 was just too much for the financially struggling nation to consider. As great as the R-7 was in its ability to launch spacecraft, its ten-hour launch preparation time precluded it from being an effective retaliation weapon, and its inaccurate guidance system made targeting problematical. Only four R-7 launch pads were built.

Although the R-7 had moved Khrushchev onto the world stage as a bona fide superpower, it had failed to provide the Soviets with a viable weapon, and he was feeling the heat from the military. As close as Khrushchev and Korolev had become as a result of their mutual needs, the R-7's inability to perform as a practical ICBM began to distance them, and the other Chief Designers saw this as an opportunity to begin making some inroads into Korolev's monopoly on missile development and space exploration. Korolev's awkward position was also to help a newcomer to the field of Soviet rocketry, Vladimir Nikolayevich Chelomey, whose design bureau had been working with cruise missiles. A new addition to Chelomey's OKB-52 was none other than the Premier's own son, Sergey Nikitich Khrushchev, who worked in guidance systems. Korolev's refusal to consider storable propellants had allowed yet another rival (OKB-586 Chief Designer Mikhail Yangel) to move his own R-16 ICBM into the development stage. This design was smaller because advances in nuclear weapons had resulted in significantly lighter warheads. If Korolev's desire to upstage the Americans was foremost in his mind, the knowledge that Yangel, Glushko, and Chelomey were hot on his heals was not a distant second.

The fear that the Sputniks had engendered in the American Congress (and the senior Senator from Texas, Lyndon Banes Johnson) had resulted in a great expansion of the American military and civilian move into space. Militant comments made by LBJ and others, often designed for political consumption within the US, were having an alarming effect on the Soviet military and political structure.

France: Veronique and Diamond

Although economically impoverished by the Second World War, France was determined to take her place among the emerging technological leaders of Europe, if not the world. Using the talents of Wolfgang Piltz, recruited from Germany following the war, the French embarked on the development of a small, liquid-fuel sounding-rocket similar in size and capability to the American WAC-Corporal. It was to be capable of launching a 130-pound payload to 40 miles. Using diesel oil and nitric acid as propellants, the rocket weighed 3,000 pounds and had a burn time of 49 seconds.

Named Veronique (a contraction of the manufacturers name, Vernon Electronique), the rocket used a unique method of achieving stabilization in the early part of its launch sequence. A set of wires attached to the ends of its fins was played out in equal measure, to assure that the rocket would not deviate from its vertical alignment during the first 300 feet of flight. The wires would then be cut, and the speed of the rocket would allow fin stabilization to provide for its vertical assent for the remainder of its flight.

Launched from Algeria in 1952, the Veronique managed only two successful flights out of eleven. By 1954 modifications had improved its performance and reliability with two of four flights being successful—one achieved an altitude of 84 miles. A parachute recovery system returned the instrumentation.

By the late 1950s, the rocket had grown to 24 feet, and telemetry had been added to allow for transmitting data back to Earth. The rocket continued to play a part in France's upper atmospheric research program with 48 Veronique AGI models flying between 1959 and 1969. The last of the family, the Veronique 61, was 31 feet long and continued to fly into the 1970s.

Building on the technology of the Veronique was the Vesta program begun in the early 1960s. It had a weight of 12,000 pounds. Ten launches between 1964 and 1970 provided the first opportunity for France to send live animals (monkeys) into space to record their biological responses to weightlessness.

The Emeraude was the successor to the Vesta, using gimbaled engines for control. Two of five flights conducted in 1964 and 1965 were successful. A solid-fuel second stage was added, and the new creation became the Saphir.

Using the technology that had been produced by these early efforts, the three stage, 41,000-pound "Diamant A" was developed as a satellite launcher using the Emeraude as the first stage. The first launch on November 26, 1965, placed a 175-pound satellite in orbit. Although its lifetime was cut short by a failure in the radio communications after only two days, France had become the third nation to compete in space. All four of the Diamant launches were successful, and the program was completed in 1967.

The three stage "Diamant B" and "B-P-4" represented the continued growth and maturity of the French space program with launch weights of up to 58,000 pounds and first stage thrust of almost 80,000 pounds. The first launch in 1975 successfully placed a 110-pound satellite in a 500-mile orbit.

While Britain and Germany had shown the desire to move into space exploration, only France, among the European nations, felt compelled by national pride to spend the large sums necessary to compete.

The United Kingdom: Black Arrow, Blue Streak and Europa

The European community had struggled to gain a foothold in the building of launch vehicles so that it could compete with the United States and USSR. The British had begun development of the single stage Blue Streak liquid-fueled, intermediate range, ballistic missile in the late 1950s that was largely dependant on American technology. However, its cost (over £500 million) and political implications ultimately led to its cancellation. Along with it went a project for using the Blue Streak as a first stage booster for a satellite vehicle named the Blue Prince.

A second satellite launch vehicle proposal, called the Black Arrow, was given approval by the British government. It consisted of an enlarged version of the liquid-fueled Black Knight, which had been developed to explore warhead reentry problems, as a first stage. To this was mated another Black Knight as a second stage and a solid-fuel third stage, and this project continued for a few years with limited funding.

When the French succeeded in launching their first satellite in 1965, additional appropriations were made available for the project. The first test, launched from Woomera, Australia, on June 28, 1969, was a failure; but on October 3, 1971, after several more attempts, the eight-engine first stage came to life, and the vehicle succeeded in propelling Britain's first satellite into orbit. However, it would be the only independent satellite launch for that country as the Black Arrow program had already been canceled— Britain could buy American rockets at a lower price than they could build them.

When the British Blue Streak was canceled in the early 1960s, another attempt was made by a group called the European Launcher Development Organization (ELDO) to create a booster that would be capable of orbiting a 3,000 pound satellite to LEO or sending 100 pounds to the Moon. The United Kingdom would provide 39 percent of the funding for the first stage, France 24 percent for the second, and West Germany 22 percent for the third stage. Italy would contribute 10 percent for satellite development, and Belgium and the Netherlands would also provide some funding. Australia would allow the organization to make use of its Woomera launching facilities.

The rocket, called ELDO Europe 1, was based on the Blue Streak for its first stage. The project experienced all the problems that beset high technology endeavors, and ultimately ELDO failed to achieve a completely successful flight after eleven attempts, and the project was canceled. A cooperative European effort would have to wait for another decade.

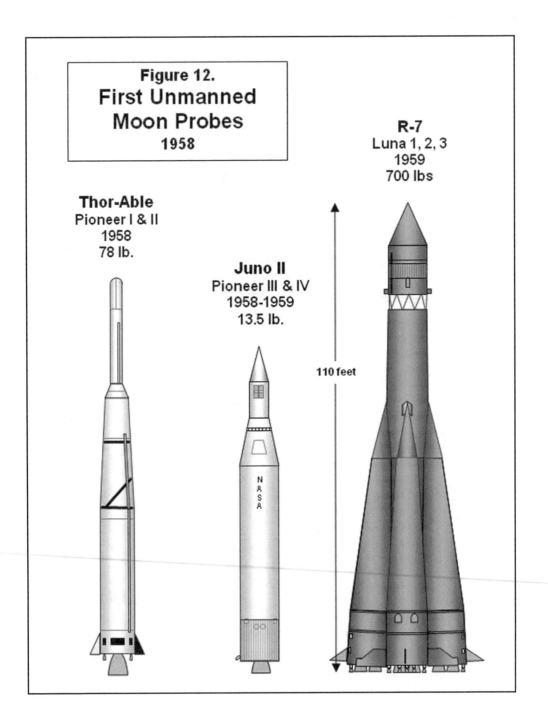

Figure 12.
First Unmanned Moon Probes
1958

R-7
Luna 1, 2, 3
1959
700 lbs

Thor-Able
Pioneer I & II
1958
78 lb.

Juno II
Pioneer III & IV
1958-1959
13.5 lb.

110 feet

NASA

A New Kind of War: A Cold War

Following World War II, Joseph Stalin made it clear to his Politburo and to the Soviet citizenry that the Soviet Union was surrounded and threatened by the United States and its allies. With numerous bases in Europe and Asia, American bombers, now armed with nuclear weapons, could attack. As he incessantly hammered about this threat, he was well aware that the United States had not pursued a militaristic program to take advantage of its nuclear superiority. In the late 1940's, only the original atomic bomb group, the 509th with a hand-full of B-29s, existed, and they were sequestered at a remote airbase in Roswell, New Mexico. The B-29 was the only plane then equipped to carry the bomb. Production of nuclear bombs was also almost at a standstill, and less than a dozen existed. Nevertheless, Stalin's power rested, in part, on his ability to use the United States as an ever present antagonist.

It was behind this façade that Stalin made his own aggressive moves. His grip on Eastern Europe, with the lowering of the "Iron Curtain" following World War II, as deftly defined in a speech by Winston Churchill, was brutal. The blockade of Berlin in 1948, the revelation of the Soviet's development of the atomic bomb in 1949, and its possession of a long-range bomber capable of delivering it (the TU-4) represented a capability that could not go unchallenged. These factors, coupled with Stalin's involvement in stirring the North Koreans to invade the South in 1950, heightened the tension between East and West.

For the remainder of his second term, President Eisenhower regretted that his lack of perceptiveness in this new frontier in space had allowed the Soviets to gain significant prestige and leverage in world power politics. However, if he had missed the opportunity to demonstrate American technology by launching the first satellite or by hitting the Moon, his support of space-based reconnaissance was his strong suit.

As a distinguished military leader during World War II, Eisenhower was acutely aware that poor intelligence had played a major role in allowing the Japanese to surprise the United States with its attack on Pearl Harbor in 1941. He also knew that good intelligence had led to the ultimate allied victory. The expression, "remember Pearl Harbor" was the rallying cry for a growing military-industrial complex that sought to plug the gaps in America's missile program—at any cost. Eisenhower recognized that there were those within his own country who would take advantage of the adversarial relationship with the Soviets to line their own pockets and had carefully crafted his military budget to thwart these individuals. Nevertheless, Sputnik began to unravel his fiscal planning.

The new paradigm in defense, as it was unfolding in the nuclear age, dictated that the next war would be fought without much warning of pending hostilities. Because of the destructive power of nuclear weapons, the first strike would likely determine the outcome. Thus, the winner of such a war would be the country who had done the best job of determining what weapons would be decisive and had those weapons on hand in quantity when hostilities began. A nation would go to war with what was in its inventory at the time.

This model of war required that a country invest in significant quantities of expensive new weaponry which, over a period of several years, would become obsolete and need replacement by more advanced machinery of destruction. Few countries could afford this approach, and most aligned themselves with one of the two superpowers (the United States or the Soviet Union) for their defense. Those who ostensibly remained uncommitted were often referred to as the Third World—usually impoverished countries under the control of dictators who liked to play both sides.

This also meant that there was little opportunity to test new and innovative weapons against an adversary before an all-out confrontation required their use. One approach to ensure a high degree of capability and reliability was to have redundant weapons. A second tenant, with the rapidly advancing technology of the 1950s, was to have new weapons in the development chain all the time. This situation of constant military preparedness was the basis for the phrase "cold war."

With the advent of the cold war, military intelligence became a critical factor. For the United States, it meant assessing the capabilities of the Soviet Union to ensure that the weapons developed would effectively counter the threat. An example of this was the long-range jet bomber fleets fielded by both sides during the 1950's. To address this threat, new manned interceptors emerged every few years. Arrays of surface-to-air missiles were positioned to defend strategic points—the number being dependent on the estimated size of the aggressor's bomber fleet.

But, how many bombers did each side possess, and where were they based? These questions were relatively easy to answer when viewed from the Soviet perspective. The defense budget of the US is public record. Defense contracts are awarded for specific numbers of bombers over a specified period. The location of US air bases could found on any of the road maps given away in gas stations during the 1950's.

However, with a closed society like the Soviet Union, how could the United States begin to answer these questions for a country with a land mass of over 8.5 million square miles—almost three times the size of the United States? Sparked by its new commander, General Curtiss E. LeMay, the Strategic Air Command embarked on a clandestine approach to gaining information on the size and capability of the Soviet military. Periodic over-flights of various portions of the Soviet Union by American reconnaissance aircraft began in the early 1950s. Although blatantly illegal by international law, RB-29s and RB-47s flew dozens of missions hundreds of miles into Soviet territory to photograph air bases and to determine the Soviet's radar frequencies and intercept-response times.

Both Presidents Truman and Eisenhower were aware of the intrusions by the Air Force into Soviet airspace. Although uneasy about the program, neither prohibited these activities, as they understood that LeMay's bombers could not be an effective deterrent unless they had valid targeting information. Every year or so, one of these aircraft would be successfully intercepted by Soviet fighters and shot down.

In another effort to penetrate the Iron Curtain, helium-filled balloons carrying cameras were launched by the United States in the mid-1950s in an effort to over-fly the Soviet Union without the risk of human life. The balloons, launched from Europe, used the prevailing westerly winds in the upper atmosphere (at 60,000 feet and higher) to carry them across Russia to be recovered over the Pacific. The obvious purpose was to photograph Soviet military installations.

Several of these balloons were recovered by the Soviets, who were impressed with the cameras and the radiation resilience of the film, as radiation in the upper atmosphere tends to fog the emulsion of film. One story that has circulated is that when the Luna 3 camera system was being engineered, Soviet technologists recognized that it had to penetrate the Van Allen radiation belts en route to the Moon and elected to use the 70 millimeter wide film from the American balloon cameras, cut down to the 35 millimeter size of the Soviet camera system.

Eisenhower needed an alternative to these provocative activities which could be construed as an act of war and which were costing American lives. He had put forth an Open Skies program in July of 1955 during his first term, in which he proposed allowing any country to over-fly and photograph any portion of US territories to confirm the size and readiness of the US military. The purpose of Open Skies was to assuage the paranoia of Soviet leaders, such as Stalin and Khrushchev, as to US intentions. However, some reciprocation by the Soviets was also expected, and to open the vast Soviet empire to the prying eyes of the United States was not something that the Soviets would even remotely consider.

Failing to get Open Skies accepted by the Soviets, Eisenhower proceeded to approve the development of a high flying spy plane that would become the U-2. He also endorsed a futuristic program called Pied Piper (project designation WS-117L), a reconnaissance satellite program.

Discoverer: Perseverance Pay-off

An interesting display was presented at the annual meeting of the American Rocket Society in Washington, D.C., in early December of 1958. The Lockheed Aircraft Company revealed a biomedical capsule in which a chimpanzee would fly into orbit and return, to help evaluate the effects of space flight on primates and ultimately man. To this point in time, no object had been returned from orbit, and to do so would be a significant "first" for the country that achieved it.

What made the display so intriguing was that the project, called Discoverer, was sponsored by the Air Force under the banner of the Advanced Research Projects Agency—not the newly formed NASA which was in the early stages of defining the national manned space program called Project Mercury. What was occurring here, however, was much more than met the eye. The biomedical research program was simply a cover for America's satellite reconnaissance effort—a spy in space.

Almost three months after the unveiling of the program, the first Discoverer satellite launched on February 28, 1959. But it did not rise from Cape Canaveral; it was launched from a new facility on the west coast on a piece of land that jutted into the Pacific Ocean north of Los Angeles named Point Arguello—Vandenberg Air Force Base. From here, a launch vehicle could be lofted to the south to enter into a polar orbit without endangering any populated areas. A polar orbit has an inclination of 90 degrees to the equator, and, as such, its ground track allows it to cover the entire globe over the course of a few days. This is the optimal orbit for a reconnaissance satellite and was the one chosen for Discoverer.

Discoverer pioneered many new and innovative concepts that were far beyond the modest Explorer and Vanguard satellites and even the Sputniks that had preceded it. The satellite itself was an integral part of a 19-foot-long, five-foot-diameter second-stage rocket that used the *Thor* IRBM as its first stage. Lockheed, the prime contractor for the *Agena* second stage (as it was named), incorporated the Bell Aircraft Company's XLR81 15,000 lb. thrust, liquid-fuel rocket engine. The engine was originally designed for a canceled weapons pod of the supersonic bomber, the Convair B-58 Hustler. The second stage Agena weighed 8,000 lb. at launch and could provide thrust for a total of 120 seconds. It was the first satellite to incorporate an attitude control system to allow it to be pointed in a specific direction (another significant "first"). This capability was required to return it to the Earth, as the small, solid-fuel retrorocket had to be precisely aligned opposite the direction of flight—another key requirement for a reconnaissance satellite.

Only a 300-pound segment of the second stage (which itself weighed 1,600-pounds in orbit) was designed to be recovered. This included a heat shield, the "biomedical" payload and the recovery parachutes. When the payload was ready to return to earth, the Agena would align itself to the proper attitude, and the reentry portion would detach and fire its retrorocket. This would slow it enough to allow it to descend into the upper reaches of the atmosphere where the drag of the air molecules would then cause it to return to Earth.

Discoverer I carried no live animals, and there were questions as to whether it had actually achieved orbit because its radio signals could not be confirmed. However, as its primary goal was to flight test the new Agena, the Air Force was satisfied with the telemetry data that confirmed its performance. Discoverer II launched into orbit in April, but it failed to produce a recoverable capsule. Four mice, who were to precede the use of primates and who were allegedly aboard Discoverer III in June, failed to make it back. A new launch occurred on an average of every six weeks, but something always managed to go wrong. One capsule appeared to re-enter over Spitzbergen near the North Pole, and it was feared that the Russians might try to recover it.

There was much soul searching among the program's management as the failures continued to mount. At total of 12 launches and 12 failures occurred. The biomedical cover story began to wear thin as those who were preparing these experimental packages were not privy to the real purpose of Discoverer, and their mice were often displaced by another payload just before launch. Nevertheless, Eisenhower stayed with the program and continued to provide the funding. In fact, an event in the skies over the Soviet Union would turn Discoverer into a critical defense program.

On May 1, 1960, a U-2 spy flight over the heart of the Soviet Union had finally been shot down. After four years and 20 over-flights, the Soviets had succeeded in stopping a U-2. Not knowing the fate of the pilot or plane, the US government issued its prepared cover story of the flight being an off-course weather observation mission. However, its pilot, Francis Gary Powers, had been captured alive, and the wreckage of the aircraft with its powerful cameras had been recovered! By skillful manipulation of the events, Khrushchev tricked the Americans into admitting that they had deceived the world. The resulting negative political ramifications were significant as Khrushchev cancelled his participation in a long awaited summit among the major powers.

Caught with his hand in the cookie jar, Eisenhower decided to come out into the open with America's intentions and reasoning for its aerial espionage program. He also promised Khrushchev that there would be no more U-2 flights over Soviet territory. He made that promise with the assurance of his advisors that the Discoverer program would soon succeed.

Discoverer XIII was launched into orbit three months later on August 10, 1960, and the sequence of events necessary to return its capsule safely finally began to work as designed. The *Agena* achieved the proper orbit and, at the appointed time and location, oriented itself for the scheduled retrorocket firing. The recoverable capsule dipped into the atmosphere and survived the scorching high temperatures of reentry. The heat shield and retrorocket were discarded at 50,000 feet, and two small parachutes unfurled. Suspended beneath the blossomed chute, the "bucket," as it was called, with its precious payload drifted lazily towards the Pacific a few hundred miles from Hawaii. A group of C-119 cargo planes waited below with a recovery trapeze suspended beneath their opened cargo doors in the rear. Guided by a radio signal emitted by the descending capsule, the planes homed in on its location. The parachute was spotted, but the planes were unable to effect a mid-air recovery as intended. A back-up procedure, with scuba divers dropped into the water where the capsule landed, successfully retrieved the bucket.

With the process now proven, the next launch just a week later was totally successful, and the bucket was recovered in mid-air. This bucket contained 1,432 frames of photos on 3,548 feet of 70-millimeter film. On the film were 1.65 million square miles of Soviet territory and the first view of a new Soviet ICBM launch site called Plesetsk. In its one day in orbit, Discoverer XIV provided more images of the Soviet Union than the entire four years of the U-2 spy plane program. Over the next year, the returned photos conclusively proved that the Soviet's ICBM threat did not number in the hundreds, as was widely feared, but was less than a dozen. Discoverer revealed that the supposed missile gap, on which Kennedy based his 1960 presidential campaign, was a myth.

Following Discoverer XXXVIII, the program was *officially* terminated, and the program became "black," meaning that no acknowledgment of its existence or results were released. However, Discoverer continued to fly under a new and classified name: CORONA.

CORONA: Opening Soviet Skies

The origins of Discoverer actually covered a winding path. In March of 1955, just months before the Vanguard satellite program was announced, the Air Force established a secret, high-priority program for developing a reconnaissance satellite designated The Advanced Reconnaissance System (ARS). The Lockheed Aircraft Company, which was contracted by the Central Intelligence Agency in 1955 to design and build the high flying U-2 spy plane, was again called on in 1956 to partner in developing the reconnaissance satellite—before Sputnik had been launched. The program, Weapon System 117L (WS-117) was originally code-named Pied Piper. This was changed to Sentry and finally to SAMOS.

The SAMOS program was to provide an orbiting camera that would transmit images to ground stations, although some missions would also provide for dropping "buckets." With the advent of Sputnik and concerns for Soviet advances in technology, it was felt that the extended time for development of SAMOS required a more immediate interim solution. The *Atlas* missile itself, which was to be the booster, was not yet ready to support the program, and the video technology needed to provide the required resolution appeared to be further into the future than the current political situation could tolerate.

With these considerations, President Eisenhower approved a Central Intelligence Agency satellite reconnaissance system code-named CORONA in February 1958. With even its name classified, CORONA was introduced to the world under the veil of Discoverer. The Air Force would continue with its announced SAMOS system for future capabilities.

Since the *Thor* intermediate-range missile was further along in its development than the Atlas, it was assigned the role of carrying the Agena, although the CORONA payload of about 500 pounds was one half the payload of the Atlas. This turned out to be a very fortuitous decision as the Thor achieved a high level of reliability and, along with the Agena second stage, had considerable growth potential.

With CORONA, a film-based camera (developed by a company called Itek) used a new, extremely thin, plastic film base, known as Mylar, developed by the Eastman Kodak Company. Similar cameras were in use in the U-2 spy plane, and the results had been excellent. Initially using a 70-millimeter, black-and-white film, resolutions of 25 feet were possible from an altitude of 100 miles and were improved to 5 feet within a few years. Later flights would also experiment with color and infrared film. While the first satellites carried only a single drop "bucket," some of the later satellites had two.

Each frame of the panoramic photos covered an area of about 12 miles by 140 miles. Some missions carried a different camera for higher resolution of points of interest rather than wide areas. The images were so good that the CIA used them to re-map virtually the entire world. In one instance the photos discovered the ancient ruins of a Roman fort. Perhaps equally important, the majority of the land mass on the earth was recorded so that later generations could see the effect of erosion and other natural and man-made events on the environment.

In 1962 the KH (KeyHole) designation was used to refer to all photo-reconnaissance satellites with the number following the KH designation indicating the type of camera system. Keyhole protocols and nuclear weapons data represented the highest levels of security in American history.

Most CORONA missions were completed within two days, while some lasted four. On a few occasions, durations up to 33 days were achieved. The cameras themselves were not recovered but were left with the bulk of the Agena, which would burn-up in the atmosphere within a few months. Subsequent spy satellites grew in size and complexity. Where the original CORONA series were "one shot" excursions into space, the intelligence satellites that followed had video systems that could transmit pictures with resolutions down to less than a foot. These 30,000-pound giants were launched by the Titan III and designed to function for years.

From 1959 through 1972, when the CORONA program was terminated, 144 missions were launched, of which 102 were successful. These flights provided images of all Soviet missile complexes and atomic weapon storage installations. They identified Soviet air defense missile batteries and command and control installations and networks. The program allowed for an accurate inventory of Soviet bombers and fighters. Moreover, when the first Strategic Arms Limitation Treaty (SALT I) was signed, it allowed for verification of compliance. Of course, other countries were monitored (including Red China) as well as activities in the Middle East during the Arab-Israeli war.

Perhaps more important was that CORONA proved the Soviets did not possess enough ICBMs to be a significant threat. This was a critical factor as there had been some suggestion in the new Kennedy administration that a preemptive American first strike should be considered before the Soviets had an opportunity to build their own first strike ICBM force. This option had been discussed in the Eisenhower administration as well. The images returned by Discoverer allowed the US to breathe easier as there was no significant ICBM capability in the Soviet forces. Through CORONA, Eisenhower had achieved his Open Skies program without Soviet cooperation. The first big return from the investment in space had been achieved.

Project CORONA established an impressive list of "firsts" that included the first photo reconnaissance satellite, the first attitude controlled satellite, and the first recovery of a vehicle returning from orbit. It represented the first mapping of the earth from space (including stereo-optical data) and led to a wide variety of commercial applications. In 1995 over 800,000 images taken from the CORONA satellites were declassified and put into the public domain. Today, a person can download a picture of virtually any neighborhood anywhere in the world from an unclassified online database.

SAMOS: A Dead End

While CORONA, under the veil of Discoverer, was finally achieving success, the first of the much larger SAMOS satellites was launched by the Air Force using the Atlas on October 11, 1960. As would be the case for most reconnaissance satellites, the launching was from Vandenberg Air Force Base. The 1,200-pound payload carried camera test equipment and a "ferret" module for recording electronic signals emanating from Soviet territory. However, an Agena umbilical cable failed to separate properly during launch, and the nitrogen attitude control gas was inadvertently vented. Without attitude control, the Agena could not maintain the proper orientation during the firing of its engine and did not achieve orbit.

It was almost three months later before SAMOS 2 finally climbed the fiery trail to orbit on January 31, 1961. It returned the first-generation photo-surveillance radio relay of images, but its photos were of disappointing quality.

With SAMOS 2 the project turned "black" (as had occurred following Discoverer XXXVIII). The Air Force Public Information Office now issued a launch schedule with only one-day notice (as opposed to five days), and pre- and post-launch briefings were discontinued. The Fact Sheets that were distributed had virtually no new information. The press could get more data from the weekly aerospace journals than directly from the Air Force.

SAMOS 3 launched in September 1961 using the newer Agena B, but the Atlas exploded on the pad when the engines shut down because of an umbilical cable that failed to release properly. SAMOS 4, with a recoverable capsule, failed to achieve orbit when the guidance system on the Atlas failed.

SAMOS 5 launched in December 1961 with a 1,200-pound payload, but the capsule was not recovered. None of the SAMOS satellite versions worked particularly well, in contrast to the success the CIA was having at the same time with CORONA. SAMOS was quietly wound up following its eleventh flight in November 1962 without producing any usable results. Portions of the SAMOS program are still considered classified, and it has been suggested that this may be due to its failure to produce, resulting in some embarrassment to the Air Force. It is also of interest to note that Sergei Khrushchev (the former Soviet Premier's son) reported that portions of several American spy satellites were recovered by the Soviets in the spring of 1961. From their descriptions, there is reason to believe these may have been parts from a SAMOS satellite.

The MIDAS Touch: Early Warning

The 1955 Rand Corporation's report that proposed a photo-intelligence satellite also stated that it might be possible to detect ICBM launches from satellites with infrared detectors. It was estimated that enhancements to the Distant Early Warning (DEW) radar, which stretched across Alaska and Canada in the mid-1950s, could detect ICBMs coming over the horizon and provide about a 15-minute warning of an attack. Because of this appraisal, procedures for getting bombers of the Strategic Air Command aloft within 15 minutes were instituted. However, if an additional 10 minutes could be gained by a satellite detection system, then more options were opened for retaliation as well as more time being allowed for the public to reach shelters.

The premise of the proposed project was that an ICBM during its boost phase (the first five minutes of its flight) emits a significant amount of heat that could be detected by infrared measuring devices. The MIDAS (Missile Defense Alarm System) program envisioned a set of 10 early-warning satellites in orbit, perhaps as soon as 1961.

As there had yet to be a satellite placed in orbit in 1955, many of these visionary programs were often discouraged for credibility as well as lack of funding. Air Force Chief of Staff Gen. Thomas D. White noted, *"Approval of the MIDAS plan is being delayed because of the doubt . . . in our ability to achieve necessary system reliability."* Another alternative provided for 84 infrared-equipped U-2s flying extended missions around the periphery of the Soviet Union at high altitudes. This option was considered impractical and was never pursued. Despite its "Buck Rogers" image, the MIDAS program was accelerated in 1960 when Khrushchev threatened to launch ICBMs at America if the United States intervened militarily in Cuba.

The first MIDAS launch occurred on February 5, 1960, from Cape Canaveral using an *Atlas Agena A*, but it failed to achieve orbit. A second launch in May was more successful, and experiments began with attempts to identify rocket launches from the Cape when the MIDAS satellite passed over. The tests were encouraging enough that seven more satellites were launched over the next two years (of which five were successful), and the first early-warning-satellite system was somewhat operational.

Although the first MIDAS was placed in a low orbit of 300 miles, the remaining satellites were much higher at just over 2,000 miles, so they could cover a larger area and remain in sight of a suspected launch for a longer time. They were also launched from Vandenberg into polar orbits to be able to over-fly the Soviet Union. A second set of three satellites with more advanced capabilities launched in 1966.

The Integrated Missile Early Warning Satellite (IMEWS) system was a follow-on project that positioned satellites in geosynchronous orbits at an altitude of 22,000 miles in the early 1970s. These satellites, while on patrol for aggressive ICBM launch scenarios, also monitored launches from the Soviet test facilities as well as those of other nations and were able to spot nuclear tests. The technology has progressed so far that today even some conventional explosives and highflying aircraft (which emit a heat signature) can be tracked. A new set of Space-Based Infrared System (SBIRS) satellites has begun replacing the older system and is able to detect and target Scud-class missiles sites.

Agena: Workhorse Upper Stage

The Lockheed Agena second stage, in addition to its pivotal work in the Discoverer, CORONA, SAMOS, and MIDAS programs, became a workhorse of early lunar and planetary space exploration. It went through a progression of upgrades from the "A" model, to the "B," which began with Discoverer XVI. This version used unsymmetrical dimethyl hydrazine (UDMH) in place of the kerosene fuel. UDMH and the oxidizer, red fuming nitric acid (RFNA), are storable, hypergolic propellants. This fuel permitted the Agena to accomplish a restart in space to change its orbit after several weeks in space.

With the "D" model ("C" was never implemented) that first flew in June of 1962, the Agena was stretched five feet to double the propellant and attitude-control fuel. This increased the burn time to a total of 240 seconds, allowing for increased payloads or significant changes to the orbital profile.

Likewise, the Thor rocket experienced growth over the early 1960s that would benefit many other space programs. A set of solid-fuel rockets attached to the periphery of the lower portion of the first stage created the "Thrust Augmented Thor." The airframe was stretched, providing for more propellant, and the burn duration was thus extended.

Zenit: Soviet Space Spy

Within a year after the United States Air Force initiated its satellite reconnaissance program in 1957, work began in the Soviet Union on a similar project. Designated Zenit (Zenith), the program called for the R-7 to orbit a 3,000-pound satellite (the size of Object D) that had all the requisite capabilities of attitude control and had a recoverable film capsule. Actual design work began even before Sputnik I was successfully orbited. However, when it was determined that a larger focal length of over 36 inches was required (the first CORONA cameras had a 24-inch focal length), the weight of the satellite more than doubled, and the need for an additional stage on top of the R-7 became apparent. The increased focal length was dictated by the need to get more detailed images since the basic topography and location of most US sites could be purchased commercially.

With the success of Sputnik and the self-criticism that was going on in the United States, Korolev recognized that he had "awakened a sleeping giant" and that the Soviets had to use their current lead in heavy lift capabilities to beat the Americans to the next prime "first" in space exploration. Manned space flight had to be achieved at the earliest possible date.

The new weight requirements for Zenit now placed more demands on the design bureaus' critical resources. The military, of course, wanted the reconnaissance satellite as soon as possible. Korolev was able to placate them with the reasoning that the same capabilities of the manned spacecraft (attitude control and re-entry) were key attributes for their spy satellites, and thus he would be pursuing a parallel and symbiotic objective. In November of 1958, the Council of Chief Designers approved the manned Vostok program with Zenit being an adjunct.

As soon as the first two cosmonauts orbited in 1961, the Zenit program became the priority. However, launch failures in December of 1961 and again in January 1962 put the program behind schedule. The Zenit-1 was finally in orbit on April 26, 1962, one year after the Soviets orbited the first man. Zenit-2, a Vostok type spacecraft carrying cameras instead of a cosmonaut, successfully returned its film from space. However, the next attempt on June 1, 1962, experienced a catastrophic failure that damaged the launch pad, which required more than a month of repair before it was operational. As with the American Discoverer program, the Soviets claimed these satellites were conducting scientific research and gave them the generic name Kosmos.

While continuing with their unmanned spy satellites, the Soviets also designed two manned space reconnaissance systems in the late 1960s. The Korolev design bureau pursued a militarized version of the Soyuz spacecraft while the Chelomey team developed the Almaz, a much larger space station. The development of these is detailed in the chapter "The Space Station."

While satellites that count cosmic rays and measure micrometeorite impacts might be of interest to the scientific community, the average man on the street in 1960 was looking for something more useful from this expensive headlong rush into space. Those who vividly recalled the exciting promises from the Walt Disney/Wernher von Braun TV production that aired in 1955 were now wanting such things as better and lower-cost overseas communications and improved weather forecasting capabilities.

President Eisenhower, in late 1960, issued an executive order for NASA to take the lead in encouraging private industry to provide space technology for commercial use. NASA was quick to react with a series of briefings on applications satellites to which more than 20 companies responded. In concert with this, the Federal Communications Commission (FCC) awarded the American Telephone and Telegraph Corporation (AT&T) licenses to operate on a series of both ground and space based frequencies.

Communications: A New Paradigm

As predicted by author Arthur C. Clarke back in the 1940s, communications were quick to assume a lead in space applications. Existing world-wide telecommunications links depended on undersea cables and a myriad of telephone interchanges to move a call from the United States to Europe. A single minute of overseas connect time could easily exceed $20 in 1960 dollars. Television was also coming into its own and needed high bandwidth to transmit images and sound to provide "live" worldwide coverage. With the capability of orbiting several hundred pounds using the Thor-Able (well beyond the 20 lb. capability of Vanguard and Explorer satellites) coupled with the miniaturization of electronics, various types of relatively simple communications satellites were now possible.

One of the first applications was the passive balloon satellite that orbited at the mid-levels of 1,000 to 2,000 miles. These 100-foot, inflatable, aluminized-Mylar spheres simply provided the ability to bounce radio signals transmitted from the earth back to line-of-sight receivers thousands of miles distant. The first of these, called *Echo I*, was launched in August 1960. The advantage of the reflective satellite was that it required no onboard electronics and thus was not limited by batteries or component lifetime. While the tests were successful and a small "constellation" of 16 satellites was envisioned, a variety of problems ranging from keeping the sphere from wrinkling and pitting (from micrometeorites) to the limited period of transmission time (line-of-sight) resulted in an early demise of this vision. Echo I and her sister Echo II continued to provide passive communications for more than five years, but the technology was a dead-end.

A second passive communications concept was called *Project Westford*. This involved the orbiting of thousands of needle-like reflectors into a belt around the earth. While it was an interesting experiment performed as an adjunct to the MIDAS 7 satellite, it failed to produce the desired result, and the active satellites soon took center stage.

Courier IB, a 500 lb., 51-inch-diameter, delayed-repeater communications satellite developed by the U.S. Army Signal Corps, was launched in October 1960 aboard a *Thor-Able-Star* rocket (the 100th firing of the Thor). A follow-on to the experiments performed by the *Atlas/SCORE* satellite orbited in 1958, *Courier IB* was the second attempt (thus the "B" suffix) to orbit the package. The first had exploded on the launch pad the preceding August. Electrical power was generated by 19,200 solar cells that covered a large portion of its spherical structure. *Courier* could handle up to 68,000 words per minute (about the size of a small paperback novel) and was an important step in developing technology for both military and commercial applications. The biggest drawback of its low 500- by 745-mile orbit was the line-of-sight limitation that allowed communications for only ten minutes from a specific station on earth. This required that messages be stored on magnetic tape and sent to the satellite at high data rates. They were then stored on-board and transmitted when the satellite passed over the destination's receiver station. If the satellite was in view of both stations, it could immediately relay the message to the destination station. *Courier* was a prototype for a more advanced military satellite communications project known as *Advent* which placed much larger satellites in geosynchronous orbits several years later.

Telstar I was the first successful active, commercial-communications satellite and was launched by an improved *Thor-Able*, called *Thor-Delta*, in July of 1962. Developed and funded entirely by AT&T at a cost of $2.7 million, the 170 lb., 35-inch-diameter satellite tested broadband microwave communications. It successfully received and retransmitted live TV pictures between America, Britain, and France from its orbit of 600 by 3,500 miles. *Telstar II* was launched into a higher orbit of 604 by 6,702 miles in May 1963 (with an inclination of 43 degrees). This allowed for 50 percent longer periods of communication (up to 40 minutes) by a single ground station.

Relay was an active repeater communications satellite designed and built by RCA to receive and retransmit radio and wideband television signals. Launched in December 1962, and powered by solar cells and batteries, the satellite was placed in a highly elliptical orbit, ranging from 700 to 40,000 miles, to evaluate the electronics as well as the effectiveness of the unusual orbital parameters. Britain, France, Brazil, and West Germany built ground stations to participate in later launchings.

Syncom-3 went aloft in August of 1964 and was the first geosynchronous satellite to orbit. Developed by the Hughes Aircraft Corporation, it was launched by a *Thor Delta*. A geosynchronous satellite orbits around the Earth's equator (zero inclination) once every 24 hours which allows it to remain stationary over the same point of the Earth. With three satellites in optimum locations, it is possible to communicate to any point on Earth with the exception of the Earth's Polar Regions. However, the launch of Syncom-1 in February 1963 failed due to an electrical system problem. Syncom-2, launched in July of 1963, entered an imperfect orbit with a 33 degree inclination and had limited usefulness.

The primary drawback to the synchronous orbit is that it requires much more energy and guidance to place the satellite into a circular orbit at 22,300 miles above the Earth. To accomplish this involves a precise, three-phased approach. The first phase uses the rocket to put the satellite into a highly elliptical orbit (called a transfer orbit) whose apogee reaches to the synchronous altitude of 22,300 miles. As the satellite coasts up to that high point, another powered segment begins using what is sometimes referred to as a "kick motor" that raises the perigee so the orbit is perfectly circular. The third phase adjusts the inclination or "plane" of the orbit to achieve a high degree of precision so the satellite remains in an apparently fixed point over the earth. This very special position is sometime referred to as the Clarke Belt (also called a geostationary orbit) in honor of Arthur C. Clarke's visionary proposals of its use. The term *belt* is used to denote that only along the equator can this orbit be achieved, and thus the positioning of geosynchronous satellites occurs along this orbital plane or belt. The amount of energy required to establish this orbital plane is essentially determined by the original latitude of the launch site.

Syncom-3 was positioned over the Pacific Ocean so that it could provide live coverage of the Tokyo Olympics in October 1964, demonstrating commercial viability. But without the availability of the other two satellites, the utility of *Syncom* was limited. Because the satellite's solar battery output was only 2 watts, the signal required a large antenna on the ground and special amplification equipment before being passed to a TV broadcast tower as opposed to the current ability of direct broadcast satellites.

In May 1965 both Syncom-2 and -3 communications were allocated to the U.S. Department of Defense because of the growing involvement of the United States in the Vietnam War where these satellites played an important role in military communications until they were replaced by other satellites in April 1969.

A prototype of an operational, commercial satellite, called *Early Bird* (HS-303) was built by Hughes Aircraft Company. It was launched by a *Thor Delta* that had solid-fuel rockets attached to its sides (called a Thrust Augment Thor or TAT) in April 1965. From a synchronous orbit over the Atlantic, it evaluated a design for a commercial system to provide communications between the United States and Europe. Funded by a new consortium called COMSAT (Communications Satellite) Corporation, it was a modified SYNCOM design that provided up to 240 two-way voice channels.

The Applications Technology Satellite 1 (ATS-1) was built by Hughes and launched into geosynchronous orbit in 1966 as the first of six satellites sponsored by NASA to test new technologies in space communications. ATS-1 experimented with the transmission of black-and-white and color TV and used a new antenna design. It also had a Spin Scan Cloud Cover camera for meteorological observations and innovations for attitude control of the spacecraft. Although designed to operate for three years, ATS-1 operated nineteen years until 1985. Much of its operational life was devoted to communications to remote areas for delivery of emergency medical services and educational programs across the Pacific and in Alaska.

Capitalist Complications

Within a decade of the launch of Sputnik I, significant advances were made in the three primary applications areas—communications, weather observation, and navigation—that had a profound effect on both military and civilian activities. But within the democratic nations, and the United States in particular, the blossoming of this new technology brought other complications.

The potential for communications satellites in particular reflected a multi-billion dollar investment as well as the potential for significant profits. As the first decade in space concluded, three giants in the international community emerged to vie for a possible monopoly. And the quandary was not as straight forward as it might appear. For example, COMSAT, who operated as an agent for a 70 nation consortium, was prepared to provide television to Alaska (which had no live capability at that time). However, RCA had recently bought the Alaska Communication System that had been built by the U.S. Air Force but had not yet been licensed by the FCC to operate. What was to become of their investment in terrestrial infrastructure? AT&T argued that their existing terrestrial facilities could be more economically tied to Alaska than the proposed satellite system—but also proposed their own satellite coverage.

Congress, in particular, was faced with the dilemma of how to sort out the problem of who should be allowed to develop a domestic communications system, and by what means should it be tied to a world wide network? This was perhaps the first time that business had been presented with the ability to replace an old technology almost overnight. How should the economic and social implications be addressed? As for the ability to get their products into orbit, all of the eager participants would have to look to the singular monopoly of NASA, who would launch satellites at cost. Thus arose the drive to perfect an inexpensive launch vehicle.

Scout: A Less Expensive Solid Solution

When it became apparent that useful applications could be packaged into a 100 to 200 lb. satellite, the need for a more cost effective launcher than the Thor-based Agena, Able, or Delta became obvious. The answer lay in the advances made in solid-propellant-rocket-motor technology in developing the Polaris submarine-launched, ballistic missile and the Pershing surface-to-surface tactical missile.

When the U.S. Navy had originally envisioned a sub-launched, ballistic rocket, similar in concept to what the von Braun team had proposed for the German Navy during World War II, it was based on liquid fuels. In fact, the Navy initially worked with the Army to partner on the *Jupiter* intermediate-range-missile program. But as that project progressed, the Navy saw that handling a large, liquid-fuel rocket in the confines of a submarine was fraught with problems and imposed many restrictions on the operational aspects.

The Navy turned to Lockheed Aircraft, who had perfected the X-17 reentry test vehicle for the Air Force, and thus began a fruitful relationship. The missile that resulted was a solid-fuel, 900-mile-range missile fired from a submarine—while it was under water. This technology would have resulted a significant impact on the development of one of the most potent weapons systems ever devised by man as well as provide another means to make the exploration of space more reliable and affordable.

There were three significant drawbacks to the solid-fuel-rocket motor that had to be overcome before it could be used as a viable weapon or as an instrument of space exploration. First, the chemical propellant mixtures that had been used were very inefficient as far as their Specific Impulse. Research into more energetic compounds by 1956 had produced far better combinations that provided an Isp of 260 seconds using a Polyurethane Ammonium Perchlorate (PU/AP) based oxidizer with aluminum as the fuel. This made the prospects of a long range missile much more achievable. These chemicals were bound together in a rubber-like polymer and poured into the rocket casing.

As with modern solid-fuel rockets, the fuel is caste in a star-shaped arrangement in the motor, and ignited along its entire length. This allows a constant surface area burn to create an even thrust during the consumption of the propellant burning from the inside out. The propellant acts as an insulator to protect the casing from combustion heat until the very end of the burn

The second shortcoming was the inability of the solid-fuel rocket to be effectively controlled in direction. To the mid-1950s most solid-fuel rockets were either spin stabilized or fin guided. The latter typically required some form of launch tower until the missile had gained enough speed for the fins to be aerodynamically effective. Techniques to swivel the exhaust nozzle (similar to gimbaling the engines of a liquid-fuel rocket) or to place thrust vanes in the exhaust were developed to provide the required control without the need to "spin or fin" the rocket. Later, more efficient methods of control provided for liquid thrust vectoring.

The final problem with the solid-fuel was that, once it starts its burn, it could not be shut down. For space applications, this was not a significant problem, but it did limit the precision with which solid-fuels could perform some applications. With respect to a ballistic missile, firing it at distances less than its maximum range required that some form of drag inducing "brake" be extended during the ascent to diminish some of its energy, and/or for the trajectory to be elevated like an artillery piece. In some applications, terminating the thrust at a precise velocity was achieved with ports in the thrust chamber that were opened in flight to reduce the chamber pressure abruptly.

With the advances in solid-fuel technology, NASA in late 1958 asked the aerospace industry to submit bids to produce an all-solid-fuel satellite launch vehicle that could orbit 130 pounds to low Earth orbit (300 miles). Vought Aircraft Company (which became Ling-Tempco-Vought—LTV) combined the Aerojet "Algol" first stage of the *Polaris* with the Thiokol "Castor" (from the *Sergeant* missile) as a second stage, and the Allegany Ballistics Laboratory's "Antares" and "Altair" (derivatives of the *Vanguard* third stage) for third and fourth stages to create a new low cost alternative.

Initial testing began on April 18, 1960, and by February 16, 1962, the *Scout*, as it was called, launched its first successful satellite, Explorer IX. The 40-inch-diameter, 72- foot-long missile produced 100,000 lb. of thrust from the first stage for 78 seconds. The vehicle then coasted upward for about 44 seconds to rise above the denser levels of the atmosphere before second-stage ignition. During the coast period, triangular fins on the first stage kept the vehicle properly oriented. The second and third stages used small peroxide thrusters to maintain pitch, yaw and roll control. The fourth stage employed spin stabilization. *Scout* also introduced a third satellite launch facility for the United States—Wallops Island, Virginia.

The price tag for *Scout* was less than one million dollars per launch, a cost which compared favorably to that of the *Thor Able Star* at over $3 million (1960 dollars).

Weather: Art to Science

One prospect of the new movement into space was the fascinating possibility of being able to look down and observe the earth and its cloud cover using the new technology of television to transmit the image in real-time. Weather not only caused people great inconvenience but it could be a deadly killer, especially if there were insufficient warning of a severe storm. The ability to predict weather was still a "black art" in 1960; therefore, tracking hurricanes was of particular interest.

The first satellite to address the cloud cover issue was the small *Vanguard II*, launched on February 17, 1959. (The first attempt to launch this particular application had failed the previous September.) Designed for the light weight capability of the Vanguard rocket, the satellite weighed a mere 25.7 pounds, and was the classic 20-inch, gold-plated-magnesium sphere. It contained two optical telescopes with two photocells to measure the global distribution and movement of cloud cover and contributed to the basic knowledge of the earth's energy budget over the daylight portion of its orbit.

Radio communication was provided by two transmitters. One sent a continuous signal for tracking purposes while the other was used as a command receiver to activate a tape recorder that relayed telescope experimental data via the telemetry transmitter. Both transmitters, powered by mercury batteries, functioned normally for 19 days. The initial orbit had an apogee of 2,063 miles and a perigee of 346 miles, providing a period of 126 minutes and a lifetime of at least 200 years. The satellite was spin stabilized at 50 rpm, but optical data was poor because of an unsatisfactory orientation of the spin axis (wobble).

Television Infrared Observation Satellite (TIROS 1), launched on April 1, 1960, by a *Thor Able,* was the first true meteorological satellite. The 42-inch-diameter cylinder, designed and built by RCA, was 19 inches tall and weighed 270 pounds. Power for the electronics, which included two TV cameras (wide and narrow angle lenses), was provided by 9,200 solar cells that recharged nickel-cadmium batteries.

Because the upper stage of the *Thor Able* was spin stabilized to 136 revolutions per minute during its firing, a means to slow the rotation (so the pictures would not be blurred) consisted of two small weights on long cables that were released 10 minutes after orbital insertion. As these weights spun off on either side of the satellite, their inertia slowed the spin to 12 RPM before they slipped off their hooks into space. Shortly after this process was completed, two additional weights within the satellite were freed to shift along a set of rods to serve as precession dampers. This "de-spin" technique was widely used with satellites launched with spin-stabilized, solid-fuel, upper stages. Because some of the residual rotation was needed to maintain its basic orientation, a set of small spin thrusters mounted on the lower rim were available on ground command to increase the spin should the rotation drop to less than nine RPM.

To avoid the problems encountered by Vanguard II, a set of infrared detectors were used to locate the horizon for timing the activation of the cameras. Despite all of these precautions, there was still some wobble in the spin of the satellite, but the scientists were able to adjust the picture quality to achieve good results. The cameras produced an image consisting of 250 pixels (dots) per line and 500 lines per frame over a period of two seconds. Magnetic tape recorded the video scan at a rate of 50 inches per second. The tape recorder, with 400 feet of tape, held up to 32 pictures until commanded to transmit them to earth. This system pioneered many advances in video tape technology.

With TIROS, mankind could, for the first time, look down (electronically) from a nearly circular orbital perspective 473 miles above the Earth and view the cloud formations that covered much of its surface. The swirl of low pressure systems and the formation of frontal conditions could be readily seen with the wide angle camera that viewed about 800 square miles. The tall build-ups of cumulonimbus were identifiable with the narrow angle camera that covered about 80 square miles. In its 78-day lifetime, more than 22,000 pictures of good quality were relayed back to earth.

TIROS II, which operated for seven months, was launched on November 23, 1960, and TIROS III followed in July of 1961. These had advanced infrared and radiation sensors that were used to determine how much of the Sun's energy is reflected and re-radiated back into space. These measurements also helped establish methods for determining the amount of water vapor present in various areas around the world to aid in predicting the intensity and amount of rain. An intensive 9-week period saw an international study team, made up of experts from many countries, examine the data and correlate it with ground and airborne observations and measurements.

The initial TIROS program was not meant to provide current weather predictions but to develop the appropriate instrumentation for future operational satellites and to establish algorithms for predicting atmospheric changes based on the data received. However, its success led to the launching of nine TIROS satellites by 1965 with some in special polar sun synchronous orbits. This type of orbit allows the solar powered satellites to remain in constant sunlight for extended periods of time.

NIMBUS was a second generation spacecraft that provided a fully operational meteorological system beginning in 1964. It was placed into a sun synchronous orbit with an inclination of 98 degrees that had a perigee of 281 miles and an apogee of 577 miles. NIMBUS carried advanced TV cameras including a sophisticated cloud mapping system and an infrared radiometer which allowed weather pictures to be taken at night. Seven Nimbus satellites were placed into orbit by 1978 and have since been replaced by the GOES and NOAA satellites.

GOES is a Geostationary Operational Environment Satellite that takes a picture of the entire western hemisphere's weather from a geostationary orbit. Its primary mission is to locate and track large destructive storms such as hurricanes. Working in pairs, these spacecraft are located at 75°W and 135°W Longitude along the Clarke Belt and are known as GOES East and GOES West respectively. The spacecraft provides visible and infrared imagery of an entire hemisphere 24 hours per day. The first seven GOES spacecraft were built by Hughes and were spin stabilized with body mounted solar arrays. The two newest spacecraft, GOES 8 and GOES 9, were built by Loral and have a three-axis stabilization system.

NOAA spacecraft can view the entire Earth's surface from their polar orbit twice in 24 hours, providing morning and evening observations by working in pairs with one being six hours behind the other in the same orbit. In addition to weather imagery with a resolution of one-half mile, these spacecraft monitor atmospheric humidity and temperature, snow and ice cover, total ozone content, atmospheric aerosol content, and COSPAS/SARSAT distress signals. The current NOAA satellites are numbers 12, 14, and 15.

Navigation: Unbelievable Accuracy

Navigation at sea was a formidable problem that became a critical issue at the time of Columbus. For the first time, voyages were being taken that were out of contact with land for weeks and even months at a time. Determining Latitude (North and South of the equator) had been possible for centuries by using the position of the Sun relative to the horizon at high noon. High noon was determined approximately by the shortest shadow cast by the sun on any given day.

But the early mariners could not locate themselves accurately with respect to Longitude (East and West) until technology provided an accurate timepiece (thank you, John Harrison, 1693-1776). By measuring the angle of the sun (or a designated star) above the horizon at a specific time, locations could be determined within 10 miles. With today's demands on accurate positioning, this level of accuracy had to be improved a thousand fold. Electronic LORAN (LOng RANge) navigation was developed during World War Two, but this was only effective within 1,000 miles of a set of transmitters, and its accuracy provided a position to within a mile.

The Navy needed much more precise navigation capability for its Polaris submarines that were just entering service in 1960, and it looked to developing a satellite system to provide an answer that could be used in any weather. The Polaris missile guidance system had to know its exact location within a fraction of a mile in order to establish its starting position relative to its intended target.

TRANSIT IB, launched on April 13, 1960, (by the *Thor Able Star* rocket) was the first navigational satellite to attempt to provide a solution to the problem. (*TRANSIT IA* failed to orbit in September 1959.) Using a constellation of multiple satellites launched into 600-mile-high orbits, *TRANSIT* employed an electronic clock memory system, a magnetic stabilization system, and four transmitters powered by nickel-cadmium batteries recharged by solar cells. This first 265 lb., 36-inch-diameter sphere operated for 15 years.

Designed under a contract with the Johns Hopkins Applied Physics Laboratory (APL), initial accuracies of less than ½ mile were achieved; overcoming the problem of atmospheric refraction to the radio signal allowed later accuracies to one-tenth of a mile. An interesting aspect to the project was that accuracies were ultimately achieved that challenged existing knowledge of geodesy (the shape and size of the Earth), and considerable research was required before the projected accuracies could be fully used and appreciated. This included the exact shape of the earth, gravitational influences, and the drifting distances between major land masses.

With *TRANSIT IIA*, the *Thor Able Star* (with its ability to restart its second-stage engine) delivered a second satellite piggyback into orbit that was said to measure solar radiation. It was the first such dual delivery which has since become more common. In later years it was learned that these second satellites were actually small electronic intelligence gathering applications whose role was initially kept secret.

The *TRANSIT 4* series were drum shaped and had the first radioisotope, thermoelectric generator (RTG) called the SNAP-3 (Space Nuclear Auxiliary Power) which provided 3 watts of power. The *TRANSIT 5BN* series were also atomic powered, but solar power was less expensive and more "politically correct," so the SNAP units were discontinued on navigational satellites. After the experimental *TRANSIT* satellites validated the concept, the operational satellites were launched from Vandenberg Air Force Base in California into Polar orbits. More than 24 *TRANSIT* satellites were launched over a period of 28 years, most by the new, low-cost, solid-fuel *Scout*, which required the satellite to be "ruggedized" because the Scout gave more vibration and G force during launch than the *Thor Able-Star*.

Attitude control of some applications satellites was obviously critical. A variety of methods were used depending on the accuracy of attitude and the types of maneuvers that might be required. In the case of the operational *TRANSIT* satellites (known as Transit-O), only a basic orientation was required (for antenna positioning) called a gravity-gradient-stabilization system. It consisting of a long rod extending from the center of the satellite and using the Earth's gravitational field to keep the spacecraft's long axis aligned with the radial direction of the Earth's center. The technique uses the decreasing strength of the Earth's gravity field (the gradient) at increasing distances. The largest mass of the spacecraft nearest the Earth experiences a slightly stronger gravitational attraction than the smaller mass at the end of the boom.

TRANSIT-Os weighed 143 lb. and were octagonal in shape with four paddle-like, solar arrays. They used an up-rated *Scout G* booster that could put two satellites into orbit at one time. The final *TRANSIT-O* launch occurred in August 1988, and the system actually remained in service until December 1996. Those satellites still operational are used for measurements of the ionosphere. The current navigational system is the Global Positioning System (GPS) that uses a constellation of 24 satellites and can achieve accuracies to within inches.

Soviet Activities

The Soviets were eager to proceed in all three application areas but did not progress as rapidly as the United States for a variety of reasons. Most prominently, they did not possess the depth and breadth of an aerospace and electronics industry that could apply itself to these multiple problems simultaneously and an economy that could support and effectively use the results. All three applications were approved for development by 1960 and used a similar spacecraft structure or "bus" whose acronym was KAUR.

The first Soviet communications satellite used a highly elliptical, twelve-hour orbit known as a Molniya orbit (typically 350 miles by 24,000 miles) and was identified by that name to the West when it was launched in October 1965. This type of orbit (similar to that used by the American Relay satellite) was chosen because it requires less rocket power to achieve than a geosynchronous orbit, and with an inclination of 65 degrees is more suited to communications in northern latitudes. Since satellites in this type of orbit move very slowly at apogee, they appear to hover for hours near that point. The disadvantage is that the sending/receiving unit must track the satellite, whereas a geosynchronous satellite is essentially stationary in space as viewed from the earth.

Molniya-1 series of communications satellites was developed by Korolev's OKB-1 team and was initially used to test the utility of such a satellite for the military. Preliminary work on the satellite began in 1960, but it was not ready for launch until 1964. The first two launches failed in January and June of that year. With the success of the October 1965 launch, Molniya-1 quickly proved the new technologies of automatic satellite control with its three-axis, gyroscopic stabilizers that aligned the spacecraft to within 10 degrees of the desired orientation for the solar panels, while its optical sensor pointed the antenna at the earth during communications sessions. The massive (for its time), 3,600-pound spacecraft had a length of 14 feet, a diameter of over four feet, and a span of 25 feet across its solar panels. The last communications session occurred in February 1966 before the solar cells deteriorated.

Operational use of the *Molniya-1* series began in 1968 with three satellites providing long-distance communications coverage that included television and two-way multi-channel telephone and telegraph service for the entire Soviet Union with an expected lifetime of two years. Later versions also included cameras for cloud coverage video and were launched several times each year under the *Molniya* name through 1975.

Development of the first Soviet weather satellite known as *Meteor* began in 1960 but (as with the communications satellite) was not ready for launch until 1964. The first of four launches used the R-7 with an upper stage similar to the Vostok to an inclination of at 81 degrees and weighed 10,400 pounds. The system went into operation in 1969.

Using the common KAUR-2 bus, a follow-on project was approved in 1968 for a strategic communications system. With the name Kristal, it was launched beginning in 1972 as the *Molniya-2* into both highly elliptical orbits and Raduga (Stationary) geosynchronous orbit.

Tsiklon, the first prototype Soviet navigation satellite system, used the basic KAUR-1 bus and the Kosmos-3M launch vehicle. Like the U.S. *TRANSIT* system, the 1,800-pound satellite provided Soviet ballistic missile submarines with accurate positional fixes so that acceptable targeting accuracy could be achieved. It used the Doppler navigation method with a constellation of satellites placed in 2,000-mile orbits using a single-axis magneto-gravitational (gravity gradient boom) passive orientation system. Initial flights began in 1967 and showed a position error of 1.6 miles, which was not acceptable. The basic problem was the inability of computer software to create an accurate spacecraft ephemeredes (location table). Significant improvements were subsequently made that also provided for gravitational perturbations and the geodesy of the earth so that an average error of only 300 feet over a five-day period was achieved. Testing continued through 1972 before an operational system was deployed.

Kosmos: A Small Soviet Launch Vehicle

Just as the United States recognized the need for a more economical launch vehicle for comparatively small applications satellites, so it was with the Soviet Union. Of course the word "small" is relative. Yangel's R-12 intermediate range missile was initially the best candidate, although it's rather low first-stage Specific Impulse (some 20% lower than that of the R-7) had to be addressed when work began in April 1960. To compensate for this lack of first-stage performance, a more efficient second stage was developed using the RD-119. This had a small, fixed, single combustion chamber with a large nozzle that allowed exhaust gas expansion of 1,350 times at the nozzle exit (about four times the sea-level expansion ratio). The main engine turbo-pumps were driven by bleeding off combustion gas, which was then directed into a set of four non-movable thrusters arranged at the base of the stage. Steering control was achieved by changing the rate of exhaust through the corresponding thruster. The second stage used liquid oxygen as an oxidizer and unsymmetrical dimethyl hydrazine as a fuel (UDMH).

The first launch of *Kosmos 1*, as the rocket was called, with a satellite designated DS-2 (from the Russian *Dneprovskiy Sputnik*) took place on October 26, 1961, but failed because of flight control problems. The second attempt on December 21, 1961, also failed when the second stage prematurely shut down. The cause was traced to excessive evaporation of the oxidizer from engine heat that was higher than anticipated, resulting in greater propellant consumption. This was resolved, and on March 16, 1962, the first successful launch orbited a DS-2, which the Soviets announced as *Cosmos-1*—a catch-all name that would be applied to most military and a variety of scientific satellites and spelled with either a 'K' or 'C'.

The *Kosmos 3* launch vehicle followed, based on the R-14 intermediate-range ballistic missile that used storable propellants (UDMH and Nitric Acid). *Kosmos 3* was eight feet in diameter, 105 feet tall, and weighed 240,000 pounds at liftoff. Its first stage was powered by a 390,000-pound-thrust RD-216 engine composed of a two-dual-thrust-chamber RD-215 engine. Four graphite vanes extend into the exhaust, one on each of the four fixed thrust chambers, to provide steering. In size and capability, it was the equivalent of America's *Atlas*.

The second stage used an RD-219 fixed, restartable main engine and a low-thrust on-orbit-propulsion system using four thrusters. This feature allowed the vehicle to place multiple satellites in different orbits during a single mission. Later versions were able to launch up to 3,000 pounds into a low earth orbit. The first R-14 derived vehicles were launched from Baykonur in 1964. By 1977 the *Kosmos 3M* had replaced the smaller R-12-based *Kosmos 1* and 2 launch vehicles and had orbited the *Tselina-O* electronic intelligence satellites, the *Strela* and *Tsyklon* military navigation and communications vehicles, and *Tsikada* navigation satellites.

The *Kosmos 3* launcher, with a 1980 cost of $15 million per vehicle, ranks third in number of launches with about 450 orbital attempts, trailing only the R-7 and Thor/Delta, before production ended in 1995. As with virtually all Soviet launchers, *Kosmos 3M* is mated to its payload in a processing building and transported horizontally to its launch pad on a railroad carrier about two days before launch.

When the Soviet Union announced the orbiting of Sputnik II in November 1957, they gave strong hints that a new power source had been used. As the Sputnik weighed over half a ton and dwarfed the proposed American Vanguard, many were more than willing to believe that indeed the Soviets had tapped into a new technology. Although most authorities in the West recognized that the weight of Sputnik II was well within the capabilities of conventional chemical propellants, the Soviet's motive was to keep the opposition guessing as to the real state of their technology in several areas. Wild speculation abounded with nuclear energy heading the list of possibilities.

America had not been lax in its own search for new methods of rocket power. Since the end of World War II, there were modest research efforts underway to explore innovative sources of reactive force before Sputnik. These ranged from higher-energy chemical propellants to perhaps the most daring and technically challenging idea posed by the use of nuclear power.

Exploring Alternatives

The most bizarre concept during the period was called Project Orion, which proposed using the principle of nuclear pulse propulsion: a series of nuclear explosions, approximately one every second or two, to propel extremely large spaceships from the Earth.

Research conducted by General Atomics of San Diego, in the late 1950s and early 1960s, proposed using small nuclear bombs to react against a large plate, called a "pusher," that was attached to the bottom of the spacecraft with a shock absorbing system. A special explosive design of the bomb maximized the momentum transfer, which, it was theorized, would lead to Specific Impulses in the range of 2,000 seconds (about five times that available from the best chemical propellants) to a maximum of 100,000 seconds. Thrust was to be measured in the millions of tons, allowing for a very large spacecraft in the order of thousands of tons. Several hundred bombs, perhaps as many as one thousand, would be ejected one at a time through an opening in the pusher to accelerate the ship into orbit. Significant engineering problems were resolved, including crew shielding and pusher-plate lifetime. A three-foot scale model (using chemical explosives) flew a somewhat controlled flight for 23 seconds in 1964, to a height of 185 feet at Point Loma north of San Diego. An obvious problem that was never really resolved was the significant radiation fallout during launch and the initial fireball that would overtake the ship itself. One solution was to use conventional explosives to get the ship off the ground before going nuclear.

The cost for the largest size of Orion was estimated at five cents per pound into Earth orbit in 1958 dollars. This was in contrast to optimistic estimates of $1,000 per pound for conventional propellants. The largest of the Orion designs, called the "super," weighed eight million tons. While the system appeared to be feasible, it was cancelled in 1965, ostensibly because the nuclear-weapon-test-ban treaty made it illegal to explode bombs in space.

While Orion used the explosive power of nuclear bombs, there were other less extreme uses of atomic energy, such as nuclear reactors. The first study of a nuclear rocket in the United States was contained in a report by North American Aviation in 1947 and again in a 1953 report by the Oak Ridge National Laboratory. However, because of the nature of the problems that confronted the effort, funding was not forthcoming.

Theodore von Karman, of CalTech/GALCIT fame, finally got the ball rolling in 1954. Within a year project NERVA (Nuclear Engine for Rocket Vehicle Applications) was underway. It envisioned both air-breathing, nuclear-propulsion systems for manned applications (the AEC-Air Force Pluto project) and the nuclear rocket *Rover* program. Both were based on plans of the Los Alamos and Radiation Laboratories of the University of California. However, by 1957, the budget had been reduced, and Los Alamos was chosen to continue the research alone. Westinghouse Electrical and Aerojet General were the prime contractors for the prototype reactors based on Los Alamos designs.

Project *Rover* developed a series of reactors that took advantage of the heat generated by controlled fission to super-heat liquids such as hydrogen. Instead of chemical combustion to produce the high-pressure gas, the hydrogen expanded dramatically from the reactor heat and was routed through a rocket nozzle to produce thrust. It was estimated that this process could produce about twice the Specific

Impulse of the best chemical rockets.

By July 1959 the first in a series of experimental nuclear rockets designated KIWI, was operated successfully at Jackass Flats, Nevada. This series ultimately included PHOEBUS, PEWEE and NF-1 (nuclear furnace) at a cost of $1.5 billion dollars. Phoebus produced a Specific Impulse of 825 seconds over a period of 12 minutes. By 1973 the experiments had shown that the next step was a flight rated engine. However, *Rover* was terminated at that point, in part because the Russians had essentially lost the Moon race, and there were no longer any dramatic firsts (except perhaps a manned mission to Mars). It was felt that this expensive and potentially environmentally risky venture should be discontinued.

The Case for Liquid Hydrogen

The use of liquid hydrogen (LH2) as a fuel had been a dream of rocket scientists since Tsiolkovsky, but domesticating the -423 degree liquid would take considerable research and technology development. Cryogenics, the term applied to extremely cold liquids, presents a bewildering array of problems. Exposing materials to intense cold results in different expansion ratios, and joining two dissimilar materials can be virtually impossible. Likewise, materials (including alloys of aluminum and steel) may become brittle and easily fractured when stress or pressures are applied at these cold temperatures. Traditional lubricants become useless gummy substances. The tanks and pipes carrying cryogenics, unless insulated, become encrusted with frost from the ambient moisture in the air, and LH2 actually causes the surrounding components of air (nitrogen and oxygen) to turn to liquid and flow into areas where they can cause even more problems. To the average person the mention of hydrogen often stirs the memory of the dirigible Hindenburg bursting into flame in the famous 1937 newsreel. This highly flammable substance requires considerable care in handling as a gas, and as a liquid, it would require an entirely new technology to make use of its potential.

The U.S. Air Force had sponsored work at Ohio State University in the late 1940s in the use of hydrogen for rockets. A miniature engine had been tested with liquid hydrogen and liquid oxygen in 1949, but the research had not progressed any further. It was recognized that the use of hydrogen with its theoretical Specific Impulse of 450 seconds (as opposed to kerosene's 300 seconds) could cut the gross mass of a kerosene/oxygen rocket in half, and the use of hydrogen in the second stage could increase the payload 50 percent without increasing the gross mass.

Abe Silverstein, then of NACA's Lewis Laboratory, organized a meeting of government and industry rocket experts in 1950 on the subject of high-energy rocket propellants for long-range missiles. Liquid hydrogen was recommended for the fuel, with hydrazine and ammonia as alternatives. Fluorine was the preferred oxidizer over oxygen. Bell Aircraft (where the former German military commander of Peenemünde, General Walter Dornberger, was employed) would ultimately pursue experiments with a 35,000 lb. thrust engine that used ammonia-fluorine. Hydrogen-fluorine uses a smaller proportion of hydrogen for combustion than the hydrogen-oxygen combination, and when the greater density of fluorine over oxygen is considered, a smaller size stage can produce the same thrust and duration.

The Air Force Scientific Advisory Board subsequently recommended a hydrogen-fueled rocket in November 1956. It proposed that two engines be developed in the 25,000 to 50,000 lb. thrust range using high-energy chemical propellants of which liquid hydrogen-oxygen was highlighted. The Air Force had some experience with the production of large amounts of liquid hydrogen because of the "Sun Tan" project. This was an effort to produce an air-breathing super-fuel reconnaissance jet, a project that Silverstein had directed in 1955. It would ultimately be developed as the A-11 but was better known to the world as the SR-71.

During this time the Pratt & Whitney Aircraft division of United Aircraft Corporation was operating a hydrogen-fueled version of their famous J-57 jet engine as a part of the Sun Tan project. Pratt & Whitney had become a premier provider of large reciprocating engines during World War II and had moved aggressively into jet engines following the war. Now, with rockets beginning to show promise, they were considering entry into that field. However, with the lead that Rocketdyne and Aerojet General had, Pratt & Whitney knew they would have to be innovative to be competitive. Using their own expertise from Sun Tan and information about producing and handling the super cold liquid from Los Alamos, Pratt & Whitney decided that there was a future for them in hydrogen-powered rockets.

With the technological impetus of Sputnik, Pratt & Whitney held discussions with the Air Force, and by March 1958, submitted a proposal for a 15,000 lb. thrust rocket engine using liquid hydrogen and liquid oxygen as propellants. In July 1958 the Air Force decided that a hydrogen-powered upper stage would be built by Convair for their Atlas and that Pratt & Whitney would supply the engines. A highly optimistic schedule called for the first delivery in 18 months. Six flight articles were to be built with the first ready to launch in 24 months.

Krafft Ehricke's Centaur

The choice of the Convair *Atlas* as a first stage of the rocket was an almost forgone conclusion. It was the only large booster then available that could take advantage of the second stage being proposed. However, it was also a direct result of the efforts of yet another former member of the Peenemünde team, Krafft Ehricke. Even before the momentous October 4th event, Ehricke had envisioned the Atlas as a major steppingstone to space exploration, and with Sputnik's arrival he was ready with several proposals, the first being a hydrogen-oxygen second stage.

Like many young boys growing up in Germany during the late 1920's, Ehricke was entranced by the possibility of space travel after having seen Fritz Lang's silent movie *Girl in the Moon* in 1928. He graduated with a degree in aeronautical engineering from the Technical University in Berlin and took postgraduate courses in celestial mechanics and nuclear physics. During World War II Ehricke served on the Russian front as a tank commander but was one of the fortunate few whose talents were needed at Peenemünde. In June 1942 he transferred there to work on rocket engine development under Walter Thiel, another proponent of liquid hydrogen.

Following the war, Ehricke, as a part of the von Braun team, traveled to the United States under Operation Paperclip and moved to Huntsville, Alabama, with the team in 1950. However, he felt constrained in that conservative engineering environment. He moved to Bell Aircraft and worked under Walter Dornberger in 1952, but it was not until 1954, when Karel Bossart was able to bring him to work on the Atlas ICBM at Convair, that he finally felt he had found a creative home. Although somewhat restricted by the mandates of the Atlas program under Air Force General Bernard Schriever, who was overseeing the project, Ehricke was able to conduct in-house studies for orbiting satellites by 1956 but could not induce the government to fund his proposals until Sputnik.

By December 1957, barely two months after the first Sputnik, General Dynamics' Convair Astronautics Division (the new name inspired by the coming space age and a change in ownership in 1954) submitted Ehricke's proposal to the Air Force entitled "A Satellite and Space Development Plan." A four-engine, hydrogen-oxygen stage was envisioned that had a total of 30,000 lb. of thrust and was called *Centaur*. The Air Force did not accept the proposal, but ARPA in early 1958, acting for the military interests, initiated a high-energy hydrogen/oxygen stage program, specifying a total of 30,000 pounds of thrust in single or multiple chambers based on Ehricke's plan.

Practical experience with liquid hydrogen in rockets at that time, however, was still very small, and the handling problems very large. The ability to produce, pipe, pump and propel with liquid hydrogen would present many problems—some of which were total unknowns in 1958.

NASA Assumes Control

In 1958 Abe Silverstein, a member of the NACA organization that was in the process of morphing into NASA, established a committee to coordinate government plans for propulsion-and-launch vehicles. During August of that year, high-energy upper stages were evaluated in ranges from 6,000 to 20,000 lb. of thrust. Both tank pressurization and turbo-pumps were considered to deliver the propellants into the engine. Also, an evaluation of the various trade-offs between hydrazine/fluorine and hydrogen/oxygen was made. It was established that pumping liquid hydrogen was not a major obstacle to the development of a hydrogen-oxygen rocket engine. The committee established a "stable" of satellite launch vehicles that consisted of *Vega* (an *Atlas* with a second stage derived from the first stage of the *Vanguard*), the *Centaur*, the *Juno V* (renamed *Saturn I*) and the *Nova*. *Vega* would soon be dropped in favor of the *Agena* and the *Nova* would evolve into the *Saturn V*.

NASA's Lewis Center conducted an extensive series of tests with a variety of high-energy fuels including fluorine and hydrazine. These resulted in a maximum exhaust velocity of 11,300 fps, which was 97 percent of the maximum theoretical performance. It was also determined that a major consideration in the design was the exhaust-nozzle-expansion ratio, which had to be optimized for the near vacuum conditions of space to achieve the impressive efficiencies.

In the final analysis, it was decided that the performance gained by using the denser hydrogen-fluorine combination over hydrogen-oxygen was not worth the additional effort. The decision to proceed with the hydrogen/oxygen engine would have a profound impact on high-energy propulsion for the next 50 years.

In October NASA Administrator Keith Glennan requested that ARPA transfer management of *Centaur* to NASA. The Air Force wanted to retain management control, but the Eisenhower administration preferred a strong civilian space effort, and in July 1959 *Centaur* was transferred to NASA with a close military management liaison to ensure its effective application to the required defense programs. The importance of the liquid-hydrogen project was emphasized during the 1960 congressional budget hearings when Silverstein stated that *Centaur* was *"the kind of thing upon which our whole technology future...rests."* Initially the concept was considered such a high risk that no specific missions were assigned to *Centaur*. Nevertheless, as the USSR continued to raise the stakes in the space race, planetary, military, and geosynchronous applications began to migrate to the vehicle.

The RL-10

Design of the RL-10 engine began in October 1958 by Pratt & Whitney. Producing 15,000 pounds thrust at sea level and weighing just 350 pounds, the engine had a mixture ratio of about 5:1 (hydrogen to oxygen), a Specific Impulse of 450 seconds, and a total burn time of 430 seconds. A unique feature of the engine was its turbo-pump. Existing turbo-pumps used either the old V-2 method of decomposing hydrogen peroxide to create steam to drive the pump, or of burning a portion of the kerosene fuel. Both systems required the overhead of more components and less efficiency. With the RL-10, Pratt & Whitney used the heat exchange properties of the liquid hydrogen itself to accomplish two of the most important tasks in rocket propulsion. A portion of the liquid hydrogen was circulated around the thrust chamber to provide regenerative cooling. The expanding hydrogen then provided the energy to drive the turbo-pump before being delivered into the combustion chamber through the injectors. With this method there was no toxic auxiliary fuel to complicate the innards, and there was no need to waste any of the precious fuel by burning it for a turbo-pump.

The original *Centaur* stage into which the RL-10 would be mounted measured 30 feet long and 10 feet in diameter. (The diameter would match the contour of the *Atlas*.) At launch it weighed more than 35,000 lb. and provided for payloads weighing as much as 8,500 lb. to low earth orbit and 2,200 lb. on lunar missions.

Because of the propellant temperatures, the tanks required special construction. Liquid oxygen, at -297 degrees Fahrenheit, was significantly warmer than the -423 degree liquid hydrogen that was carried adjacent to it. In an attempt to minimize the heat transfer which causes the super-cold liquid to "boil off," a double walled bulkhead was designed to serve as an insulator to separate the two compartments. A major effort was eventually required to resolve minute leaks in the bulkhead at the point of the welds.

In addition, the liquid-hydrogen tank was covered with lightweight external insulation to protect the tank from aerodynamic heating during the rocket's flight through the Earth's atmosphere and to prevent further boil-off of the fuel inside the tank. This "blanket" was held in place by a series of metal bands that were released by explosive bolts after the rocket had passed through the denser layers of the atmosphere. Like the Atlas, the propellant tanks were made of very thin stainless steel, less than two-hundredths of an inch thick, and were pressurized to maintain their structural integrity.

One of the most important features of the Centaur was its ability to re-start in space, allowing for the use of "parking" orbits for lunar flight and geosynchronous orbits. A Reaction Control System was employed to provide roll control during powered flight and attitude control during coast. This system consisted of small thrusters to maneuver the rocket in response to the guidance system. It was also used to perform an *ullage* maneuver to settle the propellant in the bottom of the tank just prior to restarting the engine in a weightless environment.

Because not much was known about how fluids (liquid hydrogen in particular) behaved in weightless conditions, an Aerobee-Hi rocket was launched in February of 1961 from Wallops Island Station to examine the effect of zero gravity on liquid hydrogen. In addition to pressure and temperature sensors, the liquid was filmed to observe its response.

The first successful static test of the RL-10 occurred in July of 1959, but there were several explosions during the trials, and progress was slow as a result of the many problems that had to be overcome.

The optimistic schedule was slipped repeatedly until one of the missions, the Mariner spacecraft to Venus, whose schedule could not be postponed, was downsized so that an *Atlas Agena* could launch it. Aside from military satellite missions assigned to *Centaur*, which were to be considerable, NASA planned to launch one operational Centaur every month for a period extending well into the 1970s and beyond.

Trials and Tribulations

From an organizational viewpoint, shifting *Centaur* to the Marshall Space Flight Center in July of 1960 seemed like an appropriate move. However, as it now was under the direction of the von Braun team, the entire concept of its thin-wall pressure shell came under constant criticism. Von Braun, called to testify regarding the slow progress, highlighted the advanced technologies being employed. He stated, *"In order to save a few pounds, they* [Convair] *have elected to use some rather, shall we say marginal, solutions where you are bound to buy a few headaches before you get it over with. Ultimately when you are successful you have a real advanced solution."* He referred to the thin-walled pressurized tankage as *"... a continuous pain in the neck."* Ehricke's own management skills had been challenged, and he had been replaced as well. Homer E. Newell, then Director of Space Sciences at NASA, testified to Congress that *"Taming liquid hydrogen to the point where expensive operation space missions can be committed to it has turned out to be more difficult than anyone supposed at the outset."*

Both von Braun and Brian Sparks, the Deputy Director of the Jet Propulsion Laboratory, recommended canceling *Centaur* and using the *Saturn* I with the *Agena* upper stage. Sparks wrote, *"I feel compelled to emphasize... the necessity for gearing our lunar planetary programs to be strongly competitive with the Soviet Union. Timeliness and weightlifting capability are now well established advantages of the Soviet program...Our international position and our national pride can be significantly enhanced by the earliest possible exploitation of the* [Saturn]*C-1/Agena D."* However, NASA headquarters (and Silverstein in particular) believed that liquid hydrogen was the key to the future and that its development must be aggressively pursued regardless of the cost (which had reached almost three-quarters of a billion).

On May 8, 1962, the first *Atlas Centaur* AC-1 finally rose from its launch pad, but after 54 seconds of flight, as the rocket encountered Max Q, the insulation surrounding the *Centaur* tore loose and ripped into the thin tank walls. The failure was an omen of another insulation problem that would occur 41 years later with another liquid hydrogen powered rocket. AC-2 was scheduled for October 1962, but major changes in the organizational structure would occur before the next flight took place.

In light of all the technical and management problems that beset the program, *Centaur* was transferred from the Marshall Space Flight Center (the von Braun team) to Lewis Research Center in October 1962. There was too much at stake to allow the project to continue in an environment where the relationship between the various factions in the program had become very strained.

The second flight would not occur until November 27, 1963, when the first successful launch of an *Atlas/Centaur* occurred, and the first in-flight burn of a liquid-hydrogen/liquid-oxygen engine took place. The 10,000-pound payload, a dummy Surveyor spacecraft, was placed in a high "parking" orbit. Of the eight engineering test flights over a period of five years, only four would be termed successful. The most destructive failure occurred in March 1965 with the explosion on the pad of AC-5. The resulting damage required a major rebuild of the launch pad itself. The original two-year schedule had taken almost seven years to perfect a hydrogen powered upper stage. However, when it was deemed operational, the results and reliability were exceptional.

Beginning in May of 1966, with the first Surveyor unmanned lunar spacecraft, the *Centaur* achieved a perfect record of seven consecutive successful launches. By the 1970s *Centaur* was teamed with the Air Force Titan III to provide a capability to launch spacecraft weighing up to 30,000 pounds. As the precursor of all liquid-hydrogen/liquid-oxygen high-energy stages, the technology was incorporated into the J-2 engines of the upper stage of the *Saturn* IB and *Saturn* 5 rockets for the Apollo program and the Space Shuttle Main Engines. However, the pressurized thin skin tanks would not find their way into any future vehicles.

As the early years of the space race continued to unfold, the *Agena* upper stage had proven itself in several American programs, including Discoverer (Corona), SAMOS, and MIDAS, and was planned for the next round of lunar probes of the Ranger series in the early 1960s. Nevertheless, the higher-thrust upper stage of *Centaur* was needed for much larger spacecraft being envisioned for lunar and deep space missions to the planets.

Soviet Progress

The Russians continued to make public pronouncements regarding advanced rocket propulsion, but in truth there was little research and development being done in that direction. While America forged ahead hoping it would not again be surprised by Soviet progress in this important aspect of the space race, the Soviets themselves were remarkably negligent. Even as late as 1961 (two years after the RL-10 had undergone extensive static testing by Pratt & Whitney), the Soviets had yet to even provide facilities for the production of quantities of liquid hydrogen; the refrigeration techniques had not been adequately developed.

While the military was not particularly interested because liquid hydrogen did not lend itself to weapons of quick response and field operations, Chief Designer Sergey Korolev had argued adamantly for expanded funding of the low-key efforts that were underway in the early 1960s. As for nuclear propulsion, there had been several studies and preliminary proposals in the various design bureaus but no real development activities in either nuclear rockets or liquid hydrogen until the mid-1960s, and it would be decades before high-energy rocket engines would be developed.

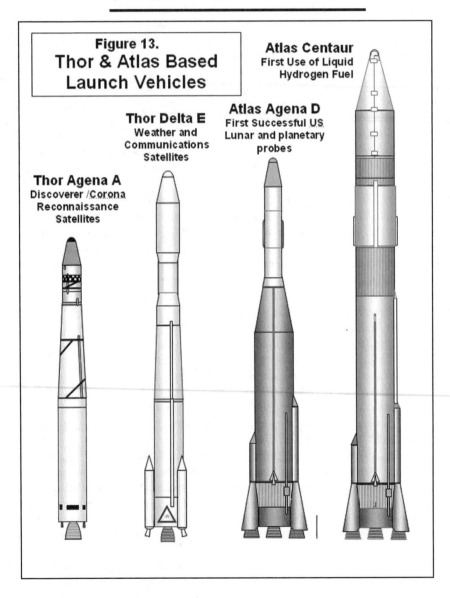

Figure 13.
Thor & Atlas Based Launch Vehicles

Atlas Centaur
First Use of Liquid Hydrogen Fuel

Thor Delta E
Weather and Communications Satellites

Atlas Agena D
First Successful U.S. Lunar and planetary probes

Thor Agena A
Discoverer /Corona Reconnaissance Satellites

Even before Sputnik, both the Soviets and Americans had made tentative paper studies on a variety of possible paths for sending a man into space. The 1946 proposal by Mikhail Tikhonravov, designer of the first liquid-fuel rocket in the Soviet Union, had looked at using the technology of the V-2 to develop a "vertical rocket" for carrying two passengers to an altitude of 190 km.; thus the project's designation VR-190. Korolev did some tentative work in 1956 that projected a manned sub-orbital flight occurring in the 1964-1967 period.

The U.S. Air Force, in February 1956, began examining the next step after the hypersonic X-15 rocket plane (only then beginning to take shape) that included a manned hypersonic glider launched by a conventional rocket booster, similar in concept to the Eügen Sanger *Silverbird* of 1933. It also examined a possible Manned Ballistic Rocket Research program.

However, none of these paper projects received development funding. That would have to wait for the impetus of Sputnik.

The Winged Messenger

The exhaust plume of Sputnik I had hardly cleared when the rush to catch up with the Russians began in earnest in dozens of military and corporate research centers in America. Combinations of existing boosters were lashed together (on paper) in an effort to find a possible arrangement that looked feasible as a "heavy lift" rocket. However, it was one thing to play the game with paper and quite another to get funding to cut metal.

The National Advisory Committee for Aeronautics, which would become NASA within a year, convened a Special Committee on November 21, 1957, (just two weeks after Sputnik II was launched) in which critical aspects of a manned space flight were assigned to working groups. These included spacecraft systems, tracking, reentry, and human factors. Each of the various NACA centers around the country participated, with the Virginia based Langley Research Center being represented by Maxime A. Faget, who headed the Performance Aerodynamics Branch. Over the months that followed, Faget and his team eliminated various approaches and by March of 1958 had settled on the "ballistic capsule" as being the path that would lead to the highest probability of success in the shortest time for the least money.

The Air Force initiated its own three-day seminar in January of 1958 at Wright Patterson AFB in Ohio, to examine possible projects and technologies to place a military man into space in the shortest period. The new Advanced Research Projects Agency (ARPA), under Roy Johnson, assumed that manned space flight would be a responsibility of the Air Force, although no mandate had been issued by the Department of Defense to establish a need for such a project. A wide variety of spacecraft configurations were described, among them a 2,300 lb. ballistic capsule proposed by the McDonnell Aircraft Company, who inferred that a mission could be flown within 24 months from the start of the project. Optimism was rampant and paper studies were still cheap.

By mid-March the Air Force, based on its analysis, had gained approval for $133 million to initiate a Man-in-Space-Soonest project (MISS). Its first objective would be a simple ballistic ride into space with follow-on goals that included orbital flights of up to two weeks. Deeper in the paper work were vague projections for circumlunar flights and lunar landings by 1965. The initial price tag for MISS was about $1.5 billion.

The Army Ballistic Missile Agency in Huntsville had its own idea of getting man into space with Project Adam, using the venerable *Redstone* booster for a sub-orbital flight of 115 miles into space and 200 miles down range. However, by July someone high in governmental circles had determined that the Army had no business in space. Once again, the von Braun team, who had the most depth and breadth in rocketry and had saved some of America's stature by orbiting its first satellite, was frozen out of the space race.

With the pending reorganization of NACA into NASA set for October of 1958, a joint NASA-ARPA Manned Satellite panel was formed in September. Administrator designee Dr. Keith Glennan had been given the specific mission of manned space flight just a few weeks earlier by President

Eisenhower. The basic layout established by Faget (and arrived at independently by McDonnell Aircraft) provided for a twelve-foot cone-shaped craft with a seven-foot diameter heat shield on the broad end. The astronaut lay reclined, and the capsule reentered the Earth's atmosphere backwards so the G loads would remain in the same direction as at launch. Small retrorockets initiated reentry, while a simple attitude control system of thrusters stabilized the craft during retro-fire and reentry.

The various representatives on the panel were not unanimous on the configuration, but simple economics, time, and technology provided the greatest convincing arguments for the ballistic approach. There were, as yet, many unknowns that included a determination of how many Gs a man could endure without incapacitation or injury. The reentry dynamics were also somewhat vague with some studies suggesting that, unless the capsule had extensive heat protection, the reentry angle had to be very precise or the spacecraft might be incinerated.

Although the Atlas was the largest rocket undergoing flight test at the time, its reliability and lifting capacity left much to be desired. Convair estimated that it could deliver 2,700 lb. to low earth orbit, and early optimistic estimates indicated that a spacecraft could be built to these weight limitations.

With the reliability of the launch vehicle itself being a major concern, Faget realized that some method had to be devised to allow the occupant (or the capsule itself) to escape from the rocket should a major malfunction occur during launch. This escape system would have to function at high mach numbers and extreme altitudes. Protecting the astronaut from the debris and flame of a catastrophic launch pad explosion as well as the windblast of a 3,000 mph abort at 20 miles altitude meant that an ejection seat had to be ruled out.

Faget's team examined the escape rocket on the Air Force proposal which positioned it under the capsule as might be expected. However, several factors encouraged the Faget team to place the escape rocket on top of the capsule in a "tractor" configuration. The first of these considerations was weight. As the escape rocket would no longer be needed in the later stages of the boost phase (the capsule could simply separate from the booster in an abort), it could be jettisoned, and its 300 lb. weight would not have to be carried all the way to orbit. A second consideration was that if the escape system were carried to orbit, its volatile propellants and pyrotechnics would present a hazard to the spacecraft, and they would eventually have to be discarded.

By November of 1958 more than three-dozen potential bidders attended a briefing in which key members of the manned spacecraft design team made presentations that described the systems and their rational for selection. As the project began to take on its own character, it had yet to be given a name. Most missiles of that era were named for Greek or Roman mythological figures, so when Abe Silverstein (who became head of NASA's Office of Space Flight Programs) proposed the name *Mercury*, it was readily accepted. *Mercury*, the messenger of the gods, was the grandson of Atlas!

"Money is not a substitute for time"

McDonnell Aircraft had been an early proponent of a manned space-flight program, and as of October 1957 (immediately following Sputnik I) the company had assigned several engineers to begin working on some of the problems that had to be resolved to put a man into space. It came as no surprise, then, when they were awarded an 18.3 million dollar contract (with a 1.5 million dollar fee) by NASA in February of 1959 for construction of 12 Mercury capsules.

While Faget's team had established the basic configuration, engineering reliable systems to conform to the weight restrictions became the critical aspect of McDonnell's responsibility. Some of the systems were an outgrowth of existing X-plane programs such as the X-15, which was nearing the flight test stage. But these systems, such as the thrusters and environmental system, required significant weight reduction. Even something as simple as the communication package had to be re-engineered to fit the small space available.

The reentry process was initiated by a set of three small, solid-fuel retrorockets attached to the base of the heat shield. An independent circuit fired each so that it was almost a guarantee that one would ignite, and only one would be needed to drop *Mercury* from orbit. Attitude control was important during retro-fire to insure that the correct thrust vector was established for maximum effect of the small retrorockets.

During the reentry itself, the thrusters would maintain the correct spacecraft attitude as the heat shield was designed for a precise angle for optimal weight savings. When the capsule descended into the more dense layers of the atmosphere and slowed to subsonic speed, a drogue chute would deploy to stabilize the craft, preventing any oscillations due to its unstable aerodynamic qualities. At 10,000 feet the main conventional parachute would open, but its descent rate of 32 feet per second would result in a momentary 40 G impact.

To soften the touchdown, the heat shield would detach from the main body of the spacecraft and extend down about four feet revealing a fabric landing bag between the heat shield and the base of the spacecraft's pressure bulkhead. This bag had a series of holes that allowed air to be drawn in as the bag was extended by the weight of the heat shield. At touchdown, the air within the bag provided a compression effect as it escaped back out through the holes, softening the impact on the capsule itself.

Because the Earth's surface is predominantly covered by water, it was decided that the craft had to survive a landing at sea in case of an early retrieval from orbit due to an emergency. This led to consideration of the ocean as the primary recovery area.

With the *Atlas D* being the only real option available as a launch vehicle, several modifications had to be made to it. The first was an abort sensing package that received input from a variety of sensors monitoring critical aspects of the vehicles "health." Any one of these sensors could initiate an abort signal that would ignite the escape rocket and allow the spacecraft to be pulled safely from the launch vehicle during the boost phase.

Because of the added structure that extended the height of the rocket almost another 20 feet, the load factor on the *Atlas* would ultimately require that its skin thickness be increased. The added length also necessitated that the rate gyros (that sensed the flight path) be relocated to allow for the change in the control dynamics of the rocket.

On reaching orbit, the spacecraft would be separated from the *Atlas* by a set of small solid-fuel posigrade rockets whose exhaust would have to be shielded from the top of the *Atlas* propellant tank by a fiberglass faring. The name posigrade relates to a separation process whereby the spacecraft receives a slight increase or "positive" velocity as opposed to "retrograde" motion.

The idea of going straight to orbital flight with the capsule, without an intermediary step of a suborbital flight, was actually never contemplated. With the entire project open to world scrutiny, the risk of such a giant step was too much for those who held the reins of responsibility. Incremental testing, and lots of it, would precede an actual manned orbital attempt. The capsule itself would be tested in a variety of unmanned modes to validate its structural integrity and the proper functioning of its various systems—including all possible abort scenarios.

To accomplish this, and to attempt to do it within a reasonable budget and time frame, four launch vehicles were proposed including the *Atlas*. First, a relatively inexpensive means of testing the escape system had to be found that could simulate the acceleration of the *Atlas* under various segments of its launch profile. This was accomplished by creating a fin guided solid-fuel rocket called *Little Joe*. This 27-foot-tall, 6.8-foot-diameter casing allowed for a total of eight different-sized solid-fuel-rocket motors to be inserted, based on the specific type of test. The order in which each set of rockets was fired allowed for different altitudes and velocities to be explored with the 23-foot assembly of the capsule and its escape tower riding on top. North American Aviation received the contract to build eight *Little Joes*.

The second rocket on the test schedule was the *Redstone*. The capsule would be boosted to an altitude of 115 miles to verify all of its systems under high-G load as well as the weightless conditions of space. This was the mission that von Braun had suggested with Project Adam. Although Adam would not have been capable of orbital flight, it would have meant beating the Russians into space, even though it was just up and back down. NASA ordered eight *Redstones* from its manufacturer, the Chrysler Corporation.

The third rocket was to have been the *Jupiter* IRBM to qualify the heat shield. However, the Army wanted as much money for its *Jupiter* as the Air Force wanted for an *Atlas*. *Jupiter* was cancelled and the *Atlas* assumed its role. Of course the *Atlas* price had increased 32% to $3.3 million per article.

Cost over-runs began almost immediately in the *Mercury* program. The McDonnell spacecraft contract grew over 30% within a few months to $41 million. Money was not the only optimistic aspect of the program that quickly fell by the wayside. The first schedule called for a manned *Redstone* flight by the end of April 1960 and the first orbital flight by the following September. A span of nineteen months was projected between the laying out of the schedule and the first manned orbital flight.

With respect to Project Mercury's prospects of orbiting a man before the Soviets, NASA Deputy Hugh Dryden said, *"...that race was lost before the space agency was founded."* He then added, *"Money is not a substitute for time."*

Selecting the Astronauts

As the hardware began to come together, the selection process of who would ride it into space began to take center stage. Originally, NASA's Space Task Group had thought that an "open" selection procedure would provide about 150 applicants from which 36 would go through a winnowing down process, and a final six would be made offers to become astronauts. There was no shortage of volunteers from virtually every walk of life. However, shortly before the end of 1958, President Eisenhower decided that that course was fraught with uncertainty, and he preferred to limit the application process to existing military test pilots. The reasoning was quite simple and pragmatic. These men had already proved their ability to operate in a flight environment under pressure and were already on the government payroll.

The basic requirements were the same as for the original Space Task Group selection criteria in that each had to be between 25 and 40 years of age, less than six feet tall, in excellent physical condition, and hold a degree in science or engineering. To this was added the necessity of having completed one of the military test pilot schools and having logged 1,500 hours of jet fighter time.

There was an interesting divergence of opinion over what some felt were excessive qualifications. Some viewed the astronaut as "spam in a can," simply doing what a monkey would have already accomplished. Others felt that the presence of man would add immeasurably to the utility of the mission with his ability to reason and make judgment evaluations. In fact, right from this first year of the space age, the debate began as to man's usefulness in space as opposed to the cost and risk of getting him there.

Of the 508 military records provided to NASA by the Defense Department, 110 men were selected by a committee for further consideration. By the time the first 69 of these were reviewed, it was determined that there was no need to consider any more of the 110 as it was going to be a relatively easy task to find the six qualified candidates. Several were eliminated by the requirements such as height, thus 56 were given written and psychiatric examinations that reduced the group to 36 who were then asked to volunteer. Four declined to be further considered, and it was at this point that the career aspect began to become a factor. What lay beyond Project Mercury was totally unknown. These military pilots had worked hard to position themselves in their career. To be diverted for an unknown period of time into a project that had questionable use for their skills caused most of them to have second thoughts.

In groups of six, the 32 prospective astronauts, who were then considered volunteers, began the final and somewhat tortuous path to the ultimate selection. Each group visited the Lovelace Clinic in Albuquerque, New Mexico, for a week-long series of intrusive physical and psychological tests. Dr. Randy Lovelace, a pioneer in aerospace medicine, was called on to create a physical examination that would ensure that each candidate did not harbor any problem that could possibly interfere with the demanding flight that lay ahead. Several of these tests were widely criticized by the astronauts in their memoirs as being ludicrous by any standard. If there was one redeeming value to the Lovelace analysis, it may have been to determine if these physical and academic specimens of manhood had the right motivation to see the process through. Only one applicant failed to move on to the next selection step.

At the Wright Air Development Center in Dayton Ohio, another set of physical tests awaited the candidates. If the Lovelace ordeal stretched the limits of humility, this new series explored the extremes of heat, cold, isolation, noise, and physical endurance. Centrifuge tests required the men to endure high forces of gravity (G loads) and a spinning chair determined the stability of their equilibrium. Another psychological analysis was performed that included motivational and peer evaluation scrutiny.

The records of the final 18 candidates were reviewed by the NASA committee, who passed their recommendations on to the Assistant Director for Manned Satellites, Robert Gilruth. The legend that has grown up around the selection process is that the number could not be reduced to the desired six because there were just too many outstanding candidates. One account that may have some credibility is that the Air Force and Navy were equally represented with three each, and it was decided to add a

Marine to the list to round out the service's participation. Oddly enough, the Marine selected was the only one who had not completed a four-year degree.

The new NASA Administrator, Keith Glennan, approved the selection, and the Mercury Seven, as they were to be called, were introduced to the world at a press conference in Washington, D.C., on April 9, 1959. The spectacle of the conference took on a surreal twist from the moment that the presenter said, *"Gentlemen, these are the Astronaut volunteers."* Amongst the popping flashbulbs and ecstatic applause, the candidates themselves were astounded by the reaction. They had not expected this kind of immediate adulation, if any. After all, they had yet to do anything heroic—except endure Randy Lovelace's probing.

Perfecting the Spacecraft

Each of the new astronauts was assigned to a part of the Mercury program to follow the progress being made and to assist the engineers with any aspect that might involve the pilot. Every week the astronauts would meet to discuss the problems that each had uncovered and arrive at a recommendation to be passed on to the engineers. It was obvious from the start that a spacecraft designed by engineers for an environment that they had never experienced was going to be a real challenge. Likewise, the desires of the astronauts themselves played a key role in such things as control and instrument placement and even window size and location.

One of the first systems to be defined was the environment of the spacecraft. If it were to hold ambient sea level pressure (14.7 lb./sq./in. or 760 millimeters of mercury) with the normal mixture of 20% oxygen and 78% nitrogen, the weight of the pressure vessel necessary to contain it would be too heavy for *Atlas*. In addition, the expected leakage through welds and openings for electrical cabling would be significantly greater, and increased amounts of air would have to be provided to make-up for the loss. A mixed gas environment would also be more complex in maintaining the proper ratio of gases.

It was determined that if the capsule's interior pressure was maintained at about one-third that of standard atmospheric pressure and used 100% oxygen, the weight and the complexity of the system could be noticeably improved. The original design called for the capsule to contain an ambient atmosphere at launch, and as the vehicle climbed, the air within would vent into space until the desired 5.5 lb./sq./in. was achieved at which time the vent would be closed. During this period, oxygen would flow into the craft to enrich the mixture until the 100% level was reached. The astronaut himself would be in the space suit with the visor down, breathing 100% oxygen during launch. If a fire occurred in the spacecraft while in orbit, the atmosphere in the craft would simply be vented into space creating a vacuum that would extinguish the fire from lack of oxygen.

However, during one test of the Environmental Control System (ECS), a McDonnell engineer lost consciousness from lack of oxygen. It was determined that there was a problem with the regulation of pressure between the spacecraft and the pressure suit. In an effort to avoid this problem, which could have been fixed, it was decided to fill the spacecraft with 100% oxygen prior to launch. It was recognized that this was a major hazard, as a fire occurring anytime before the capsule got into space would not be controllable. However, the decision was made to proceed with the "easy" solution—time was of the essence. That decision would have major repercussions in the years to come.

Another decision was to place most of the systems within the small craft's pressurized hull to minimize the number of holes that would have to be sealed against loss of the cabin's environmental pressure, which would have to be continually made up by the limited supply of oxygen. This would prove to be an unfortunate decision with respect to the ability to maintain and modify systems in assembled spacecraft.

Astronaut Navy Lieutenant Commander Wally Schirra, age 36, was assigned to monitor the Environmental Control System and the space suit. Existing suit technology was called upon to handle protecting the astronaut should the cockpit environmental support system fail at any point in the mission. Most of the changes to the space suit involved mobility issues as the astronauts sought to be able to reach any switch or control easily.

The layout of the instrumentation and development of the spacecraft simulators was the responsibility of Marine Lieutenant Colonel John H. Glenn, age 37 and the oldest. The youngest, Air Force Captain Gordon "Gordo" Cooper, age 32, was given the task to track the progress in preparing the

Redstone booster. Air Force Captain Donald "Deke" Slayton, age 35, was assigned the Atlas.

Communications was another critical system. Because of the unknowns involving the astronaut's abilities to function in space, it was desired to maintain virtually continuous communications. However, with the spacecraft being above the ionosphere, some radio communications technologies would not work. High frequency and ultra-high frequency communications bands had to be used that limited communication to "line-of-sight." This meant that a whole network of tracking stations had to be established around the world in several foreign countries. Even then, there would be periods of time when the capsule was out of sight. A basic rule was established that no more than ten minutes of "dead time" be allowed between contacts, nor would any period of contact be less than four minutes.

Mexico, Nigeria, Spain, Australia, and the United Kingdom became key partners whose countries and territories were supplemented by the use of special tracking ships: *Rose Knot Victor* and *Coastal Sentry Quebec*. Navy Lieutenant Malcolm Scott Carpenter, age 33, and Navy Lieutenant Commander Alan B. Shepard, age 35, monitored the global communications network and recovery operations, while Air Force Captain Virgil "Gus" Grissom, age 33, provided the liaison for the spacecraft electrical systems.

Two of the first changes instituted by the astronauts were the size and placement of the window and the escape hatch. The original design called for a relatively small port hole off to the side of the cockpit. However, pilots need a large and unencumbered view of the horizon directly in front of them. This, and the fact that the hatch could not be opened quickly from the inside, caused some tension between the astronauts and the engineers. Both of these features were changed to the satisfaction of the astronauts.

Thousands of changes to drawings, hundreds of minor changes to systems, and a few dozen major changes resulted in no two Mercury spacecraft being the same. This in itself caused much time to be lost. Arbitrating between the engineers and the Mercury astronauts, NASA's Faget and McDonnell's John Yardley were able not only to maintain a good working relationship but also to move the required changes through with a minimum of delay.

Object K

In February 1958, at the same time as the Mercury spacecraft was being pursued by the Americans, Soviet Chief Designer Sergey Korolev directed a design team headed by Tikhonravov to concentrate on a manned spaceship. Like their American counterparts, time and technology would play a major role in their decisions. However, unlike the Americans, the Soviets had the R-7 (with an upper stage) that provided a payload weight advantage by a factor of three! An upgraded upper stage that developed 12,000 lb. of thrust would replace the RD-0105 LOX/kerosene engine developed for the Luna program.

Within Tikhonravov's organization was a young man of 32. As a 16 year old, he had been captured by the Nazis during the war. Shot and left for dead, Konstantin P. Feoktistov, had survived and went on to become head for planning Piloted Space Apparatus. He was the Soviet equivalent of NASA's Maxime Faget, and he would prove an important ingredient in Korolev's future in space.

The first problem to be addressed by the Soviet team was reentry. Several configurations were examined including the cone shape being pursued by the Americans. However, the simplest shape that provided maximum volume with least surface area was a sphere. By selectively placing the center of gravity behind and below the reclined pilot, it would be possible for the craft to assume a stable orientation during a ballistic reentry that would avoid having to provide for attitude control during that critical period. However, the entire sphere would have to be coated with an ablation material to protect it during the initial reentry stages before the self-orientation was completed. One of the drawbacks with a sphere was that it would induce higher G loads than other configurations, but these would be tolerable.

Unlike Mercury, the Soviet craft used a sea level atmosphere with 79% nitrogen and 21 % oxygen. This required a thicker pressurized container and subjected the pilot to possible nitrogen embolism (the "bends") if the capsule experienced rapid decompression and the pressure suit (with 100% oxygen) had to be inflated. On the other side, however, the fire danger was greatly minimized. With respect to the space suit, some believed that the suit was not necessary and should be omitted to save weight. This suggestion would be accepted for a later flight and, with the second-generation spacecraft, would result in disastrous consequences.

It was also decided that only a portion of the total satellite would be returned to earth to simplify the reentry problem still further. A double truncated instrumentation cone attached to the descent apparatus would provide for the electronics and control system as well as the 3,500 lb. thrust storable-liquid-propellant "braking engine" that fired for 45 seconds to decelerate the satellite for return to earth. While a solid-fuel system would have been much simpler, the available thrust-to-weight ratio dictated a more efficient engine. This segment would be discarded after retro-fire. Like Mercury, orbital altitudes would initially be selected to allow the spacecraft to reenter within the period of the available supply of life support consumables (ten days for the Soviet craft) if retro-fire failed to occur.

As with Mercury, a parachute would slow the spacecraft after reentry but would present a momentary high G impact on landing. To avoid any injury at touchdown, it was decided that the pilot would use the ejection seat that was also provided for emergency escape during the powered ascent. A recovery from orbit within the Soviet Union on dry land would permit the entire flight envelope to be kept secret, although provisions were made for the pilot should an unplanned landing at sea occur.

The use of an ejection seat for emergency exit was not the first option explored. Korolev had wanted an escape tower topped by a solid-fuel rocket similar to Mercury, but the weight of the system was excessive, and an ejection seat was chosen that would provide for the first 40 seconds of flight. Should a malfunction occur after that period, the spacecraft would separate from the rocket and assume a normal reentry profile.

The crew cabin had three portholes, one of which had a special optical "viewer" that had precise lines engraved on it to permit manual orientation of the spacecraft to the horizon for retro-fire. A miniature globe showed the pilot his current position and where he would land if the retro-fire sequence were initiated at any particular time.

Again, like Mercury, the attitude control system was fully automated but provided a manual back up. However, unlike Mercury, the manual back up was seen as just that—a back up. The pilots chosen to ride the Soviet spacecraft would be more passenger than pilot, while their American counterparts became an integral part of the spacecraft development process and the flight plan profile.

There were two primary reasons for this view. The first was that, as in America, there was a strong feeling that the psychological impact of being in space might prove an overwhelming experience that could incapacitate the pilot. The affects of high G forces at launch followed by weightlessness and isolation were unknown. A second reason was that the spacecraft designers were not aviation people and did not know or understand the piloting environment. To them the occupant was a passenger. This view would hinder the development and utility of first- and second-generation Soviet spacecraft. Thus, the role of the cosmonaut reflected the social philosophy of the Communist State: tight control, and little autonomy.

With the preliminary design complete, a meeting of the Chief Designers, held in November of 1958, reviewed the prospects for orbital and sub-orbital manned flight as well as the reconnaissance satellite. It did not take long to dismiss the intermediate step of sub-orbital flight—the Soviets would move directly to a manned orbiting satellite.

Moreover, as the resulting manned space flight technology would be able to transfer to a reconnaissance satellite as soon as another "first" in space was achieved, the military was satisfied. But, not everyone was. There were voices in the Politburo who felt that the risk was too great, and that, should a failure become public, it would do immeasurable damage to Soviet prestige. Nevertheless, Korolev's Sputnik aura was still preeminent, and his confidence, and that of the team, was high.

The spacecraft went through several names during its design phase that related to its origin as a reconnaissance satellite. However, when it emerged with its own identity, it was designated "Object K," with the K being an abbreviation of *Korabl*—Russian for ship. The initial prototype that appeared in the spring of 1959 had a total weight of 10,400 pounds, of which the descent module represented about 6,000 pounds. The craft had a maximum diameter of 8 feet and a length of 14 feet. The total height of the assembled R-7 was now a towering 123 feet.

The Cosmonauts

The Soviets arrived at the same conclusion as their American counterparts: military pilots provided the most logical source for occupants of the new spacecraft that was starting to take shape in the summer of 1959. The fact that the Americans had already chosen seven military test pilots gave a renewed sense of urgency to the Soviet selection process. The criteria were similar in all respects but two: the Soviets were not looking for engineering test pilots. A bachelor's degree in engineering or science was not a prerequisite nor was high performance jet time. Korolev's philosophy reflected the automated environment, and in this respect he was looking for men who could execute a specific script laid down for the cosmonaut by the "controller." Men who could, and would, follow directions to the letter would be selected.

The records of several thousand potential candidates were reviewed, and those who appeared to posses the desired qualifications were interviewed and tested physically and academically. The purpose of the exams was not disclosed to the pilots at this point. By the end of 1959, a pool of candidates was available from which Korolev decided to select twenty. It was reported in later years that the number 20 was decided on simply to be larger than NASA's seven. The twenty reported to the new Cosmonaut Training Center (TsPK) in Moscow, under the direction of Yevgeniy A. Karpov, in February 1960.

This first group of twenty (and their ages) consisted of Ivan N. Anikeyev (27), Pavel I. Belyayev (34), Valentin V. Bonderenko (23), Valeriy F. Bykovskiy (25), Valentin I. Filatev (30), Yuriy A. Gagarin (25), ViktorV. Gorbatko (25), Anatoliy Y. Kartashov (27), Yevgeniy V. Krhunov (25), Vladimir M. Komarov (32), Aleksey A. Leonov (25), Grigoriy G.Nelyubov (25), Andrain G. Nikolayev (30), Pavel R. Popovich (29), Mars Z. Rafikov (26), Georgiy S. Shonin (24), Gherman S. Titov (24), Valentin S. Varlamov (25), and Dimitriy A. Zaykin (27). The Soviet cosmonaut's average age (26) was almost 10 years younger than their American counterparts.

The training regimen consisted of lectures on various subjects including rocket fundamentals, geophysics, astronomy, and radio communication. A demanding physical conditioning program and intensive parachute training required that each candidate make 40 to 50 jumps.

As the time to launch came closer, more advanced training was provided for a subset of six candidates that was, oddly enough, called "The Vanguard Six." In some respects, this training program was similar to that of the Mercury astronauts. It was at about this period (the summer of 1960) that the cosmonauts (as they were to be known) finally came face-to-face with the Chief Designer, Korolev, and were also introduced to the spacecraft they were to fly. Unlike the Mercury astronauts, the cosmonauts were not a part of the spacecraft development team.

Korabl Sputnik

With the published (and much revised) Mercury schedule in the summer of 1960 showing a possible sub-orbital flight by December of that year, Korolev had his work cut out for him. It was vital that the Soviet spacecraft be in orbit before the Americans. If NASA's schedule had been optimistic, Korolev's was as well.

Many of the components for the spacecraft had already undergone tests using R-2A and R-5A missiles, when the first unmanned version of Object K was launched at Baykonur on May 15, 1960. This prototype did not have all of the systems and lacked the ablative coating as it was not to be recovered. The primary objective was a test of the attitude control system. A 195- by 230-mile orbit was achieved with a 65-degree inclination. The Soviets announced the successful launch and used the name *Korabl-Sputnik 1*, which loosely translated to satellite ship, although the TASS New Agency referred to it as a spaceship. As there was no indication that it was a manned prototype, the Western press referred to the satellite as Sputnik IV.

The flight plan called for it to remain aloft for three or four days while a variety of tests were run on the various systems. On May 19th it was determined that there were some problems with the attitude control system, and when the retrorocket was fired, the spacecraft was pointed 180 degrees from that which was intended. This caused the spacecraft to be boosted into a higher orbit where it stayed for five years before disintegrating on reentry.

A second Korabl-Sputnik was launched on July 28th and contained two dogs, Chayka and Lisichka, and more advanced attitude control sensors. However, 19 seconds into the launch, a fire developed in one of the boosters causing the vehicle to be destroyed 10 seconds later. As there was no escape system in these test vehicles, the dogs were killed.

A third Korabl-Sputnik (labeled "Korabl-Sputnik 2" by the Soviets, thus not acknowledging the earlier failure) was launched on August 19th with two more dogs, Belka and Strelka, and a variety of mice, rats, insects, and plant seeds. This launch was successful, and the spacecraft attained an orbit of 191 by 211 miles. For the first time, two TV cameras recorded the responses of the dogs in a weightless environment, startling the doctors observing them from the earth. While the vital signs of the two animals were within tolerable limits, the two appeared lifeless. After several orbits, they began to show some movement which appeared convulsive, and Belka vomited.

Telemetry from the spacecraft indicated that the primary attitude control system had again malfunctioned, but the back up (using the sun as a reference) was successful, and the satellite reentered properly. At the designated altitude, the dogs were safely ejected and landed only a few miles from their intended point of touchdown—becoming the second object ever recovered from space. The American Discoverer XIII had accomplished that "first" only a week earlier. The flight was an important milestone in the spacecraft development as it essentially validated all of the critical systems.

Although both dogs were found to be in excellent condition, because of the dogs' alarming responses in space, it was recommended that the first manned flight be limited to a single orbit. Two additional flights of the prototype were recommended before the more advanced manned version was to be scheduled. There was still time to beat the Americans. Then a tragedy occurred that delayed further testing for several weeks, but it had nothing to do with Korolev's manned spaceflight program.

Yangel's R-16 was poised on its launch pad on October 24th for the first test of this new storable-fuel ICBM. A propellant leak was discovered, and repairs were completed without draining the fuels or recycling the countdown. Almost 200 technicians and managers including the Chief of the Soviet Strategic Missile Forces, Marshal Mitrofan Nedelin, were standing in the area immediately around the base of the rocket. As the countdown was set to continue at T-30 minutes, and the assembled group prepared to leave for the blockhouse and safer viewing points, the second stage of the missile unexpectedly roared to life. The destructive force of the exhaust ripped into the first stage spilling its toxic and flammable content down upon the men below. An enormous fire ensued, and most of those in the area were incinerated including Marshal Nedelin.

Operations were brought to a standstill at the test facility while the carnage and destruction were analyzed and cleaned up. Fortunately for Korolev, the Americans were having their own set of technical problems, and the December launch date for the first manned Mercury sub-orbital launch had long since slipped into 1961.

The next Korabl-Sputnik soared into the sky on December 1, 1960, and achieved the precise low orbit planned for the manned launch (112 by 155 miles). Aboard were the dogs Pchelka and Mushka, and improved biomedical sensors to record more accurately and completely the condition of the dogs while in the space environment. After a full day in orbit, the retrorocket fired, but the burn was much less than planned. The new perigee was low enough that the spacecraft began its reentry after completing another orbit and a half—the initial calculation showed that it would land beyond Soviet territory.

Now an automated sequence of events aboard the spacecraft was put into action that called for explosives to destroy the ship if on-board sensors determined that it was not on its planned flight profile. The objective was to assure that no useful components that could lead to determining the technology employed in the spacecraft systems could be recovered by the Americans. The ship was destroyed and its passengers killed. The Soviet News Agency TASS simply announced that an incorrect reentry profile had resulted in the craft being destroyed on reentry—a rare admission of failure for the Communist regime.

The trouble with the braking engine was quickly isolated and remedied, and the last of the prototype spacecraft was erected within three weeks. The first flight of the R-7 with the improved RD-0109 upper-stage-unit thundered aloft on December 22, 1960, with the dogs Kometa and Shutka. All went well until ignition of the new upper stage. A failure in the gas generator for the pump that supplied the propellants to the engine caused a premature shutdown. The emergency escape system was activated and the spacecraft separated from its carrier, arching more than 133 miles into space and reentering 2,200 miles down range in one of the most inaccessible regions of Siberia.

Rescue forces were immediately sent but did not arrive in the vicinity until almost two days had elapsed. Moving through waist deep snow, the technicians carefully examined the spacecraft which still had its destruct charge set to detonate 60 hours after landing (to destroy any evidence)—and it had been more than 60 hours since launch! They disarmed the charge and found the dogs in the spacecraft alive but undoubtedly very cold in the -40 degree C temperatures. The ejection seat to which their enclosure was attached had failed to exit successfully due to a series of malfunctions. The escape system would be redesigned. There was no mention of the aborted flight in the Soviet press.

With the two failures it was recognized that the next flight of a fully man-rated and equipped space-craft could not be made in the time frame scheduled, and that two more unmanned flights were need-ed. The pressure on the Soviet team was now intense, as a successful American sub-orbital effort would dim their accomplishment. The first man-rated spacecraft was launched on March 9th (called Korabl-Sputnik 4 by the Soviet press). The flight was perfect in all respects. Korabl-Sputnik 5 went aloft on March 25th carrying the dog Zvezdochka (Starlet). It was another complete success.

At this point there was some discussion whether the manned launch should be timed for the annu-al May Day celebration. Khrushchev actually preferred that it occur before or after as he was getting somewhat anxious that a failure resulting in loss of life would undermine the propaganda success that had thus far been achieved. However, the imminent launch of a manned Mercury spacecraft dictated that the attempt be accomplished as soon as possible—and a life had already been lost in the fledgling space program.

Cosmonaut-trainee Valentin V. Bonderenko had been undergoing a training session in an isolation chamber filled with 50% oxygen at reduced pressure to simulate the spacecraft environment. On the tenth day of the planned 15-day exercise, Bonderenko removed a biomedical sensor attached to his body and cleaned the skin with an alcohol wipe. He tossed the moist pad towards the garbage bag in the corner but it landed short on a hot plate and immediately ignited. Bonderenko tried to extinguish the small flame but it quickly grew and ignited his flight suit, engulfing him in flames that resulted in severe burns to virtually his entire body. He died within hours. The incident became a state secret that was not revealed until 1986.

Nevertheless, the stage was now set for the first manned Soviet launch.

Mercury Beset With Problems

If the Soviets had encountered problems along the way with the development of their manned spacecraft, the Americans could well empathize. On the first test of the Mercury escape system using the *Little Joe* solid-fuel rocket in August of 1959, an errant electrical signal caused the escape rocket to fire 35 minutes before the *Little Joe* flight was scheduled to launch. This test, designed to evaluate the escape system under maximum dynamic pressure (Max Q) would have to be repeated.

A few weeks later, the first Atlas with a Mercury capsule launched from the Cape on September 9, 1959. The objective of the test called "*Big Joe*" was to prove the integrity of the heat shield. As the missile arched over in its trajectory, the planned separation of the two booster engines at BECO (Booster Engine Cut-Off) failed to occur, and the skirt unit remained attached. The added weight did not allow the Atlas to achieve the desired velocity. Adding to the problems, the capsule did not sepa-rate at the planned time, and its thrusters attempted to wrestle the entire Atlas into a 180-degree atti-tude change to present the blunt end of the capsule forward to the line of flight for the reentry. This exercise in futility resulted in the thrusters running out of fuel. However, when the capsule did sepa-rate, it was able to withstand the unanticipated heating until its center of gravity performed a self-align-ment with the atmosphere. Although not a complete success, the test was encouraging.

Little Joe 1A (a repeat of the first August test) validated the Max Q requirement. The Little Joe tests were conducted at a new firing range off the coast of Virginia called Wallops Island. Although not near-ly as large and sophisticated as the Cape, the facility was close to the Langley Research Center and pro-vided the required range safety consideration out over the Atlantic Ocean. Another Little Joe test in November provided for further low altitude escape sequences, but the escape rocket fired late, and the test objectives were not fully accomplished. The next Little Joe test in December carried a monkey safely through the abort sequence, and another in January of 1960 was also successful.

Delays in the production of the spacecraft continued to plague the test schedule. One aspect that had not been considered in the schedule was that the Mercury capsule was so small that only one or two technicians could be working in the tight confines at any one time. In addition, the production workers still lacked the sensitivity for a "clean room" environment. Small pieces of wiring, loose nuts and other debris were left in the craft during their manufacture and modification. This could not be tolerated in the weightless environment of space where anything not fastened down would float around and could lodge in critical places. The first production capsule was delivered in April of 1960 and was fired in an "off the pad" abort test at Wallops Island in May. It was clear at this point that the December 1960 manned sub-orbital launch schedule would not be met.

However, the big test of a production spacecraft with an *Atlas D*, termed MA-1 (Mercury Atlas) occurred in July of 1960. This capsule carried no escape tower and was to qualify the production spacecraft structure under maximum aerodynamic loads and after-body heating during the powered ascent. A heavy rain shower that passed over the Cape delayed the launch. At ignition, the booster rose cleanly into the sky, but, as it began entering the area of maximum dynamic pressure at 40,000 feet, the sides of the Atlas buckled, and the entire vehicle was destroyed.

The failure was heavily covered by the press and played up in the aerospace media. Only 4 days earlier, the Apollo follow-on program had been proclaimed by NASA, but press announcements did

The launch of a Mercury Redstone

not accomplish milestones in space exploration. The 1960 Presidential race was in full swing, and the young candidate from Massachusetts, John Kennedy, remarked that if a man was to be placed in space before the end of the year *"his name will be Ivan"* (an obvious reference to Russia's superior capability). Although Kennedy had no way of knowing, one of the Soviet cosmonauts then in training was named Ivan!

On Election Day, November 8th, another test with the Little Joe was made to verify the escape system at the most critical portion of Atlas flight profile: where the G forces were at their peak. However, early separation of the spacecraft occurred, and the test objectives were not met.

The first test of an unmanned Mercury Redstone (MR-1) occurred on November 21st. The Redstone was touted as the most reliable missile in America's arsenal. Nevertheless, a failure in the launch sequence caused the Redstone to shut down just as it was lifting off, and it settled back on the pad. Fortunately, its structural integrity allowed it to remain upright. However, a portion of the escape sequence initiated, but the capsule remained firmly attached to the Redstone as the tower itself shot into the air. Three seconds later the parachute deployment began from the nose of the capsule as it sat on the missile amid the smoke. As the self-destruct system of the booster was armed, it was not prudent for anyone to venture near to cut away the parachute and begin the clean-up process until the next morning when the batteries that powered the explosives would have exhausted their charge.

The damage to the Redstone was minimal and quickly repaired, and the source of the erroneous sequencing was located for both the booster and the escape system. On December 19th, MR-1A (as the revised shot was designated) lifted off to execute a flawless flight. One more successful test of the Mercury Redstone and then a man would fly. That next test would carry a chimpanzee named Ham. As the test pilots who derided the spam-in-a-can approach to manned space flight had decreed two years earlier, man would fly only after a monkey had shown the way.

MR-2 launched on January 31, 1961 after an agonizing series of "holds" to the countdown causing the poor chimp to be sequestered in the capsule for more than 4 hours before ignition. As the booster climbed into the morning sky, the cabin inflow valve erroneously opened, literally sucking all the air from the capsule. The spacecraft environmental control system sensed the potentially lethal situation and inflated Ham's custom made space suit. To add to his discomfort, the trajectory was more vertical than the profile called for, and the liquid oxygen depleted at about the time the engine was scheduled to shut down. The abort sensor recognized the low pressure in the LOX line and was quicker than the shutdown sequencer, resulting in the triggering of the escape system. Poor Ham had to endure another sudden high G load as the spacecraft was lifted an additional 30 miles into space.

Because the escape system had activated, the retrorocket firing was not carried out, and the capsule descended at a faster velocity than planned, resulting in a 15 G load for Ham instead of the planned 12 Gs. But Ham's experiences were not over yet. As the capsule descended into the water, it landed at a slight angle, and the heavy Beryllium heat shield that was hanging down to extend the floatation bag slammed into

An American Mercury Atlas launch.

the side of the pressure vessel that was now providing buoyancy, and shoved a bolt through the titanium bulkhead allowing water to begin entering the capsule. The heat shield was quickly torn away by the wave action, and the capsule tipped on its side allowing more water to enter through the open cabin-pressure-relief valve. Ham was in deep trouble, but the recovery forces were unaware of his plight. As the capsule had overshot its intended landing area, it was not hoisted out of the Atlantic until almost two hours had elapsed. Ham survived the ordeal and lived a long and quiet life thereafter.

The straps on the landing bag were increased in number and size, and a fiberglass bulkhead was inserted between the heat shield and the titanium pressure bulkhead. However, von Braun was adamant that another test of the Mercury Redstone was needed before he would consider it "man-rated."

As the debate began concerning the need for another Mercury Redstone test before a manned flight, the next Mercury Atlas, MA-2, was scheduled for launch on February 21, 1961. This event was also contentious, as MA-1 had experienced a structural failure. The D model Atlas had undergone a weight reduction program, and its already thin skin had been shaved even more. The redistribution of weight that occurred with the addition of the Mercury capsule on top of the Atlas had caused the MA-1 failure. NASA had ordered the thicker skinned Atlas, and the first would arrive within a few weeks. The Atlas on hand was the thin-skinned version. A controversial decision was made to place an eight-inch wide steel band around the critical location on the body of the Atlas instead of waiting.

The launch occurred on schedule, and the rocket climbed into the morning sky. Everyone who was aware of the steel band was watching intently as the Atlas began to generate the "frozen lightening" contrail indicating that the Max Q region of aerodynamic stress was being encountered at 30,000 feet. MA-2 continued on its way—the "fix" had worked, and the capsule, driven to 114 miles above the earth, sped over 1,400 miles down range. The instrumentation revealed that it had endured a Max Q of 991 pounds per square feet and 15.9 Gs of deceleration on its 18-minute flight.

As for the Mercury Redstone, NASA was ready to fly an astronaut, but the ground rules for a manned space flight dictated that at least one full-scale rehearsal covering all the key elements had to be successful, and according to the von Braun team, the Redstone needed that one more test. However, the program had run out of production Mercury capsules, so an old "boilerplate" used in the Little Joe 1B flight was brought out and refurbished. On March 24th, MR-BD, which incorporated some eight modifications, flew a near perfect trajectory. The next MR flight would carry an astronaut. The following day, March 25, 1961, the Soviets flew their last unmanned test with Korabl-Sputnik 5. Even Hollywood could not have scripted a more exciting cliffhanger as April 1961 moved onto the calendar. Only a few in the Soviet Union really knew how close it would be.

Vostok and the First Man Into Space

It was 5:30 A.M., April 12, 1961, when cosmonauts Yuri A. Gagarin and Gherman S. Titov were awakened from a sound sleep at a small house in the secret Baykonur Launch Complex deep in the Soviet Union. It was the same dwelling formerly occupied by Marshall Nedelin, before his untimely demise the previous October. A sensor placed surreptitiously beneath their mattresses to monitor movement, confirmed the restfulness of their slumber to the doctors who were hovering over their every move. After a sparse breakfast of meat paste, marmalade, and coffee, they underwent a brief medical examination and then donned a light blue pressure suit covered by an orange jumpsuit.

The two, selected from the short list of six by a series of reviews, had to pass the final approval of both Chief Designer Sergey Korolev and Soviet Premier Nikita Khrushchev. Gagarin had only received the nod for being the first cosmonaut four days before and was told the following day. Titov was prepared to assume the primary roll right up to the moment of launch should anything inhibit Gagarin from completing the preparations.

Gagarin had been a favorite choice of the selection committee as his easy-going personality and intellect complimented his personal commitment and ability to focus on a task at hand. In peer votes for whom they would like to see be the first in space (cast by the cosmonauts themselves), he had scored higher than any of the others. He had also satisfied the Communist Party in that he came from a working class family (having grown up on a collective farm), was a devout atheist, and his ethnic background was Russian.

Following a short ride in the transport van to the launch pad, the two were greeted by a bevy of officials including Korolev—who reportedly spent a restless night himself. He had concerns for the reliability of the upper stage of the rocket. At the base of the service structure to one side of the giant R-7, Gagarin and Titov parted, the former taking the elevator to the top level and the waiting spacecraft, while the latter returned to the transport van to continue his stand-by role. The R-7 itself had been erected just two days earlier on the 10th.

Gagarin was helped into the spacecraft and secured by 7:10 A.M.—T-120 minutes. All through the procedure, various technical and party officials had been observant for any signs that the stress of the momentous occasion might be too much for the young 26-year-old. However, Gagarin was quite composed considering the history-making event of which he was the center. The spacecraft

The Soviet R-7 Vostok carrying Yuri Gagarin

hatch was closed at T-80 minutes, but incorrect seating caused it to be removed and laboriously re-installed. The count continued, and at T-30 the technicians left the service tower which was then lowered out of the way.

With fifteen minutes to launch time, Gagarin put on his gloves as the spacecraft communicator in the control room reported to him that his pulse was 64 and his respiration was 24, good signs of his controlled physiological disposition. At T-5 minutes, Gagarin lowered and locked the helmet visor. Only he knew the thoughts crowding his young mind as the final seconds ticked off. Perhaps he had already reconciled the odds of completing the next 90 minutes successfully, as he sat atop a combination of thousands of parts that had only flown 16 times in their present configuration, and only half of those had completed their mission.

As the countdown reached zero, the 32 rocket engines beneath him came to life, and a half million pounds of volatile propellants began to burn, hopefully in a controlled manner. Here was the first man to ride the fire into the cosmos, the first to experience the sound and vibration of sitting atop the most powerful creation of man, the first to sense the full measure of this new adventure.

The forces of gravity experienced by the cosmonaut during launch were in direct proportion to the current weight of the rocket relative to the thrust being generated. Thus, at lift off the R-7 produced about 50% more thrust than its weight, and Gagarin experienced 1.5 Gs. As fuel was consumed, the weight of the rocket decreased, and the G forces increased. At T+119 seconds, the four boosters had devoured virtually all of their fuel, and the thrust to weight ratio pinned Gagarin to his seat with 8 Gs.

Just then, the boosters shut down and separated, falling off to each side as if in a carefully choreographed, slow motion ballet. The central core unit continued to thrust onward as the G forces reduced back to three, and the rocket arched over to become parallel to the earth's surface at over 100 miles above the frozen wasteland of Siberia. The aerodynamic shroud that covered the spacecraft was released, and it too fell quickly behind the accelerating rocket which now subjected its occupant to more than 5Gs. Gagarin experienced difficulty in enunciating words into the radio as the skin on his cheeks was drawn tightly backwards, and his body weighed more than 1,000 pounds. His heart rate had risen to 150 beats per minute.

At T + 300 seconds, the core unit of the R-7 shut down, and the Gs again reduced as the upper stage RD-0109 came to life. In the control room at Baykonur, those who observed Korolev reported that he was visibly shaking during the ascent as he agonized over the possibility of something going wrong during the eleven-minute powered period. However, the telemetry showed all three stages had given the expected performance.

Suddenly, at T+676 seconds, the acceleration ceased, and Gagarin perceived he was being thrown forward against the restraining seat harness. In reality, it was the suddenness of going from a high G environment to weightlessness which provided the illusion. With a perigee of 110 miles and an apogee of 188 miles, somewhat higher than planned—but acceptable, man was in space for the first time.

Gagarin reported that he felt fine with no physical problems and was making an initial adaptation to his weightless environment. He was immediately drawn to the view from the small portholes—the muted colors of the earth contrasting with the bright white clouds a hundred miles below and the deep blue of the vast ocean.

The ground controllers could hear from his reports that there appeared to be no psychological impairment. Like the *Mercury* program, the Soviets had stationed tracking ships (Sibir, Suchan, Sakalin, and Chutkotka) along the ocean path of the spacecraft for communication and telemetry reception.

Although the Soviets gave no advanced notice of the launch to the world, sealed press packets had been prepared for the Soviet News Agency TASS and the state radio and TV news organizations. Once Gagarin reached orbit, they were telephoned and authorized to open the packets and make the flight public using the carefully scripted information provided. As was the case with many important announcements, it was preceded by the patriotic Soviet anthem, "How Spacious is My Country." In addition, a sealed packet contained the notice of a fatal malfunction if it occurred after the initial announcement was made. A third packet contained information relative to an emergency landing made outside the Soviet Union. The final revelation in the press packets was the name of the spacecraft—*Vostok* (the Russian word for East), which had been a closely guarded secret to that point. Soviet schoolchildren and factory workers were given the day off to celebrate.

The time passed quickly as Gagarin completed the rather simplistic chores assigned, and soon it was time to prepare for reentry. The mission was designed to be flown automatically by the spacecraft without any intervention by the cosmonaut. Nevertheless, it was recognized that should something go wrong with the automated sequence, the pilot had to be allowed to manually control the ship. As a result, the last three digits of the combination to unlock the manual control system were in a sealed envelope that the cosmonaut could open in an emergency.

The spacecraft's attitude was set to align the braking rocket opposite to the line of flight based on the position of the sun. The time of day for the flight was chosen specifically to allow the sun to be used for this purpose. At T+75 minutes the retrorocket fired for 40 seconds. The spherical descent module was then supposed to separate from the instrument section where it had been held in place by four metal straps. But something went wrong, and Gagarin felt a sharp jolt and observed that the craft had begun to rotate slowly around its longitudinal axis at about 6 revolutions per minute. He could see the continent of Africa periodically passing the window with each revolution. He reported through the voice channel that the braking rocket had functioned properly, and he should be landing in the designated zone.

Apparently, the metal straps that fastened the two segments of the spaceship together malfunctioned. Only some of the electrical connections between the two units were disconnected while the reminder were holding the two units together. It was conceivable that conflicting signals could have interfered with the remaining reentry sequence. Gagarin perceived he had a potentially serious problem but retained his composure and even reported through his telegraph key the code *VN* indicating "all is well"—apparently believing there was no need to excite those on the ground with a potential problem over which they had no control. It may be fortuitous that the destruct package, which had accompanied all the unmanned missions, was not aboard as it might have detected a faulty reentry and destroyed the spacecraft. The KGB representative on the flight planning committee had argued for its inclusion, but was over-ruled.

Gagarin felt another bump about 10 minutes after separation was to have occurred, and it is now assumed that the two units finally parted at this point, perhaps because of the building aerodynamic loads. It is also possible that, had the two segments not separated, the heat on the descent sphere would have exceeded critical values.

Gagarin now witnessed through his small openings to the world outside a bright purple light that began to surround the spacecraft as the heat pulse created an ionized sheath of air, inhibiting the ability to communicate by radio. The flaming reentry had begun.

The cosmonaut felt the increase in temperature radiating through the hull and noted the sphere rotating about all three axes as the spacecraft entered the more dense layers of the atmosphere, and the G forces began to build. The oscillations gradually dampened as the aerodynamic forces and the center of gravity of the spaceship began to reach equilibrium. The G forces reached 10, and Gagarin later reported that his vision began to blur and he "grayed out" for a few seconds. He had experienced as many as 13 Gs in the centrifuge at the astronaut-training center.

A large contingent of more than two-dozen aircraft and seven recovery teams were spread across the expanse of the Soviet Union should the spacecraft fall outside the intended landing zone. At about 20,000 feet, the main parachutes opened, and the hatch covering the ejection seat jettisoned. Within a few seconds, a small rocket charge ignited, and the seat shot away from the descent module. Two seconds later the seat restraints released, and Gagarin tumbled free to open his own parachute for the final phase of the landing.

It was a beautiful spring day, and the farmland of the Saratov region over which he was swinging gently in the parachute harness, southeast of Moscow and north of the Caspian Sea, was familiar territory to him. He landed gently in a field just 1 hour and 45 minutes after launch. Yuri Gagarin had become the first man into space. The Soviet Union had once again beaten the Americans to another important "first" in the space race. The "backward" country continued to lead the United States more than three years after the first Sputnik had jolted the Americans into action.

"We're All Asleep Down Here"

It was still dark on the morning of April 12th in the United States when the telephone rang on the nightstand next to where Lt. Colonel John "Shorty" Powers was sleeping soundly in Florida.

Answering the phone, he heard an excited New York news reporter on the other end inquire if Powers had any comment on the launching of the first Russian Cosmonaut. Powers was the astronauts' Public Affairs Officer, and, along with others in the NASA hierarchy, had already prepared a statement for the expected announcement of a successful Russian manned flight into space. The preparation was a result of the UPI press agency reporting a rumor that the Soviets were about to send a man into space and the CIA confirmation.

But, being caught at 4 A.M. was just a little too much for the Eighth Astronaut and Voice of Mission Control. *"We're all asleep down here,"* he responded in reference to Florida's location south of New York. The reporter duly noted his comment, which was subsequently published with its double-entendre, and it became an embarrassing icon of the position of the United States in its race with the Soviets.

James Webb, President Kennedy's new NASA chief, appeared on nationwide television at 7:45 A.M. EST to congratulate the Soviets. He also wanted to reassure the nation that *Project Mercury* would achieve the same goal "soon." The following day Webb and several of his NASA team were called before the House Space Committee where strong disappointment was expressed that America had again come in second best.

At a presidential press conference, a reporter asked when the United States would catch up. Kennedy's stressed response was, *"However tired anybody may be, and no one is more tired than I am, it is a fact that it is going to take some time. We are, I hope, going to go in other areas where we can be first, and which will bring perhaps more long-range benefits to mankind."* This was a strong indicator that Kennedy had been doing some soul searching.

The first of the thick-skinned Atlas missiles, 100-D, was launched two weeks later on April 25th in the first attempt to actually orbit an unmanned Mercury capsule. The decision to orbit rather than perform another sub-orbital flight was made only after the Gagarin flight as a means of providing some lift for America's morale. However, the Atlas guidance system failed to pitch the rocket over into a trajectory to the southeast and instead continued its vertical flight. The destruct signal was sent and yet another failure was etched into the American psyche. For the escape system engineers, the unplanned test resulted in a successful exercise of their systems. The Mercury capsule was pulled from the booster and descended safely under parachute. However, that was little consolation for the nation.

On the 28th of April a Little Joe rocket made its third attempt to simulate a Max Q escape, but one of the solid-fuel rockets in the cluster failed to ignite. However, the test conditions exposed the capsule to double the aerodynamic loads and three times the G-load of a worst case Atlas abort. The capsule escape system was considered "qualified."

Just as with the Soviet program, Project Mercury had moved three of its astronauts to a short list for the first flight the previous February. Alan Shepard, John Glenn, and Gus Grissom went through the final intensive training program with the actual capsule, production spacecraft No. 7. The original launch date of March 6th had been slipped to the end of April and then into early May. The press had begun the guessing game as to who the first American astronaut would be, when in fact Robert Gilruth had already told "The Seven" back in January that, barring any change in his physical or psychological profile, Alan Shepard would make the flight with Glenn as the back-up.

Unlike the anonymity of the Soviet cosmonauts, the identity of the American astronauts was well known to the world, as was the façade of their character thanks to a contract with Life magazine. The contract was in itself somewhat controversial as it provided a $25,000 one-time payment to each of the astronauts to allow their lives to be revealed to the world. While the Life series appeared to convey the homogeneity of the astronauts, the depth of these individuals and their often-superhuman egos would not come out for some time to come. Alan Shepard was a complex man with an inner drive to be the best. His quick wit to whomever he chose to befriend was often hidden by a veneer of cold steel to those he chose to shut out.

As the flight date drew closer, the press grew more demanding for information and access, and there was a point at which NASA felt that the pressure might be too much for not just the astronauts but the technical team as well. As the Russians had been so secretive, there was some derision of the validity of their flight in the world press. So NASA and the Kennedy administration felt obligated to open as much of the launch preparations as possible without interfering with the flight itself. Kennedy felt that the U.S. was leaving itself exposed should a fatal flight occur and, like Khrushchev a month earlier, was apprehensive about the negative propaganda that would result.

Mercury Redstone MR-3: Alan Shepard

With the arrival of the first week of May, more than 300 press representatives from virtually all of the major countries of the world descended on the Cape to cover the unfolding story. On May 2nd Shepard went through the first stages of the countdown and had gone as far as suiting up; but the weather didn't cooperate, and the flight was "scrubbed" (another word in the lexicon of the new space jargon that means canceled) two hours and 20 minutes before launch. It was at this point that NASA decided to reveal the identity of the astronaut. A second attempt on May 4th was also canceled.

Shepard was awakened at 1:30 A.M. on the 5th of May to again begin the process of preparing for the flight. The medical people and NASA management observed Shepard carefully to see if the strain of the delays was showing. However, Shepard was well prepared in all respects and had, as Gagarin undoubtedly had, reconciled the risks with the steps taken to mitigate them. While the concept of fear, as most would define it, was not a factor in the astronaut's thought process, the phrase "heightened awareness" was.

Shepard ate what would become the traditional pre-flight breakfast with filet mignon wrapped in bacon with eggs and orange juice. The diet for the three days before a flight would concentrate on high protein, low residue foods.

The suit-up process then began with a brief physical followed by the attachment (or insertion, as the case may be) of the three primary bio-medical sensors for temperature, respiration and heart rate. Blood pressure would be a fourth monitor for later orbital flights. It was almost 4 A.M. when Shepard emerged from his quarters at Hanger S to begin the van ride to the launch pad. It was the same pad from which America's first satellite, Explorer I, had been launched. All of the other astronauts were engaged in some aspect of the launch, with Glenn having run through the spacecraft systems test in the early hours of pre-dawn.

Shepard began the elevator ride to the top of the Redstone at 5:15 A.M. and was helped into the tight confines of the capsule by the technicians. There he found a sign left by Glenn that read, *"No handball Playing here"*—a reference to Shepard's exceptionally good athletic prowess at the sport. Again, another tradition was being established with humor and camaraderie playing an important role in preparing the astronaut for flight.

When the space suit sensors were plugged into the telemetry system, Shepard's heart reported 80 beats per minute. The Mercury cockpit was filled with a wide variety of switches and instrumentation, far more than the rather sparse Vostok, and reflected a cockpit designed for a pilot. The hatch was bolted shut by 6 A.M. and Shepard began to breathe 100% oxygen to purge his system of nitrogen.

Low clouds began to accumulate over the Cape, and the countdown was "held" several times and then "recycled" back to T-35 minutes. For more than four hours, Shepard lay on his back waiting for an opening in the overcast. At one point he related that his bladder was in need of being relieved, and there was a flurry of discussions on what to do. Because the flight was only to be a quick 15-minute "up and down," no provision was made for the pilot's waste products. Finally, the decision was made to allow him simply to urinate in the space suit. Although this had some negative affect on the electrical biomedical sensors, it resolved the problem. With some irritation over several other glitches that caused added delays, Shepard at one point commented, *""Why don't you fix your little problem and let's light this candle."*

As capsule No. 7 had been mated with Redstone No.7, Shepard named his spacecraft "Freedom 7" to reflect the accumulated sevens that had come together in the project, and another precedent was set: each astronaut was allowed to name his spacecraft. Several of the changes the astronauts had requested (some would say demanded), such as the larger window in front of the pilot and the new lightweight quick-release hatch, were not in this version.

As the countdown neared zero, Shepard's heart rate increased to 100 beats per minute, and people all across the nation, many of whom had delayed going to work or school that morning, leaned toward their TV sets… 3—2—1—Zero! The Redstone rose rapidly into the morning sky with Shepard providing a portion of his own narrative, *"Ahh, roger, lift-off and the clock has started."* This was an expression that many of the other astronauts would repeat as a reassurance to those on the ground that the astronaut had his act together and was coherent. It also was recognition that the timing of the events that would occur as dictated by the on-board clock would be in synchronization with the planned schedule. And then, *"This is Freedom 7, the fuel is go, 1.2 G, cabin 14 psi, oxygen is go."*

At T+45 seconds, the area of Max Q presented Shepard with vibrations and buffeting that soon smoothed a bit as the outside sky quickly changed from blue to black. *"Cabin pressure holding at 5.5"* was an important call at just after T+60 seconds that his predecessor Ham (the Chimpanzee) had been unable to make; it had caused his space suit to inflate. However, with Shepard, failure of the cabin vent valve to close would initiate the abort sequence.

But all continued to go as planned, and the Redstone shut down at T+142 seconds after imparting a velocity of 5,200 mph. The escape tower jettisoned, and the small posigrade rockets separated the Mercury spacecraft from the Redstone. Five seconds later the Automatic Stabilization Control System turned the capsule around. Shepard then changed the spacecraft attitude with the manual system essentially to prove that man could perform critical tasks in the weightless environment. He then performed a quick evaluation of his ability to observe the earth and reported these to mission control.

The precious seconds ticked off as the spacecraft was realigned for retrorocket firing. Of course, this was not necessary since Shepard had achieved less than a third of the speed needed to go into orbit, but the functioning of all the critical systems had to be tested and evaluated. Retro-fire was completed, the pack jettisoned, and the spacecraft aligned for the correct reentry attitude. At about 38 miles above the earth in the descent, Shepard noted the .05G indication—the atmosphere was beginning to show its affect on the spacecraft. The G forces increased to a maximum of 11 as the atmospheric friction slowed the descent to a subsonic velocity over the next few minutes. Sub-orbital spacecraft were equipped with beryllium heat-sinks rather than the ablation type heat shields because they would not experience the extreme heat of an orbital reentry.

The drogue chute deployed on schedule at 20,000 feet followed by the main chute at 12,000 feet. At 9:49 A.M., after a 15-minute 22-second flight, Shepard's Freedom 7 splashed down in the Atlantic in sight of the recovery forces. Within 15 minutes he emerged from the recovery helicopter onto the deck of the aircraft carrier USS Lake Champlain. Shepard was in excellent physical and psychological condition, and except for a few minor glitches, the flight has been perfect. This was reflected in the news media that reported the expression "A-OK" which was used by the NASA press spokesman "Shorty" Powers. It was assumed that Powers was relaying Shepard's exact words when in fact he was simply using his own jargon. However, "A-OK" had been indelibly impressed on the American lexicon of speech that day.

A few days later, President Kennedy invited all the astronauts to the White House and awarded Shepard NASA's Distinguished Service Metal, and there was a big parade down Pennsylvania Avenue. Not only was the nation excited about the flight but also the world responded most generously. Khrushchev was at first bewildered and then angered by the response. While the Soviets received gracious compliments on Yuri Gagarin's flight, there was not the same outpouring of emotion. But Shepard only went up and down—a short 15-minute flight, not around the world as Gagarin did—so, why all the fuss? The big distinction between Gagarin and Shepard was that, through the media, the world had become a part of the frustrating preparations as well as riding along with Shepard by way of Shorty Powers' real-time voice of Mission Control. They had suffered through the failures and postponements over the past two years. The American people (and to some extent the world) felt as though they had been a part of the drama. That was the difference!

However, the success that the Americans and the Soviets had achieved in their first flights would be sorely tested in subsequent ventures as the capability and reliability of the various spacecraft systems were pitted against the unforgiving extremes of the environment of outer space.

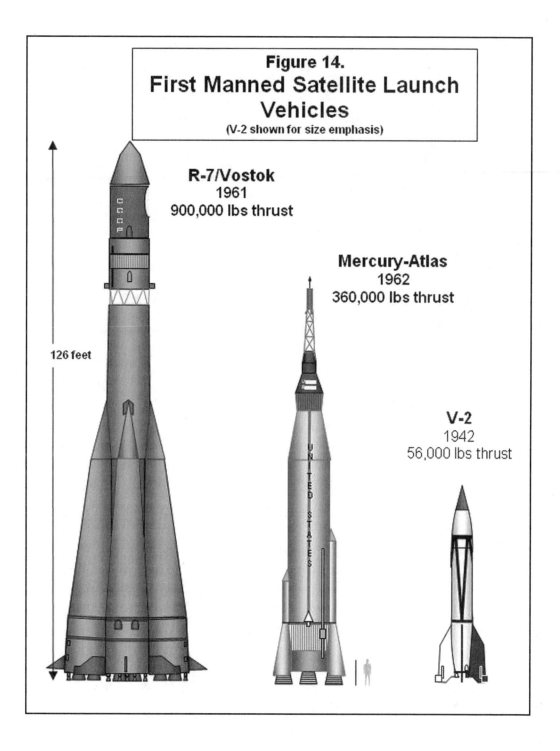

Figure 14.
First Manned Satellite Launch Vehicles
(V-2 shown for size emphasis)

R-7/Vostok
1961
900,000 lbs thrust

Mercury-Atlas
1962
360,000 lbs thrust

V-2
1942
56,000 lbs thrust

126 feet

As the space race moved into the summer of 1961, the Soviets still had a commanding lead in terms of spectacular accomplishments and weight lifting advantage. Gagarin had orbited once around the earth, and Shepard had flown the brief 15-minute suborbital hop. Neither flight had truly exposed the pilot to the full rigors of the space environment nor challenged the ability of man to function effectively in space for more than 90 minutes. However, the results to that point were encouraging. This perception would change.

Grissom Escapes Drowning: MR-4

Captain Virgil "Gus" Grissom was given the nod for the second Mercury Redstone flight (MR-4). The spacecraft, serial number 11, had several significant changes from Shepard's. The large window was now installed as was the new quick-release hatch. The 70 bolts that secured the hatch each had a small hole drilled into them that would permit the bolt to fracture when a thin explosive fuse that was wrapped around them was detonated. Activated by a plunger close to the pilot's right hand and requiring a force of 7 pounds to initiate, the new hatch saved 40 pounds.

Grissom had selected the name "Liberty Bell 7" as a patriotic symbol akin to the capsule's shape and the oneness of the seven astronauts. As was too often the case for launches at the Cape, the first scheduled firing time on July 19, 1961, had to be scrubbed just 10 minutes short of launch due to weather. The Redstone had to be recycled back to the start of a two-day countdown cycle because it had to be drained of propellants.

Grissom was the smallest of the astronauts and more reserved. Perhaps because of his diminutive stature, his personality had been formed to reflect a high degree of professionalism, and he was said to be the most sensitive and yet the most prone to use extreme profanity. Grissom entered the small spacecraft at 4 A.M. on the 21st for the scheduled 6:20 A.M. launch. Because of Shepard's experience, Grissom's spacesuit was configured with a urine collection device that had been devised just the day before.

A few short delays, including one for letting some clouds move through, resulted in ignition occurring at 7:20 A.M. *"Ah Roger, this is Liberty Bell 7. The clock is operating,"* Grissom responded to the call of *"liftoff"* from Mission Control. Shepard, who was the CapCom for the flight (Capsule Communicator), responded with, *"Loud and Clear Jose, don't cry too much."* A comedian of the day had created a nightclub routine that used a reluctant Spanish accented astronaut to parody various aspects of a flight into space. It had been a great hit with the astronauts as they had gone through training and many of the lines had become a part of their communications vocabulary.

The flight was textbook perfect with a few changes in the procedures from Shepard's journey just 10 weeks earlier. As the spacecraft splashed down into the Atlantic, Grissom began the process of disconnecting himself in preparation for egress—except for the oxygen hose inlet and the helmet communication lead. He began recording some of the instrument readings while the recovery helicopters positioned themselves for the pick-up. Grissom indicated to the recovery team that he would need another five minutes before he was ready to exit. A few minutes later he called, *"OK, latch on, then give me a call and I'll power-down and blow the hatch."* Grissom had removed the cover from the hatch safety pin and had removed the pin itself. All that was needed was a firm punch on the plunger.

But before the helicopter could secure the capsule, Grissom heard a loud "bang," and the hatch disappeared leaving blue sky and green seawater pouring into the cockpit. Grissom removed his helmet and literally dove through the narrow opening into the ocean. The helicopter pilot and his recovery crew observed the untimely hatch activation and saw that Grissom was apparently moving clear. They proceeded to snag the sinking spacecraft, as Grissom looked on from a few yards away. When he saw the helicopter had hoisted the capsule from the water, and the horse collar that would pull him up being lowered, he began to swim towards it when suddenly it started back up.

The helicopter pilot had observed an "over temperature" warning condition of the engine on the instrument panel. Decisions had to be made quickly now. The spacecraft had filled with water, and along with the impact bag that was slowly draining its content, it represented about double the weight of a normal capsule at this point. The helicopter pilot was reluctant to release the valuable cargo, and as he pondered his next course of action, the helicopter sagged under the weight of the capsule toward the water, its landing gear awash.

The pilot released the spacecraft, advised the standby helicopter of the situation, and proceeded back to the carrier while the second helicopter moved in towards Grissom, whose plight had suddenly worsened. Grissom had failed to close the oxygen inlet valve on the suit which would normally provide buoyancy. The suit was filling with water, and Grissom could barely keep his head above the encroaching swells that were now breaking over and threatened to submerge him.

The second helicopter hovered over Grissom who, with only seconds to spare before he would have sunk beneath the surface, was able to slide into the recovery sling. In minutes he was hoisted aboard the chopper and taken to the Carrier Randolph where his condition was determined to be excellent. The capsule sank in 15,000 feet of water but was recovered—40 years later.

NASA convened an investigation board headed by astronaut Wally Schirra. There was no conclusive proof as to what had happened. Grissom affirms that he did not intentionally hit the plunger for the hatch release, but the stigma of having lost his spacecraft hung heavily around his neck. Unlike Shepard, he was not invited to the White House for presentation of his NASA award, and one of the stories that circulated was that he had panicked. There was never any indication of that type of behavior from his communications or biosensors. The experience embittered him (and his wife), turned him inward, and hardened him professionally. He would continue in the astronaut corps and became respected for his resolve and engineering capability. But fate would deal him yet another blow before his career would come to an untimely end.

The original plan was for each of the seven astronauts to fly a suborbital Redstone to give them a taste of rocket flight before the main event of going into orbit and experiencing the rigors of the Atlas. However, NASA management had decided, only a month after Shepard's flight, that they would cancel MR-6 and await the results of MR-4 before deciding on MR-5. With the exception of the hatch problem, Grissom's flight had again proved all the capsule's systems. And while the lethal prospects of a hatch blowing while in space caused a detailed analysis to be made of the design and circuitry, NASA needed only a bit more prodding by the Russians to move on to orbital flight.

Vostok II: Titov Overcome by Space Sickness

Although from the outside the Soviet space program appeared to be a well-orchestrated agenda, the actual flights proceeded on a very ad hoc basis. Following Gagarin's flight, Chief Designer Sergey Korolev wanted to move directly to a 24-hour mission. However, the medical team and the cosmonauts themselves including Gagarin, who was now the commander of the cosmonaut group, preferred a less ambitious 3-orbit flight. Nevertheless, Korolev took his argument to the State Committee for Defense Technology and forced the issue in his favor.

While the 25-year-old Major Gherman S. Titov was elated to be selected to fly the second mission, he was still disappointed that he had not been given the first. He was considered by most of his peers to be of superior intellect and more urbane in many ways than Gagarin. However, that same sophistication alienated him from his contemporaries and the professional staff. His individualism almost caused him to be dropped from consideration for the second flight, which had been tentatively scheduled for August of 1961. However, the exact schedule of the flight would be determined by other considerations.

At a meeting in Mid-July with Khrushchev, Korolev had indicated that the next flight would take place about mid-August. Khrushchev then told Korolev that in fact, the flight would occur before August 10th. Although Korolev was given no reason for this rather arbitrary requirement, Khrushchev had planned to isolate West Berlin behind a brick wall, and construction of that wall was set to begin on the 13th. Khrushchev was again looking to a "space spectacular" to strengthen his hand in dealing with the West.

The launch procedures moved ahead smoothly for Vostok II, and the call from the block house of *"She's off and running"* came at 0900 hours on August 6, 1961. The powered portion of the flight went as planned, but shortly after the spacecraft achieved its 110-mile by 160-mile orbit Titov began to feel disoriented. He later recalled being in a *"strange fog,"* unable to read his instruments or to orient himself with respect to the earth that was visible from the porthole. He tried several head and body movements to shake-off the symptoms that were much like vertigo a pilot might experience following a series of maneuvers.

The spatial disorientation Titov was experiencing had been predicted by both Soviet and American medical specialists as an affect of weightlessness on the sensory structure for a person's balance. The lack of gravity on the small bones in the inner ear cause conflicting nerve impulses to be sent to the brain. Both the American and Soviet cosmonauts had undergone training to familiarize themselves with the weightless phenomena, but only periods of up to 30 seconds at a time could be achieved by putting an aircraft in a parabolic arc (a gradual pitch-over from a steep climb). The average person perceives only a slight sensation of weightlessness when in an elevator that begins a descent or on an amusement ride.

By the second orbit, Titov considered requesting a return to Earth, but at the time he was not in communication with mission control, and when communications was re-established, he decided to hold off. The biosensors relayed some indications to the medical staff on Earth that all was not well with Titov, and by the third orbit they decided to make an inquiry. But Titov was not about to give in to what he felt certain was a condition that would clear if only he gave it enough time. He reported, *"Everything is in order."*

The television pictures beamed down to those on the ground were not of sufficient quality to perceive the anguish that Titov was experiencing, and they were available only during brief periods of the flight. At the start of the sixth orbit, Titov elected to eat a meal per the schedule. However, after sampling some of the various gourmet foods prepared as a paste and squeezed from tubes, he became nauseous and vomited.

After a short rest period, he resumed the experimental schedule exercising attitude control for the first time with the manual thrusters. The seventh orbit started his sleep period, but since he was still fighting the spatial disorientation problem, his symptoms now included a severe headache that seemed, as he would later report, to permeate his eye sockets. At his next pass over Moscow he somewhat surprised the controllers with the comment, *"Now I am going to lie down and sleep, you can think what you want, but I'm going to sleep."* For the most part, his sleep period, which lasted more than six hours, was restful, but when he finally awoke and decided to resume his activity, he still felt very poorly. He performed some head and eye movements as well as some experiments to determine his mental acuity and was surprised to find that he was more oriented than he had been at the start of the flight. However, an attempt to take in some fluids resulted in another bout of vomiting.

As Vostok over flew selected third world countries, Titov transmitted greetings and read a statement extolling the virtues and technical superiority of the Communist state. As the exact schedule was not provided to these countries, few were actually able to hear the transmission. However, the Soviet consulate in each country ensured that the text of the prepared statement was made available to the local media.

By the twelfth orbit, Titov finally began to feel better, and his condition continued to improve as the flight moved to its termination at the start of the seventeenth orbit. Following retro-fire, the separation of the descent module occurred with a loud "crack." However, unusual sounds and the movement of the spacecraft led Titov to believe that, like Gagarin's flight, the separation was not clean. During the early phases of reentry, the instrument section was again torn from the descent module, and the remainder of the return was normal. Titov ejected and landed after 25 hours and 11 minutes in space.

Titov was truthful during the debriefing about his physical and mental state but was also elated about the opportunity to fly and his ability to "tough it out." There was a high level of anxiety in Korolev and the medical team about the possible causes and effects of what was called "space sickness." As the U.S. and Russia gained more experience with man-in-space, it was established that good physical conditioning and exposure to the weightless environment through the use of numerous aircraft flights performing parabolic arcs would diminish the onset of space sickness. It was also concluded that the astronaut should avoid rapid head movements.

To certify the missions of both Gagarin and Titov as new world records, an application was made by the Soviets to the International Aeronautical Federation (known internationally by its French name Fédération Aéronautique Internationale—FAI). The FAI was organized in 1905 because of the early claims being made by the pioneer aviators. It defined its principal aims as being to *"methodically catalogue the best performances achieved, so that they be known to everybody; to identify their distinguishing features so as to permit comparisons to be made; and to verify evidence and thus ensure that record-holders have undisputed claims to their titles."*

However, the FAI requires the pilot to return from a flight in his craft in order for the record to be valid. The Soviets were not about to reveal any aspect of the flight that might compromise security, but they were not about to miss holding the record for the first space flight by a technicality. So Gagarin was forced to be creative with his responses to the press regarding those aspects of the flight such as the use of the ejection seat during its final phase. It was not until ten years later in 1971 that the truth came out, and by then no one cared. Another interesting anomaly in "squeezing the truth" occurred when the site of the launch had to be identified to the FAI. As the real location of Tyura-Tam was still a state secret, the name of the small obscure and far away town of Baykonur, some 230 miles to the northeast, was picked. That is how the Tyura-Tam launch site became known as Baykonur. As there were no photos released of the spaceship or the R-7, there was continued distrust around the world as to Soviet claims.

The fame that followed both Gagarin and Titov, as they toured the world to spread the Communist gospel in the months that followed their epic flights was also the preverbal curse. The once straight-laced and easily embarrassed young Gagarin became a womanizer and liberally imbibed Vodka. Both he and Titov were reprimanded by the Communist Party for their behavior, and these would be their only flights into space. As for Gagarin, it was the desire by the Soviets to keep the first spaceman as a safeguarded national treasure, which unfortunately would be cut short by his untimely death while flying a Mig-15 during a proficiency flight in 1968. For Titov, it may have been his personality that alienated him from his peers and the Communist ideology. There was, of course, no fame or recognition for the Chief Designer, Korolev, who was protected by anonymity lest the United States find out his identity and attempt to have him assassinated.

Final Preparations for Mercury

With the daylong flight of Vostok II, NASA recognized that the Soviets had already exceeded the basic capabilities of the Mercury spacecraft. The pressure was on to "get an American in orbit," and any tentative plans for another Redstone following Grissom's flight were discarded. The failure of MA-3 in April had put the orbital schedule way behind. MA-4, the second thick-skinned Atlas D, was erected at the Cape and was mated with capsule serial number 8 that had been recovered from the MA-3 flight. Then a significant problem with some of the transistors in both the capsule and the booster electronics required their de-mating, and several weeks were lost changing out the suspect components.

Atlas 88-D finally erupted to life on September 13, 1961, and quickly accelerated the spacecraft and its mechanical astronaut into orbit. This was the first time the Atlas had ever lifted that weight to orbital velocity and only the second time it had been used to orbit a payload without an upper stage (the first being Project Score almost three years earlier). The Booster Engine Cut-Off (BECO) had occurred slightly earlier than programmed as did the Sustainer Engine Cut-Off (SECO); however, the high levels of performance of the engines provided the required energy.

The spacecraft itself used an automatic sequencer to provide all of the control and timing of the various events. A single orbit would essentially prove all of the systems in the hostile space environment without inducing a prolonged opportunity for problems to occur. While the most obvious tests centered on the spacecraft and the performance of the booster, the entire manned spacecraft tracking and data transmission network was exercised for the first time. A more comprehensive test of the tracking system had been planned using the new solid-fuel *Scout* rocket to orbit a small satellite that would mimic the signals from a Mercury spacecraft. However, this rocket, with a mis-wired control system, was destroyed by a range safety officer seconds after it was launched.

For NASA, it was critical that an "almost real-time" monitoring of the spacecraft and its astronaut take place to provide the highest levels of safety. All of the seven astronauts were involved in the MA-4 test, with four being situated at tracking sites around the world and three working at the Cape itself. As the spacecraft flew over each of the stations, the data streaming down was retransmitted via terrestrial links back to the Goddard Spacecraft Center in Greenbelt, Maryland, at data rates of 1,000 bites per second (pitifully slow by today's standards).

All of the scheduled orbital events went off at the required intervals, and the spacecraft was soon back on earth following its 90 minute, 25,000 mile journey. Of greatest interest for many was the fact that the ablative heat shield had survived the reentry; the other tests with Big Joe had been suborbital and had not induced the full duration of heat as this orbital test did. NASA was elated. The American public (and the rest of the world), while thankful, saw the test as simply another hurdle to catching up with the Russians and not a space spectacular. One more test separated America from a manned orbital flight, and it was still possible that an American would orbit the Earth in the same calendar year as a Russian.

Using a chimpanzee, MA-5 was a final dress rehearsal for a manned flight and was scheduled for three orbits. Atlas 93-D arrived at the Cape in early October and was delayed as the same transistors were changed out. Spacecraft 9 had been on hand since February, and it was quickly mated for a launch scheduled for November 7th. However, problems continued to plague the project, and it was not until November 29, 1961, that Mercury Atlas-5 sat fueled and ready for launch.

The occupant was a chimpanzee named Enos. Like Ham of MR-2 fame, Enos had gone through an intensive preparation period that consisted of more than 1,200 hours of training over a period of 16 months. Chimps were used in America's space program because they had the ability to learn a variety of tasks that could be performed at critical times during a flight to determine the ability of a primate to make cognitive functions under the same conditions that the astronaut would face. Four distinct activities were planned for Enos. One required that he pull a lever exactly 50 times to be rewarded with banana pellets. It was observed during training that he would quickly pull the handle 45 times before slowing and making the last five pulls more deliberately until the fiftieth, when he placed his hand under the dispenser in anticipation.

Those who worked with the chimps felt the animals enjoyed the challenge and tolerated the uncomfortable aspects such as the centrifuge rides in anticipation of the rewards that followed. Unlike humans, the Chimps showed no physiological or emotional indications of anxiety even while being prepared for an uncomfortable test they had previously encountered.

The Atlas lifted into the morning sky and powered itself into orbit over the next five minutes. Enos proceeded through the series of planned exercises with his actions monitored by telemetry and recorded by an on-board 16-mm movie camera—one of four that recorded various aspects of the flight.

President Kennedy was in the middle of a press conference when a note was passed to him of the launch. His quick wit was displayed to the pleasure of all when he looked up after reading the note and announced, *"The chimpanzee took off at eight past ten."* He paused just slightly and then added, *"And he reports that everything is perfect and is working well."*

A variety of problems began to plague the mission after the first orbit. The device designed to give a mild shock to the Chimp malfunctioned and buzzed the slightly confused primate even when he was doing his tasks correctly. Enos took it all in stride and continued to perform. An inverter that converts direct current to alternating current for the electrical system showed indications of overheating—but there was a back-up if the first one failed.

More critical was a drift observed with respect to the spacecraft's attitude; it was consuming excessive thruster fuel and could deplete the supply if all three orbits were flown. At the very last minute, the decision was made to bring Enos back at the end of the second orbit. As the decision was being relayed to begin reentry, a farmer in Arizona ran his plow through the buried communication line. But that is why back-up capabilities were provided, and within seconds an alternate link reestablished communications, and MA-5 recovery was made in the Atlantic without mishap.

In the post-flight analysis, all of the problems were carefully evaluated, and the most serious, the attitude control, was traced to a small metal shaving that blocked a thruster fuel line. In a humorous political cartoon that appeared in a newspaper the following day, a chimp in a space suit was depicted walking away from his spacecraft with the caption reading, *"We're a little behind the Russians but a little ahead of the Americans."* It was time for America to orbit a man.

"A Real Fireball Out There": MA-6 Glenn

The schedule had called for the launch of MA-6 on December 19th, but with only one Atlas launch pad allocated to Project Mercury, the time required to refurbish the pad after MA-5 and erect MA-6 would run the mission into the Christmas holidays. Mercury Project Manager, Robert Gilruth, after consultation with key members of NASA and his management staff, felt that keeping the thousands of government employees and contractors working through Christmas would be inappropriate. Moreover, if the past was any indicator of the future, there was no guarantee that the launch would come in calendar year 1961 in any case.

If a lack of broad based technology slowed the Soviets, America's need for a miniaturized high-tech environment and the changes that resulted was their impediment. Meticulous tracking of any change, no matter how small, required paperwork, approvals, and confirmations; and the number of changes that transpired for the Mercury spacecraft were considerable. Not only was a new type of flight vehicle being engineered but also it had to be reconfigured depending on the specific requirements of each mission. And the fact that only one technician could be working in the capsule at any one time meant that the installation of each change was very time consuming.

Spacecraft 13 was built in May of 1960, then reworked to the new specifications to include the larger window and quick opening hatch. When, in April of 1961, it was determined that it might be the first manned mission then set for late that year, it went through another set of changes before finally being delivered to the Cape in August of 1961, after which 200 more changes were made before it would finally fly. Many of these were based on the early flight experiences, while others were a result of more exhaustive testing and the desire to provide the safest, most reliable vehicle.

As Marine Lt. Col. John H. Glenn had been among the three finalists for the Mercury Redstone Flights selected the previous February and would have been the pilot for MR-5 had there been one, the first orbital flight of Mercury officially became his mission as soon as Enos had returned from orbit. Glenn was the oldest astronaut at the age of 40. He had a long and distinguished career beginning with his Naval Aviation Cadet Training in March of 1942. He flew 59 combat missions in the Pacific Theater and another 63 during the Korean War, and shot down three MiGs in the last nine days of that war. He returned to the United States to attend the Test Pilot School at the Naval Air Test Center, Patuxent River, Maryland.

Glenn came into the public eye in July 1957 when he flew the Navy's new F8U Crusader fighter/reconnaissance jet from Los Angeles to New York in 3 hours and 23 minutes—the first transcontinental flight to average supersonic speed. He was well liked by his fellow astronauts, although he sometimes rubbed them the wrong way because he was very straight laced and was known to preach about good moral character to his more lusty fellow astronauts—he was not the typical Marine.

The 4,250-pound spacecraft serial number13, christened *Friendship 7* by Glenn, was mated to Atlas 109D on January 3, 1962, to complete the MA-6 configuration. The first flight date of January 23 slipped to the 27th because of problems with the Atlas. After lying on his back for over five hours in the spacecraft, the shot was postponed on that date because of bad weather. During the recycling process, a propellant leak discovered in the Atlas caused another two weeks to slip by. Then, still more problems and bad weather moved the launch date out another few days. Because of the need for acceptable weather at three primary locations—the Cape, the abort area, and the recovery area—weather played a major role in the early American space flights.

Glenn was awakened at 2:20 A.M. on February 20, 1962. There had been much speculation in the press that the lengthy delays might have put too much psychological stress on Glenn and that the anticipation would be too much for any man to bear. However, the press and the public had much to learn about the emotional make-up of the Mercury seven astronauts. There was no problem on Glenn's part as he ate breakfast, had his vital signs examined, and suited up for the flight. Only the addition of the blood pressure cuff on his right arm differentiated Glenn's preparation from that of Grissom and Shepard.

When Glenn arrived at the launch pad, a problem encountered with the launch vehicle delayed by one hour his elevator ride to the top of the giant gantry for the squeeze into the spacecraft. Then, with the various connections made between the spacesuit and the spacecraft, the "closeout" procedures were completed. However, a broken bolt, one of the 70 that fastened the hatch, had to be replaced—more delay.

Finally, at 9:47 A.M., the missile came to life when the countdown reached zero! From launch pad 14, the great silver missile, with its band of pure white frost defining the location of the liquid oxygen tank, began to rise, balanced on its orange-sheathed flame. Leaving its colorful backdrop of the red service tower and surrounding green shrubs, it accelerated into the blue morning sky. *"Roger, the clock is operating. We're underway,"* Glenn announced. *"God's speed John Glenn,"* replied the voice of CapCom (Capsule communicator) Alan Shepard.

Riding the Atlas was a new experience as it was unlike the Redstone; indeed, it was unlike the Russian R-7. Glenn would recall that he could feel the small corrections that the powerful engines induced as they moved to balance the 250,000-pound rocket. He described it as being at the end of a long swaying pole. But this pole was also vibrating as it surged into the atmosphere building G forces on Glenn and creating an aerodynamic shock wave as it pushed the air out of its way. *"A little bumpy along about here,"* Glenn reported in the understated language of a test pilot. Through Max Q, thirty seconds into the flight the vibrations smoothed a bit.

At T+129 seconds BECO occurred and the two booster engines slid away as the single sustainer continued to power the spacecraft into the sky, which had now turned from a bright blue through dark blue and quickly into the blackness of space. The G forces dropped from six back to three with the change in thrust and then began to build again as the rocket consumed it propellants at an astonishing rate.

The escape tower appeared to jettison early as Glenn reported, *"I saw the smoke go by the window,"* but that was a visual queue from some other unnamed event, and the tower finally left on schedule and was plainly visible. *"There, the tower went right then! Have the tower in sight way out."* This was an important event because it indicated that a malfunction at this point would be handled by simply separating the spacecraft from the Atlas. It was also a critical event because the weight of the tower had to be shed in order for the Atlas to reach orbital speed.

Glenn continued to report to Mission Control not only his observations of the launch phenomena but readouts from the various systems. At T+280 seconds Mission Control responded *"Roger, twenty seconds to SECO;"* it was as much a wish that all that was now needed to achieve a milestone was within their grasp. SECO, sustainer engine cut-off, was an indisputable event as the 7.7 Gs and the sound of the engine suddenly disappeared. Glenn then commented, *"Roger, the capsule is turning around and I can see the booster doing turnarounds a couple hundred [yards] behind me. It was beautiful!"* The experience of sight and sound has to be one of the most enthralling that man has ever witnessed.

Mission control now related another important item. *"Roger Seven, You have a go, at least seven orbits."* This particular communiqué has been often misinterpreted as the intended length of the mission. This was not the case. It had always been planned for three orbits. The statement was to advise Glenn that the computer at Goddard Space Flight Center had analyzed the velocity vectors imparted by the Atlas, concluded that the 100-mile by 163-mile orbit was stable, and would result in at least seven orbits— assuring a premature reentry would not occur.

From the Bermuda tracking station, Grissom took up the CapCom duties as Glenn continued to comment on the beauty and visibility. More instrument readouts followed as the capsule maintained its rearward facing, 34-degree-pitch-down attitude with the automated control system until a series of telemetry readings to Goddard confirmed the integrity of all the spacecraft systems. Over Africa Glenn switched to Fly-by-wire attitude control and reported that he had no problem positioning the spacecraft using outside references, especially when the pitch down was increased to about 60 degrees.

Moving out across the Indian Ocean, and now communicating through one of the tracking ships, he moved into his first sunset. Glenn again was in awe of the visual impact of his high-flying perspective *"...the sunset was beautiful. It went down very rapidly,"* and he then added, *"I have no trouble seeing the night horizon,"* and speculated that Moonshine from behind him was probably illuminating that portion of the earth.

Halfway around the earth from the start of his journey he continued to make observations while out of range of the tracking ship, but they were recorded by an on-board tape unit: *"Friendship Seven broadcasting in the blind, making observations on night outside...I can identify Aries and Triangulum* (constellations of stars)."

Over Australia Glenn observed the lights of the city of Perth as its residents had turned on as many as possible in their early morning hours for his viewing. He asked CapCom Gordon Cooper at the Muchea tracking station to thank the people of Perth. Then his first sunrise in space evoked the comment, *"...Oh the Sun is coming up behind me in the periscope, a brilliant, brilliant red..."* He then proceeded to describe thousands of brilliant luminescent particles (that became known as "Glenn's fireflies") that seemed to swirl around the spacecraft but quickly disappeared when full sunlight bathed Friendship 7.

Glenn now encountered the first of several problems. The spacecraft was not holding its position relative to the "yaw" axis (left and right), and he came to the conclusion the low power (one pound) left thruster was not functioning. There were two sets of thrusters for each axis, and the low powered units were used for small adjustments. However, with the left thruster out, the automated control system would use the larger thruster when the yaw passed the 20-degree tolerance point. Glenn switched to the manual "fly-by-wire" system so that he would not consume so much hydrogen peroxide attitude-control fuel.

At about this time, two hours into the flight, Goddard Space Flight Center noted a telemetry code of "Segment-51," the indication that the landing bag had deployed. As the bag was contained between the heat shield and the bulkhead of the capsule, it could only mean that the heat shield had unlatched— if the Segment-51 indication was correct. The retrorocket package with its three metal straps was probably still in place over the heat shield and secured directly to the bulkhead of the spacecraft. However, when that unit was released after retro-fire, it would separate from the bulkhead if the heat shield was lose. If this occurred, it would undoubtedly be torn away during the reentry and the spacecraft would burn up from the incinerating 1,600 degree centigrade temperatures.

Not wanting to alarm the astronaut, Mission Control proceeded to make oblique queries as to the status lights that Glenn was observing and on occasion asked if he heard any noises when the capsule attitude was changed by the thrusters. Back at mission control, queries were made to determine if the Segment-51 signal was valid and possible alternatives if it was. The validity of the signal could not be confirmed, and they thought that it was probably erroneous. However, Flight Director Chris Kraft could not make that assumption, and a contingency plan was examined to leave the retrorocket package in place after it fired. It was felt that the straps would hold the heat shield in place until enough aerodynamic force had built that would probably keep it there after the straps for the retro-module melted through.

However, there had been no evaluation made as to what the effect of having the retro-module in place would be to the heat shield as a result of an uneven heat pulse generated by the module. And, this was only an option if all three of the retro's solid-fuel rockets fired. If one failed to fire, then the reentry heat would eventually ignite it, and its exhaust might burn through the heat shield or cause the spacecraft to assume an attitude that would allow the heat pulse to impinge on the side of the spacecraft.

Although Glenn had responded to the landing bag light queries, Mission Control had not confided in him their worst thoughts. With all of the rapid communications going on, Canton Tracking Station assumed that Glenn had been made aware and finally let the cat out of the bag. "*Friendship Seven, this is Canton. We also have no indication that your landing bag might be deployed. Over.*" Glen immediately responded, "*Did someone report landing bag could be down? Over.*" "*Negative,*" reported Canton, "*we had a request to monitor this and to ask if you heard any flapping.*" Now some of the other communications that Glenn had fielded began to make sense to him, but he thought perhaps the "fireflies" he had reported had started the queries, and he again gave more description of them.

Over Hawaii, as the time for retro-fire approached, Glenn was told of the Segment-51 indication and given some more checks to make. They came back negative. There was no confirmation that the bag had deployed. Glenn put Friendship 7 into the Retro-fire attitude (flying backward with the capsule pitched nose down by 34 degrees), and the three solid-fuel rockets fired as scheduled. Now Flight Director Chris Kraft made his decision, and it was communicated by Texas Capcom to Glenn. "*We are recommending that you leave the retro-package on through the entire reentry.*" Now Glenn wanted the complete story. "*What is the reason for this? Do you have any reason? Over.*" However, all he got was, "*Not at this time; this was the judgment of Cape Flight.*" With only about four minutes until the ionization sheath of reentry would block any more radio communications, Glenn realized any clarification would have to be swift. It was—as the capsule came into range of the Cape CapCom, "*... we are not sure whether or not your landing bag has deployed. We feel it is possible to reenter with the retro-package on.*" There was no time for Glenn to reflect on the engineering aspects of a statement that began with "we feel…" The Cape advised Glenn that he would have to perform some of the reentry tasks manually, such as the periscope retraction, to bypass the retro jettison event. The first of these occurred with the detection of the upper fringes of the atmosphere when the .05 G sensor light flashed on.

Now yet another problem confronted him—only 15% of his attitude control fuel remained, and the most critical part of the mission was ahead of him. The pitch attitude of Friendship 7 had to be held to within 10 degrees during reentry. With the realization that there might be a serious problem with the heat shield, Glenn elected to concentrate on those things that he could control and to move those that he couldn't out of his conscious thoughts.

As the G forces and ominous sounds began to build, the super-heated ionized air choked off the communications with Mission control. Glenn was on his own now, and he could see the orange glow surrounding his capsule plainly through the window. A part of the energy that the Atlas rocket had generated to get him into orbit at 17,500 miles per hour now had to be dissipated by friction with the upper atmosphere for his return. He watched as molten pieces of what he hoped were the retro-package began flashing past, and he could feel the heat building on his back—which was less than a foot from the searing temperature of the heat shield. He also kept making his observations into the tape recorder with one comment that seemed to sum up his feelings of the moment. *"That's a real fireball out there!"*

As the glow outside the window subsided, Glenn thought he had made it through the toughest part. But just then, the fuel supply became exhausted, and the capsule began to oscillate. He debated deploying the drogue chute to help stabilize the craft as he had now decelerated to less than 700 miles per hour. However, the automatic sequencer beat him to it, and the main chute soon followed. He established communication with the recovery forces and again repeated his observations of the reentry: *"My condition is good, but that was a real fireball, boy. I had great chunks of that retro-pack breaking off all the way through."*

The landing was uneventful, and within minutes the U.S. Navy Destroyer *Noah* had retrieved *Friendship 7*. Glenn was eager to exit the hot confines of the spacecraft as it sat on the deck of the ship. Rather than take the long way out through the egress hatch at the top, Glenn advised the crew to stand clear, and he blew the hatch. In striking the plunger to initiate the explosive charge, Glenn bruised the palm of his hand and cut his knuckles. This had also happened during training, and it was a clear indication that Grissom (whose hands showed no injury) had not used the plunger during his ill-fated landing in *Liberty Bell 7*.

America was ecstatic with the results of the MA-6 mission. The entire country had come to a virtual standstill during the four hour and fifty-five minute flight as many stayed home from school and work to keep up with the events as they unfolded. Even though there was no live TV of the spacecraft in orbit nor of the splashdown, the running commentary of Colonel "Shorty" Powers and the "talking heads" of such enthusiastic anchormen as Walter Cronkite kept the nation and the world appraised of each aspect of the flight.

During the debriefing, it was clear that Glenn was not pleased that Mission Control had not kept him informed of the heat shield problem. But when the various factors were considered, such as the truly unknown environment in which Glenn was flying, the decisions made by Mission Control appeared reasonable. Nevertheless, Glenn and the other astronauts made it clear that, from that point on, the pilot-in-command of the spacecraft must be given all of the facts related to the flight so that a balanced evaluation could be made. As for the Segment-51 light, it turned out to be an erroneous indication that resulted from a loose rotary switch. The thruster problem turned out to be a small metal shaving just as it had with Enos in MA-5. Better clean-room facilities and a more cautious work ethic were emphasized.

Glenn's flight had clearly demonstrated that man could function in space. He established detailed and accurate observations, cognitive evaluation, and judgment. While Titov had experienced space sickness and had been somewhat incapacitated for more than half of his mission, he too had prevailed emotionally until his physiology had stabilized.

John Glenn, accompanied by the other six astronauts, took part in a big ticker-tape parade down Broadway in the heart of New York. The cheers effectively drowned out the skeptics and those who still felt that a massive manned space program was the wrong response to the Russian threat. However, the fact remained that both the United States and the Soviets had encountered potentially serious problems. That neither had lost an astronaut was fortuitous. America would move ahead cautiously, but the Soviets needed to continue to demonstrate a commanding lead.

"I'm out of manual fuel...": MA-7 Carpenter

At the same time that Glenn had been named to MA-6 in November 1961, astronaut Deke Slayton had been assigned to MA-7. However, Slayton had displayed a mild idiopathic atrial fibrillation during a routine examination—essentially an irregularity in the heart's ability to cope with stress. NASA had sent Slayton to several specialists and the results had always come back the same; he was physically fit to fly. Nevertheless, NASA continued to elicit medical opinions until finally one summary concluded that so long as there were other astronauts equally qualified who did not have any problem, then Slayton should not fly.

Slayton did not accept the decision with grace, but when he realized he was up against a power structure that could not be moved, he resigned himself to perform in a job he could adapt to. He was assigned the role of Chief Astronaut—the man who would select, manage, and discipline the astronaut corps for more than ten years. He would be known as a tough decisive taskmaster. His opportunity to fly would eventually come, but for now, his slot for the next orbital mission would pass to Navy Lieutenant Malcolm Scott Carpenter. Carpenter had the least jet time of any of the astronauts because he had been primarily a multi-engine-patrol-bomber pilot since completing naval flight training in April 1951.

Although MA-7 was to be an identical flight profile to MA-6, Carpenter was to expand the role of man in space, and to that end the schedule of scientific and engineering experiments was impressive. America, despite the desperation race, was still pressing to do science in space. The only major change in the spacecraft configuration was the deletion of the earth-path indicator since Glenn indicated that it was relatively easy to determine position from outside references, and the communications network had worked well.

Spacecraft 18, mated with Atlas 107D, was named Aurora 7 by Carpenter for the light that this mission would shine on the dawn of a new age. The launch, originally scheduled for May 15th, was delayed by problems with the booster, the spacecraft attitude control system, and modifications to the parachute deployment mechanism until May 24, 1962. Unlike previous Mercury Atlas flights, the countdown went smoothly, and an early morning liftoff occurred at 7:45 A.M.

Carpenter was not as expressive as Glenn had been and even more understated as he passed through Max Q. As each booster had its own personality, it is possible that Carpenter was given a smoother ride than Glenn. On arrival in orbit Carpenter noted that the horizon sensor was off by about 20 degrees, a very significant amount. Also, he had to adjust the suit temperature continually, which ran on the high side. With these distractions, Carpenter set about accomplishing each of the tasks on the flight plan, and by the end of the first orbit he had kept up with the schedule.

The Russians had by this time given some indications that Titov had experienced some space sickness, but Carpenter felt completely at home in this new and strange environment just as Glenn had been. Some attribute this to the extensive weightlessness training the American astronauts received and the fact that the Mercury capsule had a much smaller interior so that the astronaut felt more an integral part of the ship rather than a passenger. Carpenter even tried to induce vertigo by performing sharp head movements. His visual acuity appeared to improve in space, perhaps because of the weightless condition, and he reported being able to see the dust trail from trucks driving along unimproved roads and the wake of ships at sea.

Carpenter was able to observe Glenn's "fireflies," and this took more of his time than he realized. He determined that, if he hit the inside of the ship, he could generate the luminous particles and that the hydrogen peroxide thrusters also generated them. *"I can rap the hatch and stir off hundreds of them,"* he reported. He soon found himself behind schedule with all of the activities (some of which were poorly planned by the experimenters). A few times he accidentally bumped the manual control system causing it and the automatic system both to be active, which was depleting the thruster fuel more quickly than anticipated.

By the time he was ready for retro-fire, there were several uncompleted tasks left to prepare the spacecraft for reentry. Moreover, when the retrorockets failed to fire by the automatic sequencer, Carpenter's manual response placed the ignition about three seconds late. This in itself was not a major problem. But the fact that the spacecraft was not properly aligned was, it being about 25 degrees off from the proper azimuth (the retros being pointed to the left of the flight path). The combination of all of the factors would put his splashdown almost 300 miles beyond the intended point where the recovery forces were located.

As he passed over the west coast of the United States, he ominously reported to CapCom Alan Shepard, *"I'm out of manual fuel, Al."* Carpenter was potentially in big trouble. He allowed the spacecraft to drift and waited for the .05G indication before attempting to align the ship to the proper reentry attitude. He realized that at any time now the fuel supply for the automated system, which he was using in the fly-by-wire mode, would also become exhausted. However, by the time that occurred, the spacecraft had driven itself into the denser layers of the atmosphere and had achieved a degree of stabilization. As the ship became subsonic, it began to oscillate as Glenn's had done, and Carpenter elected to deploy the drogue chute early.

The last communication he had heard from the Cape was that he would be landing long and that it would be an hour before recovery forces would be available. After splashdown, Carpenter elected to egress through the top hatch to avoid any possibility of losing his capsule as Grissom had. He inflated the life raft and proceeded to climb in. He noted, as his feet began to get wet, that he too had neglected to shut the suit inlet hose valve, but he quickly remedied the problem. He lay in the raft for over two hours before being retrieved by helicopter.

While his flight was essentially a complete success, Carpenter's performance came into question. That he had been given the most demanding experimental schedule, which with few exceptions he had handled quite well, did not seem to impress NASA management. That he had run out of fuel during reentry and had landed long, causing considerable distress to those whose heads would roll if he were not successfully recovered, seemed to be the primary focus. His independent spirit also played a role, as his attitude appeared to be too cavalier to suit the more conservative elements in Mission Control.

Carpenter would never again fly in space. He recognized the warning signs and subsequently took a leave of absence from NASA, to participate in the Navy's Man-in- the-Sea Project as an Aquanaut in the SEALAB II program in the summer of 1965.

Group Flight: Vostok 3 and 4

Following Titov's flight, the emphasis for the available R-7s and Vostok spacecraft (of which 18 were in various stages of construction) was switched to the Zenit reconnaissance satellite. However, Korolev continued to remain focused on manned space flight. He envisioned that each flight must be a significant step beyond the previous one—Titov's experiences not withstanding. Within a month after completing Titov's flight, Korolev began planning a group flight of three Vostoks launched on successive days, with the longest remaining in space for at least three days.

However, Korolev continued to be restrained by others who felt that two ships would be all that the current tracking and recovery facilities could effectively handle and that two days should be the limit because of the unknowns revealed by Titov's space sickness. And, although Korolev was eager to launch as soon as the Zenit schedule freed up the launch facilities, the spectacular success of Glenn's flight in February of 1962 suddenly put pressure from above to accelerate the next Vostok mission. The enthusiasm shown by the world's media towards the openness of the United States' efforts put the Soviets in a defensive position. The edict to launch by the middle of March 1962 was unrealistic, but the process was begun.

Two Zenits had been planned before the next Vostok booster could be erected on the launch pad, and these were delayed by a series of problems. Failures in December of 1961 and again in January 1962 put that program behind schedule. The first was finally in orbit on April 26, 1962, but the next, on June 1st, experienced a catastrophic failure that damaged the launch pad, and it required more than a month of rework before it was operational.

The objective of the group flight, as it was being evolved, was for the two spacecraft to be launched in such a manner that they would pass very close to each other. Because Vostok had no on-board propulsion, they could not actively rendezvous. Even the possibility of the two cosmonauts being able to observe each other's ship would be a monumental achievement. The length of the mission was also still cloudy as Korolev pushed for 3-4 days, but it was officially to be a two-day flight at that point.

The Cosmonauts were chosen less than two weeks before launch and included Captain Adrian G. Nikolayev and Major Pavel R. Popovich, who were both then 32 years old. Nikolayev was launched first in Vostok 3 on August 11th at 11:30 A.M. into an orbit of 113 miles by 147 miles and an inclination of 64.98 degrees. The medical team at Tyura-Tam closely monitored his condition, and the world was informed of the ongoing mission after he completed the first orbit. However, Nikolayev reported no problems, and on a subsequent orbit the flight plan called for him to un-strap himself so that he could float freely within the cabin (something that the much smaller Mercury capsule would not allow). In an effort to blunt some of the criticism of Soviet secrecy, live TV pictures of Nikolayev were routed for broadcast within the USSR. There was no capability to provide live coverage across the world at that time.

As soon as the rocket had left the pad, the launch crew began its restoration, and the Vostok 4 booster rolled out. Popovich was launched at 11:02 A.M. on the 12th into a 112 mile by 148-mile orbit with an inclination of 64.95 degrees. The precision with which Korolev's team was able to launch and the orbital insertion parameters achieved were impressive. The world was not told the nature of the mission nor that rendezvous was out of the question. The Soviets were not about to clarify their intentions. The media assumed that the two ships would rendezvous and gave the Soviets another dramatic and technological first. Korolev's First Deputy, Vasiliy Mishin, would note in later years: *"...with all the secrecy, we didn't tell the whole truth, the Western experts, who hadn't figured it out, thought that our Vostok was already equipped with orbital approach equipment. As they say, a sleight of hand isn't any kind of fraud. It was more like our competitors deceived themselves all by their lonesome. Of course, we didn't shatter their illusions."* The two ships never came closer than three miles and then gradually drifted farther apart.

Because Nikolayev's physical condition was not showing any signs of space sickness, the decision was finally made to allow him to go for the third day, and after the completion of that, there was more disagreement over the fourth day which he finally was allowed to complete. Khrushchev requested a fifth day, but several problems, such as a steadily decreasing temperature in the spacecraft, finally prompted the recovery on August 15th after 3 days and 22 hours. Popovich followed just six minutes later after 2 days and 22 hours. Korolev had again upstaged the United States.

"She's riding beautiful": MA-8 Schirra

The Mercury spacecraft had been engineered to simply get an American into orbit, and three orbits back in 1959, when Mercury was being designed, was a big step. But Titov's day-long, 17-orbit flight had upped the ante, and NASA carefully reviewed the Mercury design to see if it could provide for more extended operations. Simply another three-orbit flight would prove little and would be seen by the world as evidence that America's technology was still woefully behind the Russians. Thus, MA-8 sought to expand the endurance capabilities until the next generation spacecraft was available.

The obvious problem in increasing the time in space was the availability of the consumables. Most obvious was the attitude-control fuel—the hydrogen peroxide that had been totally depleted by both Glenn and Carpenter. However, the environmental supplies of oxygen, water for the cooling system, and lithium hydroxide for removing the carbon dioxide from the air were also limited. And finally, the batteries for the electrical systems could not be readily extended, especially since the Atlas was already at the limit of its ability to boost the 3,000 pound craft into orbit; the physical confines of the capsule left virtually no room for any appreciable expansion.

With respect to the hydrogen peroxide, there were several approaches to making it last longer. First was more discipline in its use and the structuring of the flight plan and the experiments. Drifting was an option for extended periods. The tolerance that the ship could be allowed to drift (called the *deadband range*), when a desired attitude was commanded, was expanded from ±3 degrees to ±6 degrees. The electrical system had several components that could be powered down when not in use, and it was determined that this technique itself would double the battery life.

The biggest problem with the oxygen was the leakage rate of the spacecraft itself which was replenished from the two cylinders. The initial design requirement of 1,000 cubic centimeters per minute in the pressurized mode (in space) was reduced to 600 cc/min., and 460 cc/min was actually achieved.

Despite all the effort, the weight of spacecraft 16 increased by almost 50 pounds as it was mated to Atlas 113D. Navy Lt. Commander Walter Schirra had been selected more than a year earlier for the flight and had served as the back up for Carpenter. He could be terse and outspoken, but like Grissom, he held a strong belief in good engineering principles. A Naval Academy graduate, class of 1945, he flew 90 missions during the Korean War and received the Distinguished Flying Cross. He was perhaps the most well prepared astronaut to date as he had spent countless hours in the simulators as well as the spacecraft itself, going through the various activities. He was determined to fly the perfect mission.

Aiming for a September launch, the Americans were once again upended by Korolev with the group flight of Vostok 3 and its four-day mission. A series of slippages finally saw Sigma 7, as Schirra had named the capsule in honor of the sum total of the American effort, take to the sky on October 3, 1962. *"Ah, she's riding beautiful, Deke,"* he commented to CapCom as he "slipped the surly bonds of earth."

The booster engines did not perform well, and the sustainer had to fire for a longer period to make up for the deficiency, using fuel that the boosters had not burned. SECO occurred 15 seconds later than in the previous missions. The velocity vector was sufficient to place Schirra into a 100 by 176 mile orbit. Other than a problem with the suit temperature (similar to what Carpenter experienced), the flight proceeded smoothly as he demonstrated discipline over the use of thruster fuel and powered down unnecessary equipment to conserve the electrical power. Schirra was the most vocal of the astronauts, continually chatting away into the recorder when he was out of range of a tracking station. Even with the addition of three more tracking ships to the Mercury communications network, the 10 minute period that had been the limit of being out of communication range had been expanded to 30 minutes as the track of the craft for the period of 6 orbits covered a much wider swath of the earth. Because of the extended flight, the splashdown shifted to the Pacific, and Schirra reentered with more than 50% of his fuel remaining. As he descended under parachute, the excess fuel was vented as it represented a hazard to recovery personnel. He landed less than 4 miles from the primary recovery ship to the cheers of the crew. He elected to remain in the craft as it was hauled aboard the carrier *Kearsarge* and then blew the hatch. He received the same injury to his hand as Glenn had, again vindicating Grissom. He had flown the perfect mission thanks to the pioneering efforts of Shepard, Grissom, Glenn and Carpenter and the tireless work by NASA and its contractors.

Lots of Problems: MA-9 Cooper

With Schirra landing with an excess of consumables, the planning for the last Mercury flight, MA-9, swung into high gear. It was to be Air Force Captain Gordon Cooper's turn, the youngest and last of the seven who had yet to fly (save for the grounded Deke Slayton). While Cooper had not been quite as outspoken with his spiritual beliefs as Glenn had, he was perhaps the most philosophical of the astronauts. He named his spacecraft *Faith 7, "as being symbolic of my firm belief in the entire Mercury team, in the spacecraft which had performed so well before, and in God."*

The flight plan was quite aggressive for the small craft as it called for more than a day in orbit—actually a total of 33 hours. One orbit more than Titov's, it reflected the growing confidence of the NASA Manned Spacecraft Center that had recently moved into the new facilities built in Houston Texas.

The heavy, 75-pound periscope was removed as the previous flight pilots felt its value was not worth the weight. The Rate Stabilization and Control System (RSCS), one of the four attitude control modes, was also removed saving another 12 pounds. In their place two of the 1,500 watt-hour batteries were replaced with 3,000 watt-hour batteries, and another auxiliary tank with 16 pounds more hydrogen peroxide was added. The oxygen supply was increased along with water for the environmental cooling and for drinking. The thrusters were replaced with a more efficient and lower powered set. In all, there were more than 183 changes to the spacecraft to make it suitable for the extended mission. Yet Cooper's capsule weighed only four pounds more than Schirra's.

Atlas 130D with spacecraft 20 in place was on pad 14 by April 22nd. Then, a series of delays occurred, including a scrubbed flight on the 14th of May, after Cooper had spent four hours on his back in the capsule—a result of problems not directly associated with the Atlas or the Mercury spacecraft. At 8:04 A.M. on May 15, 1963, the last Mercury flight lifted off, and five minutes later SECO occurred. As had happened with great regularity with previous flights, the temperature of the interior of the spacecraft began to reflect the heat that had been generated on the outside of the spacecraft by the acceleration through the lower layers of the atmosphere. Cooper's suit was unable to compensate for the 118 degree F cabin temperature. However, after several orbits, and some patience and sweat by

Cooper, the interior temperature fell to 95 degrees while the suit reflected a much more comfortable 70 degrees.

Cooper observed Glenn's "fireflies" with each sunrise and transmitted the first TV pictures of himself at two frames per second. With lots of time available, Cooper also spent considerable time evaluating the topography of the earth and, like Carpenter, claimed to be able to see extremely small details including the smoke from a steam locomotive as he passed over India.

With a much more organized flight plan, Cooper was also able to conserve fuel and powered down unneeded equipment, and by the fourteenth orbit all was going exceptionally well. In marked contrast to the propaganda broadcasts made by the cosmonauts and the anti-American rhetoric that emanated from them, Cooper transmitted a short prayer: *"Father, thank you for the success we have had in this flight. Thank you for the privilege of being in this position, to be up in this wondrous place, seeing all the many startling, wondrous things You have created. Help, guide and direct all of us that we may shape our lives to be good, that we may be much better Christians..."*

Within a few more orbits, Cooper was going to need some of that divine guidance as the .05G light came on. This was supposed to indicate that the spacecraft had sensed the upper limits of the atmosphere. Mission Control confirmed the orbit was stable, and that it was an erroneous indication. More investigation revealed the signal had initiated the reentry activities (as it should, if it had been valid) and that several events would have to be done manually to re-sequence the process. At that point it was found that both gyroscopes and the horizon scanner were without power. The attitude control for the reentry would have to be performed manually.

On the 21st orbit, the ASCS main 250 v-amp inverter blew a fuse. This unit converts the Direct Current (DC) of the batteries to Alternating Current (AC) for some of the electronics. Switching to the back up, Cooper discovered that whatever had killed the primary had also taken out the secondary unit. This essentially incapacitated all automated sequences.

Cooper now reported that the Carbon Dioxide (CO_2) level was building up. As he had exhibited such calm demeanor towards all of these problems, Mission Control advised him to take a Dexedrine tablet to arouse his awareness level. However, Cooper's response simply reflected a man who was unflappable in the face of trouble—one of the reasons they had chosen him as an astronaut!

Cooper's would be the first Mercury flight to return solely using the Manual Proportional mode with the astronaut controlling all aspects of the reentry including the firing of the retrorockets. When the parachute blossomed over the blue Pacific, *Faith 7* was just one mile from the intended splashdown point.

Although there were two more capsules and Atlas boosters available, NASA decided there was little to be gained from another flight, and the effort and dollars should be put into the next generation spacecraft being designed.

Lady Cosmonaut: Valentina Tereshkova

With the success of Titov's flight and before planning for the first group flight of Vostok 3 and 4 had been formalized, Lt. General Nikolay Kamanin, chief of the cosmonauts, put forth the notion of sending a woman into space. While he met with much dissent, his stated objectives supported the current Soviet space philosophy in that eventually a woman would fly in space, and it must not be an American. The flight of a woman under the Communist banner would strengthen the Communist party among women of the world. After a selection process that involved combing the parachute and flying clubs for possible candidates, five women were selected for cosmonaut training by April of 1962 (this was reduced to four when one dropped out).

After the Nikolayev and Popovich group flight, the plan began to come together by March of 1963. A male cosmonaut would launch for a long duration (8-day) flight. A few days after his launch, the woman would orbit for 2-3 days. Despite the rhetoric of the opportunities for women in the Soviet Union, it was generally understood that this would be the one and only flight and that the rest of the team would not have another opportunity. Thus, the competition was intense among the four remaining women.

It is interesting to note the criteria that General Kamanin used in making the selection: *"Ponomareva has the most thorough theoretical preparation and is more talented than the others... but she needs lots of reform. She is arrogant, self-centered, exaggerates her abilities and does not stay away from drinking, smoking and taking walks* [a euphemism for promiscuous activity]. *Solovyena is the most objective of all, more physically and morally sturdy but she is… insufficiently active in social work* [not an ardent Communist]. *Tereshkova…is active in society* [being a member of the Young Communist League]…*is especially well in appearance. Yerkina is prepared less than well… We must first send Tereshkova into space flight… she is a Gagarin in a skirt."*

During a visit of the female cosmonauts to Tyura-tam to witness the launch of Vostok 3/4, Tereshkova had caught the eye of Korolev and became his favorite choice as well. Although Ponomareva had cultivated a loyal following, in the end Tereshkova was selected.

None of the first four Vostoks had included any significant scientific experiments or observations. But this would change with the selection of Senior Lieutenant Valeriy F. Bykovskiy and his long duration flight that would include a series of photographic experiments.

A variety of problems confronted preparation for the mission, including the fabrication of a space suit for a female. Unlike the smoothness of the countdown that characterized other launches, the R-7 for Bykovskiy's Vostok 5 had many difficulties. Ranging from radios to gyroscopes, the problems brought Korolev and his deputies to harsh words.

As the delays moved the launch into mid-June, the estimated shelf life of the assembled Vostok spacecraft became an issue. Unless it was flown by August, it would be declared un-flightworthy. Just when it appeared that the launch would take place on June 11th, a significant Solar flare occurred causing a high level of radiation in the upper atmosphere—another delay of a few days. As the countdown proceeded towards ignition on the afternoon of June 14, 1963, numerous small problems continued to plague the rocket. When a power cable umbilical failed to eject minutes before the launch, frustration and impatience dominated Korolev's thinking. He decided to proceed; the cable was ripped from the missile as it rose into the sky.

The R-7 produced less than nominal performance, and Vostok 5's orbit of 109 by 139 miles made the likelihood of a full 8-day mission questionable. Nevertheless, Bykovskiy proceeded to work through the lengthy flight plan that included a brief period of manually orienting the spacecraft and much photography. At one point a film canister jammed in a camera, and he discovered that yet another canister had no film at all.

Tereshkova's pre-launch preparations proceeded much smoother than Bykovskiy's, and she followed him in Vostok 6 three days later on June 16, 1963. Her orbital inclination, while within .02 degrees of Vostok 5, was intentionally launched out-of-plane so that there was no possibility of a rendezvous, even if they had been capable of orbital maneuvers. They did come within several miles of each other on a few occasions, but otherwise it was somewhat anti-climactic compared to the Vostok 3/4 flight. The big news for the world, of course, was that the Soviets had orbited a woman; another "first" and a propaganda triumph had been recorded for the Russians. In another anomaly of Soviet thinking, the TASS announcement indicated that Tereshkova was a civilian, when in fact she had been given a commission in the Soviet Air Force as a Junior Lieutenant for the purpose of the flight.

Knowing that the Americans had developed very sophisticated methods of intercepting communications, the Russians prepared a series of coded phrases that the cosmonauts would use to indicate if there were problems. Using the term "feeling excellent" would indicate just that—all was well. But the expression "feeling well" would be cause for concern as there was some aspect that could cause early termination of the flight. The most ominous phrase, "feeling satisfactory," would require immediate termination at the first opportunity.

Live television of Tereshkova highlighted the flight and thrilled the Russian citizenry. However, she began to experience space sickness within the first few orbits. Consideration was given to bringing her down early, but she asked to continue the flight with the knowledge that Titov's equilibrium had returned to a more normal state after the first 12 hours. Two days into the flight there were strong indications that she was still unable to perform adequately when she failed to manually adjust the attitude of the spacecraft. With the help of Gagarin and Titov radioing up instructions, she was finally able to accomplish the basic maneuvers that would be necessary if the automatic system failed and she had to position the spacecraft.

Tereshkova was brought back to earth after three full days. During the retro-fire and reentry, she did not communicate at all. But she had logged more time in space than all six of the Mercury astronauts.

Bykovskiy's reentry was more event-filled. His spherical reentry module again failed to separate clean-ly from the instrument compartment in a situation similar to that of Gagarin and Titov. He had spent almost five days in space.

During the post-flight debriefing, Tereshkova noted of her incapacitation that, *"removing the film was very difficult. I didn't conduct any biological experiments—I was unable to reach objects."* Yet she also added, *"Weightlessness did not arouse any unpleasant sensations...I threw up once but that was because of the food, not to any vestibular disorder."* A critical review of the flight, which was kept secret for decades, revealed that virtually all concerned, including Kamanin and Korolev, considered her performance inadequate. There has been some speculation that this judgment was influenced by possible gender bias.

The completion of the Vostok 5/6 group flight effectively closed out the Vostok manned flight oper-ations. Korolev wanted to turn his attention to the development of the new super booster and a more advanced spacecraft capable of orbital rendezvous—and his health was failing. Russia, at the end of the sixth year of the space race, still appeared to have a commanding lead over the Americans. As 1963 ended, so did both the Vostok and Mercury programs after each sent six flights into space. Both coun-tries had demonstrated that man could function in space and that the spacecraft and its systems had exhibited a resiliency and reserve capability to overcome serious problems.

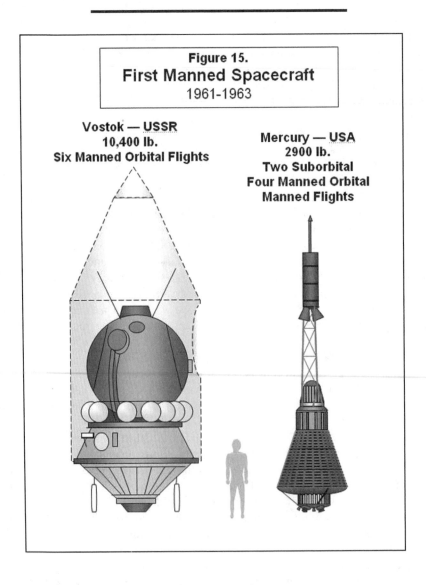

Figure 15.
First Manned Spacecraft
1961-1963

Vostok — USSR
10,400 lb.
Six Manned Orbital Flights

Mercury — USA
2900 lb.
Two Suborbital
Four Manned Orbital
Manned Flights

Early Studies

Ever since man began to have conflicts, the ability to secure the "high ground" has provided an advantage in combat. It was recognized, even before the Wright Brothers flew, that when a flying machine was invented it would play an important role in military affairs. Therefore, it was not at all surprising that studies in the late 1940s and into the 1950s indicated that a military presence in the coming space age could be a deciding factor in a nation's ability to survive. President Dwight Eisenhower was keenly aware of this and supported military satellite reconnaissance programs (discussed in an earlier chapter) even before Sputnik. His administration funded more advanced investigations into the usefulness of this new "high ground" after the surprise of Sputnik in 1957. But Eisenhower was reluctant to authorize offensive weapons systems as he realized the consequences of the arms race moving into space. It quickly became apparent that unilateral restraint would not prevent the Soviets from seeking to determine what the various offensive capabilities might be, and research into several weapons systems was subsequently authorized by Eisenhower.

But direct applications in space for combat were only conceptual in the early days of man's movement into this new frontier as it was not clearly understood what function he might serve in that hostile environment. The early post-Sputnik thinking pointed towards direct applications of weapons residing in space, and this was one of the leading arguments for allowing the military to control all operations in space. However, Eisenhower rejected that notion and placed NASA in charge of the national space program. Nevertheless, the military was far from being left out. Several military programs were begun in the late 1950s and into the early 1960s that were far more perceptive in determining what man was capable of doing in space, compared to automated satellites, than NASA was willing to concede.

The concept of the Sänger-Bredt Silverbird of the mid-1930s continued to captivate visionaries in both the United States and the Soviet Union. In the early 1950s, Dr. Walter Dornberger and Krafft Ehricke, Peenemünde alumni who had emigrated to the United States after the Second World War and who were both working for Bell Aircraft, conceived the piloted "Bomber-Missile" that was known as BoMi. Launched vertically like a rocket, the upper stage was a winged vehicle that would carry nuclear weapons.

A second version configured a manned winged bomber carried aloft on a much larger delta wing mother ship that could be reused. These were quite large vehicles with gross weights well over one-half million pounds. However, the many problems that had yet to be resolved with large liquid-fuel rocket engines, guidance, and the temperatures at reentry caused the Air Force to show little interest. In a final attempt to continue their concept to fruition, the two reworked the BoMi into a hypersonic transport in 1957.

The Air Force did provide several small contracts during the mid-1950s to study the skip-glide method. An outgrowth of this was the Rocket Bomber (RoBo). These studies essentially showed that the skip-glide (or boost-glide as it became known) was not as practical as it first appeared and that it was much more effective to use a slightly higher speed, and enter orbit or a fractional orbit trajectory.

Dyna-Soar & Raketoplan

Within a week of the launch of Sputnik I, the United States Air Force began assembling all of the previous studies on the military aspects of space into a new program. As might be expected, the Air Force perceived operations in space as being simply an adjunct to its winged aircraft. The basic flight scenario was similar to the Sänger antipodal bomber envisioned in the 1930s. The designation given to the project was WS-464L and its name was Dyna-Soar—a contraction of the words *dynamic soaring,* a term that Eugene Sänger himself had coined.

When planning for a weapons system reached the stage where a firm concept could be designed, a study contract was awarded to the Boeing Aircraft Company in March 1958 for a winged vehicle that initially would carry a crew of one. However, as the project progressed, it became obvious that the boost-glide concept has several flaws. Dyna-Soar was modified for low Earth orbit operations, and a second crew member was added. Like the ballistic spacecraft being considered for manned spaceflight, the vehicle would be launched vertically on a large rocket. However, Dyna-Soar would have short delta wings for landing like a conventional aircraft after each mission and would be capable of being reused after each flight. A contract for construction was awarded in 1961, and the vehicle itself was assigned the experimental designation of X-20 in June of 1962.

The main difference between the X-20 and previous boost-glide concepts was that this craft would achieve orbit, and its role would not be as a bomber. Instead, it would be more like a fighter capable of intercepting possible hostile satellites for inspection and destruction. It could also be used for reconnaissance in a manner similar to the U-2 spy plane. It was to have a payload bay and could place small satellites in orbit.

The 35-foot-long craft had a wingspan of just 22 feet and would weigh about 11,000 pounds. The structure used both titanium and molybdenum and had an underside surface protection of an ablative coating reapplied after each flight. As the space plane continued to grow in size, and the scope of its flight plan expanded, so did its need for higher levels of thrust. The Titan II, originally slated as the launch vehicle, required the addition of large solid rocket boosters attached to either side resulting in the Titan III.

The project was well along when a critical design review of high cost military expenditures by President Kennedy's Secretary of Defense, Robert McNamara, determined that the role it was to perform had not been adequately defined. The Corona spy satellites were producing excellent results, and the ability to intercept and destroy hostile satellites could be handled by unmanned interceptors for much less money. On December 10, 1963, (three weeks after President Kennedy's assassination) McNamara ordered the project canceled—with all the schedule slippage that had occurred it was still three years away from its first flight. The essential concepts of Dyna-Soar would be resurrected seven years later, and a much larger craft would emerge as the Space Shuttle. The Titan III booster would continue its funding and become the backbone of heavy lift for both the military and scientific payloads for the next four decades.

In the same time period as Dyna-Soar, the Soviets conceived the Raketoplan, a product by Vladimir Nikolayevich Chelomey's design group, which had a similar concept. However, it too failed to survive tight budgets and was canceled in May of 1964.

Soyuz R & Almaz

The Vostok had secured the lead in manned spaceflight for the Soviet Union, and now its designer, Sergei Korolev, who had also developed the unmanned reconnaissance version Zenit, was being pressed by the military for yet more capability. Perhaps spurred by the knowledge that the Americans were developing the X-20 Dyna-Soar and a manned orbiting laboratory for their Air Force, a manned militarized space vehicle was seen as a possible offensive weapon. Korolev proposed a design in late 1962 based on his second generation manned craft called Soyuz which would be pursued both as an intelligence gathering satellite (Soyuz R), as a possible follow-on to the Zenit, and as a satellite interceptor (Soyuz P).

But Korolev was stretched thin by this time with the Vostok (and an interim Voskhod), planetary, and lunar programs. He assigned the Soyuz R and P to OKB-1 and Chief Designer Dmitri Ilyich Kozlov. The Soyuz R would emerge in the planning as more of a small space station with separate modules launched individually by the R-7 and assembled in space. Korolev would later re-immerse himself into the Soyuz project which would develop the ability of two satellites to rendezvous and dock, but by then he would have lost the space exploration initiative to the Americans.

A second large manned spacecraft was in fact a space station begun by Vladimir Nikolayevich Chelomey, who headed the competing space design bureau OKB-52. Approval was granted to move forward with the "Almaz" in October 1964. This 40,000 pound giant required the new UR-500 Proton booster which was still in the development stage. Incorporated into the Almaz was a two man spacecraft that looked very much like the American Gemini.

While elements of both projects would eventually fly as a part of the Salyut space station, their military significance as a manned offensive weapons system was never proved.

Manned Orbiting Lab: MOL

At the same press conference in which Secretary McNamara announced the demise of Dyna-Soar, he also announced a new military initiative for placing a series of relatively small manned space stations in orbit, using the same Titan III booster being developed for the cancelled Dyna-Soar. To supply the stations and to transfer crews on what was estimated to be 30-day duty cycles, the Air Force would use a version of the two man Gemini spacecraft then being developed by NASA for the civilian manned space program. Called Blue Gemini to denote its Air Force connection (not its color), its principle role was to provide an experimental workshop for classified research in space.

With the official designation of KH-10, the 30-foot-long, ten-foot-diameter cylinder would be crewed by two men in a shirtsleeve environment. The KH designation had been used exclusively for the Keyhole reconnaissance program, so its unexpected use here was a surprise to many. Perhaps to gain support from the other services, two of the seven pilots chosen to begin astronaut training were from the Navy and one from the Marine Corps. Five of this elite group would ultimately become Space Shuttle pilots and one a NASA administrator.

Just as the hardware was reaching the flight test stage in the spring of 1969, the project was suddenly canceled. Originally scheduled to fly as early as 1968, the MOL schedule had continually slipped, and the target date had moved out to 1972. That NASA was moving forward with the Skylab space station that was almost four times the weight and more than 10 times the interior volume of the MOL may have been a contributing factor. It is believed that the CIA's successful CORONA reconnaissance project was again the survivor of the budget squeeze. Neither the United States nor the USSR had conceived a significant role for a costly manned military spacecraft.

Fractional Orbital Bombing System: FOBS

With its newfound space capability, the Soviets were able to leverage their boast of achieving a 100-megaton thermonuclear bomb. Where most ICBM deliverable weapons were in the 2-4 megaton range, Soviet Premier Khrushchev in August 1961 announced that his country had tested the unbelievably large weapon. While it was acknowledged that the size of such a weapon would be considerable, the Soviets enjoyed the credibility of their accomplishments in space to back up their rhetoric.

What made the availability of the huge bomb more terrifying was the subsequent announcement of the Fractional Orbital Bombing System—FOBS. With this technique, a nuclear warhead, placed in orbit, could be brought down on America after completing less than a full revolution—thus the term "fractional." The advantage of such a weapon was its ability to be launched from Soviet territory over the South Pole and arrive over the U.S. from a completely unexpected direction. Virtually all of America's early warning radar had been established across the northern frontier of Alaska and Canada with the expectation that it was the shortest distance between the two super-powers. Likewise, the early planning for an anti-ballistic missile defense had made the assumption of an attack from the north.

Using the two-stage R-36 rocket developed as a successor to the R-7 by Mikhail Yangel's design bureau, a third stage provided the orbital capability. Twenty-four tests were carried out between 1965 and 1972 under the guise of the Kosmos series of scientific satellites. During this time, however, the United Nations was formulating the Outer Space Treaty that would ban nuclear weapons from space. Not withstanding their participation in the treaty, the Soviet FOBS system became operational with eighteen underground silos by 1974.

The system was deactivated during the period of 1982-1984 because of the SALT II treaty with the American government. The R-36 was subsequently used as a satellite carrier into the 1990s. It is believed that the Soviets, in retrospect, honored the spirit of the Outer Space treaty in that no nuclear weapons were actually orbited although the capability existed.

Dawn and Dusk Rockets

Not much was known about our neighboring planets of Mars or Venus in the early 1960s. Fascination with Italian astronomer Giovanni Schiaparelli's elaborate drawings of a network of *canali* (made during the close opposition of the Earth with Mars in 1877) still evoked speculation about possible life on the planet. Optimistic estimates of the Martian atmosphere and areas of apparent changing color encouraged the belief that there was seasonal vegetation. Venus was our closest neighbor, but no surface features could be distinguished with Earth based telescopes through what appeared to be a shroud of clouds.

The ability to send instrumented spacecraft to the Moon by 1958 (35,000 fps) meant that, with a bit more velocity, it was possible to send a probe to either Mars or Venus (37,000 fps). Thus, both the Americans and Russians immediately sought to use their newfound capabilities to reach out to the nearby planets. However, the technique required to fly from the Earth to either Mars or Venus is considerably different from that of a lunar mission. In the case of the Moon, a spacecraft is launched into a highly elliptical orbit whose path intersects the orbit of the Moon, about 240,000 miles away. The timing is such that the Moon will be at that point of its orbit when the spacecraft arrives (similar to leading a moving target in a shooting gallery). The launch usually occurs during the Moon's first quarter to maximize the slight gravitational pull of the Sun.

To reach Mars, the problem involves accelerating the spacecraft so that it adds to the orbital velocity of the Earth and causes it to swing outward to create an elliptical orbit toward Mars. To fly to Venus requires accelerating the spacecraft in the opposite direction so that it subtracts from the Earth's orbital velocity and moves the orbit inward.

All of the planets orbit around the Sun in the same direction and close to its equatorial plane (except Pluto). Based on Kepler's work, the closer a planet is to the Sun, the greater its velocity in order to remain balanced against the Sun's gravity. This greater velocity of the inner planets, coupled with their smaller orbital circumference, means that the closer planets complete their orbits in a shorter time than those farther out. The Earth, 93 million miles from the Sun (the basis of the Astronomical Unit—AU) completes its orbit in 365 Earth days. Venus (with an average distance from the Sun of .72 AU) has an orbital period of 225 days while Mars at an average distance of 141 million miles (1.5 AU), requires 687 days.

To travel to Mars then, a spacecraft must be launched from the Earth in the same direction as the earth is orbiting the Sun so that the velocity imparted by the rocket (25,000 mph or 37,000 fps) is added to the orbital velocity of the Earth (66,000 mph or 98,000 fps). Assuming that the rocket is simply launched vertically from near the equator, this event would occur at dawn as the vertical path of the rising rocket would accelerate it in the direction of the orbital path of the Earth. An increase in velocity of about 37,000 fps will cause the spacecraft to achieve a new aphelion (the farthest point of its new elliptical orbit around the Sun) that will intersect the orbit of Mars after one-half revolution around the Sun. Walter Hohmann first described the technique of effectively changing an orbit in this manner in 1925. Referred to as the "Hohmann transfer" in his honor, the process provides the most energy efficient method of traveling from one planet to another.

However, it is not enough to intercept Mars' orbit as the traveler must do so at precisely the point at which Mars occupies that space. As the total time to travel this distance will be about 6½ months, the alignment of the two planets that will permit this intersection occurs only once every 25 months. Thus, even though the planets can be as close as a mere 34 million miles apart, when their orbits bring them in "opposition," a spacecraft must travel 220 million miles to complete the journey.

The technique to send a rocket to Venus is the same except that since Venus is inside the Earth's orbit, the velocity of the rocket has to be decreased by about 36,000 fps relative to that of the earth and so it would be launched at "dusk." The rocket would "descend" inwards towards Venus en route to a new perihelion (that point in an orbit closest to the Sun) to intercept the orbit of Venus. Again, it is necessary to time the launch so that Venus occupies the point of intercept when the spacecraft arrives. This opportunity occurs only once every 19 months and has a "window" of about 50 days during each launch opportunity.

In actuality, the rocket is not launched vertically but arcs over into a trajectory around the earth to take advantage of the earth's rotational velocity of 1,040 mph at the equator (about 1500 fps)—but the result is the same. The velocity imparted by the rocket, is either added to, or subtracted from, the orbital velocity of the Earth, depending on whether the target is outside or inside the Earth's orbit. Additional energy is required when the rocket arrives at the target planet if it is to be captured by its gravitational field for the purpose of orbiting that planet or descending to its surface. For the first attempts to reach our neighboring planets, a simple "flyby" would be sufficient to probe the basic attributes of the planet and obtain pictures of its surface. The British Interplanetary Society first used the term "probe" in 1952 to describe a robotic mission to Mars.

The outer planets require significantly larger velocity increases, with Pluto at the top of the energy list with an increased "Delta-V" of about 53,000 fps. Mercury, the planet closest to the Sun, requires the rocket to subtract a "Delta-V" of 42,000 fps from the Earth's orbital velocity.

Missions to the planets are considerably more difficult than lunar flights as they required more energy, accurate guidance, and long-term reliability to remain functioning over months and years in the hostile environment of intense heat, cold, radiation, micrometeorites, and the vacuum of space. The frustrations that had been experienced by the Americans and the Russians in achieving lunar success would be exacerbated by having to wait months after a launch to see if their efforts would be rewarded.

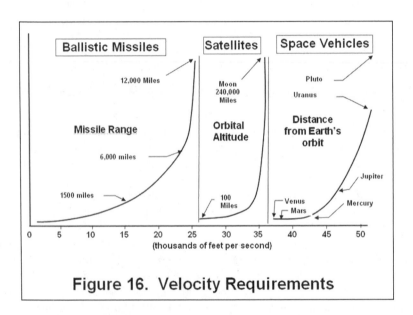

Figure 16. Velocity Requirements

Deep Space Communication and Tracking

Launching a spacecraft to another planet is just the start of a complex mission that includes communicating and navigating. Receiving the data being sent while the spacecraft is en route and sending commands to make it perform different tasks along the way involve the ability to transmit high-powered signals and receive extremely weak returns.

While the initial velocity imparted to the spacecraft by the booster is important, the precise vector (direction) of the energy is equally critical. Thus, the rocket has to have an extremely accurate guidance system. This challenge is eased somewhat by providing for mid-course adjustments. Small corrections are made early in the long journey to the planet to affect the positioning of the craft so that it arrives at the proper time and place. However, these mid-course adjustments can only be effective if a precise determination can be made of the location of the probe at a specific instant in time.

The radio signals exchanged with the spacecraft provide some of the necessary information about the position and motion of that craft. Because the transmitter has relatively low power (to save weight and electrical energy), these signals must be discriminated from the background of electromagnetic
"se" of space, amplified, and refined by appropriate signal processing and analysis.

ʻe time required for a radio signal to travel from the Earth to the spacecraft and back (round-trip light time—RTLT) can be used to measure the distance between the two. This problem is somewhat complicated by the fact that the Earth itself is both moving through space and rotating on its axis. By comparing the frequency at which the spacecraft is transmitting and measuring the shift in that frequency (the Doppler affect) as it arrives, the velocity of the craft can be determined.

A variety of factors in the signal path between the Earth and the spacecraft affect the accuracy with which the Doppler shift and RTLT can be determined. The charged ions in the plasma emanating from the Sun, known as the solar wind, distort and delay the radio signal, as will the Earth's own ionosphere and the water vapor and other gasses of the denser lower atmosphere. All of these factors are highly variable and are based on the intensity of solar activity, the season, time of day, and weather. In the early days of space exploration, these aspects had to be measured, calibrated, and modeled, to achieve the needed accuracy. In addition, precise information cannot be determined from a single site or reading of the signals, and simultaneous observations at two or more sites are needed to determine the relative position of the spacecraft much more accurately.

The Jet Propulsion Laboratory (JPL) was assigned the responsibility to track and receive the data being generated by the first American Explorer satellites. In March 1958 they chose a site in the Goldstone Dry Lake area of the Mojave Desert of California to establish their first space communication station—a three-hour drive from the JPL headquarters in Pasadena. Portable radio tracking stations were also set up in Nigeria and Singapore.

In December 1958, JPL transferred from the Army to NASA and was given responsibility for all lunar and planetary exploration programs. Thus, in addition to designing and building spacecraft, JPL had to provide for the telecommunications and tracking. All of this occurred in a relatively short span of time, and there was little experience with the problems of capturing faint radio waves, tracking a spacecraft far from Earth, and maintaining communications. This situation ultimately led to the establishment of the Deep Space Network (DSN) as a separately managed and operated communications facility that would accommodate all American deep space missions.

The first DSN Deep Space Station (DSS) facility constructed at Goldstone was the Pioneer Station (DSS 11) whose primary antenna was an 80-foot-diameter, steerable parabolic dish mounted atop a 62-foot-high tower. Its design was similar to those used in radio astronomy (such as Jodrell Bank in England) and included a closed-loop device for automatically pointing the antenna at the space probe and tracking it without human intervention. It became the prototype antenna for the entire Deep Space Network and had many of its design features incorporated into later improved antennas.

Over the ensuing years the Goldstone Complex expanded to four Deep Space Stations: Pioneer (DSS 11), Echo (DSS 12), Venus (DSS 13), and Mars (DSS 14). These stations were named for the projects in which they first participated. To provide continuous 24-hour coverage, NASA established two overseas tracking stations in Canberra, Australia, and in Madrid, Spain. Each was spaced approximately 120 degrees of longitude apart so that spacecraft would always be in communication with at least one tracking station.

The First Interplanetary Probe: Venera

Design of a reinforced Russian R-7 to carry heavier upper stages began in January 1960 with the primary purpose of launching interplanetary probes. A new second stage (the R-7 being composed of the zero and first stage) used a 66,000 lb. thrust engine, developed for the R-9 ICBM, that burned for 200 seconds. The third stage had to restart in the weightless conditions of a parking orbit. This presented the need for an attitude control system and a set of low-powered cold-gas thrusters to settle the propellants in their tanks (ullage) so that the engine could ignite. This third stage structure was based on the existing Vostok second stage. The configuration could accelerate a 2,000 lb. payload to escape velocities at a cost of $30 million dollars per launch.

As might be expected, the Soviets recognized the importance of being first in all critical aspects of space exploration, and Chief Designer Sergey Korolev had assigned a group the responsibility of preparing missions to Mars and Venus for the first available launch windows. The initial operational use of the new booster occurred on October 10, 1960, with the first attempt to send a spacecraft to Mars. The vehicle was destroyed at T+300 seconds, apparently because of high resonance vibrations. A second attempt four days later was also unsuccessful when the second stage failed to ignite because of a faulty liquid-oxygen-valve seal. Mars would have to wait.

Korolev planed three launches to Venus during the January and February window of 1961, but the emphasis for resources was on the upcoming manned Vostok flight. Object V (the Venera spacecraft) and the booster encountered several delays, and the first launch did not occur until February 4, 1961. The total weight that went into the "parking orbit" was 14,262 lb. But the third "escape stage" experienced a premature cutoff due to one of the recognized perils of restarting a liquid-fuel rocket engine in space—cavitation. Cavitation occurs when a pocket of gas (rather than the liquid) is ingested into the propellant line and upon reaching the pump causes an over speed and subsequent breakdown of the pump. This is typically caused by failure to completely settle the propellants in their tanks (ullage) before ignition.

The failure presented the Soviets with their usual dilemma—how to handle the public announcement of an unsuccessful event. The State Commission met within hours of the launch and debated the problem. Glushko wanted to emphasize that a very large satellite had been orbited and put as much positive spin on the size as possible. Others wanted to minimize any statement and prevent speculation that it was a failed manned attempt, because this was during the time that the unmanned Korabl Sputniks were being launched. Kamanin wanted to point out that the rocket demonstrated the Soviets ability to deliver an eight ton nuclear warhead anywhere on the planet. In the end, it was decided to refer to the satellite simply as Sputnik 7 with no mention of Venus being its intended objective.

As would be the case for both the Russians and the Americans, a pair of attempts was made for each available launch window. A week later on February 12, 1961, Venera 1, as officially named, was successfully propelled toward Venus from its 143-mile high transfer orbit. The actual probe weighed 1,415 lb. and consisted of a six-foot-long cylindrical body topped by a pressurized sphere. Two solar panels extended from the cylinder, and a large six-foot-diameter high-gain antenna received signals, while a long antenna arm transmitted signals back to Earth. The scientific instruments included a magnetometer, ion traps, micrometeorite detectors, and cosmic radiation counters. The pressurized sphere contained a Soviet pennant and was to float on the presumed oceans of Venus after the intended impact. Venera 1 had no on-board propulsion systems, and temperature control was achieved with thermal shutters.

A week after launch, at a distance of just over one million miles from Earth, contact with the spacecraft was unexpectedly lost. It was estimated that on May 20, 1961, Venera 1 passed within 60,000 miles of Venus and remains in a heliocentric orbit. While essentially a failure, it did permit the Soviets to gain valuable experience with launching, tracking, and communicating with an interplanetary probe.

During the next Venus launch window, in November 1963, the escape stage failed to ignite, and the payload was identified only as Cosmos 21. Yet another attempt in February 1964 experienced a second stage failure.

America's Mariner 2

America had not been able to prepare a launch for the 1961 window to Venus, but by July 22, 1962, a basic deep space bus was designed and delivered within 9 months. It provided for communications, attitude control, and electrical power (the same bus formed the basis for the later *Ranger* lunar missions as well) and was configured for the *Mariner* series to Venus. The *Atlas Centaur* was to carry the spacecraft but was way behind in its development schedule. The weight of the craft was reduced, and the *Atlas Agena* was assigned the task. The Atlas guidance system failed shortly after liftoff, and the range safety officer sent the destruct signal a few minutes into the flight. The second spacecraft, *Mariner 2,* was successfully launched a month later on August 27, 1962. The 447 lb. probe carried a flux-gate magnetometer, an ion chamber, and Geiger-Mueller counters.

The spacecraft achieved the required escape velocity (25,820 mph) 26 minutes after lift-off following the second burn of the Agena from its parking orbit. Solar panel extension occurred 44 minutes after launch. Two days into the mission the cruise science experiments were turned on, and a midcourse maneuver was initiated on September 4th. Four days later the spacecraft suddenly lost its attitude control, but this was restored within a few minutes.

On October 31, the electrical output from one solar panel dropped abruptly, and the science cruise instruments were turned off to conserve power. However, a week later the panel resumed normal output, and the instruments were turned back on. Within a few days, however, the panel failed completely, but by then *Mariner 2* was close enough to the Sun so that one panel could supply the required power. Almost four months after launch, *Mariner 2* achieved the first successful flyby of another planet on December 14, 1962, after traveling more than 182 million miles, although the spacecraft (and Venus) were "only" 36 million miles from the Earth.

The cruise experiments discovered that interplanetary dust was scarcer than predicted, and, by using a solar plasma detector, high-energy charged particles were detected coming from the Sun. The instrumentation observed several brief solar flares and detected cosmic rays originating from outside the solar system. On December 14, the radiometers were turned on as *Mariner 2* approached Venus from 30 degrees above the dark side of the planet, and passed below the planet at the closest distance of 21,000 miles.

As it flew by Venus, *Mariner 2*'s infrared and microwave radiometers revealed that Venus had a relatively cool and continuous cloud cover that was estimated to extend to an altitude of about 35 miles. The flyby was not capable of determining if Venus had oceans of water or teeming jungle life under its mantle of clouds, but it did sense extremely hot surface temperatures and high pressures in an atmosphere that was predominantly carbon dioxide. No magnetic field was observed, and improved estimates of Venus' mass and a more accurate value of its astronomical unit (distance from the sun) were made. *Mariner 2* also determined that Venus rotated opposite to all the other known planets, but the spacecraft did not carry a camera because the bright cloudy atmosphere would preclude any visible look at its surface, and the available resolution would not permit any distinguishing cloud formations. After the flyby, the cruise mode was resumed, and *Mariner 2*'s signal was tracked until January 3, 1963, when the spectacular mission formally ended.

The Russians made two more failed attempts to send a spacecraft to Venus on September 1st (designated Sputnik 20) and the 12th (Sputnik 21) during the 1962 window. A renewed attempt in March 1964 resulted in the payload remaining in Earth orbit and was designated Cosmos-27 by the Russians. A week later yet another attempt, designated Zond 1, achieved escape velocity, but communications failed shortly thereafter. Soviet success with its interplanetary program was proving very elusive.

Mars: The Mystery Unveiled

If the Soviets had experienced frustration with their Venus program, Mars proved equally difficult. In October of 1962, a Russian Mars probe was launched as a part of an ambitious program to photograph Mars on a flyby trajectory. A malfunction caused the spacecraft to be destroyed with some of the debris remaining in Earth orbit for a few days. As this occurred during the Cuban missile crisis, U.S. military radar installations were initially unsure if the launch was actually some form of space borne weapon about to be targeted on the U.S.

The effort was simply named *Sputnik 22* with no explanation by the Soviets, and another attempt was made on November 1, 1962. This shot achieved escape velocity and was named *Mars 1* and weighed almost 2,000 lb. On March 21, 1963, when the spacecraft was at a distance of 66 million miles from Earth, the communications unexpectedly stopped, perhaps because the spacecraft lost attitude control and could no longer point its antenna back towards the Earth. The silent *Mars 1* made it closest approach to the planet on June 19, 1963, at a distance of about 120,000 miles.

Four days after *Mars 1*, another companion ship launched on November 4, 1962, with the intent of making a soft landing on Mars. However, the escape stage engine could not be ignited, and it remained in Earth orbit and was identified as *Sputnik 24*.

A new Mars probe identified as *Zond 2* launched in November 1964 on a photographic flyby trajectory. *Zond 2* was unique in that it carried the first low-thrust electric-plasma-type rocket engines to provide attitude control instead of the cold-gas systems. One solar panel failed to deploy, and the communications ceased at the halfway point in May 1965. The silent spacecraft flew by Mars on August 6, 1965, at a distance estimated at only 1,000 miles.

America had high expectations for its first Martian probe, *Mariner 3*. The launch on November 5, 1964, was thwarted when the payload fairing failed to separate, and the added weight precluded the required escape velocity from being obtained. *Mariner 4* was flawless in its launch on November 28, 1964, to begin its 325-million-mile journey over the next eight months. The 575 lb. solar-cell and battery-powered spacecraft was designed to make scientific measurements of Mars and to obtain photographs of the planet's surface and transmit these to Earth.

The experimental payload included a solar instrument designed to measure the charged particles that compose the solar wind, and a trapped-radiation detector to measure the Van Allen belts of earth and possible similar formations around Mars. An ionization chamber and Geiger-Mueller tube were to measure ionization caused by charged particles. A cosmic-ray telescope to detect protons in three energy ranges, a helium magnetometer to sense if Mars had a magnetic field, and a cosmic dust detector completed the instrumentation.

Electrical power for the spacecraft was provided by 28,244 solar cells mounted on four folding panels designed to deploy in flight. The increase in panels from the two used on Ranger and Venus missions was required because of the decreased sunlight intensity as the probe moved farther out into the solar system. The cells provided 700 Watts of electrical power to run the spacecraft and recharge the battery, which would still generate 300 Watts when the Martian encounter occurred. Following separation from the Agena escape stage, the solar panels deployed and Sun acquisition (to provide attitude orientation) occurred 16 minutes later. A midcourse maneuver was successfully completed on December 5, 1964. Adjustable solar pressure vanes mounted outboard on the solar panels allowed for the use of the solar wind to provide passive stabilization of the spacecraft, saving precious attitude control gas.

After a 228-day cruise, the spacecraft flew by Mars on July 14 and 15, 1965. The camera sequence began on July 15, and twenty-one pictures were taken. The images covered a path across the Martian landscape that revealed only about 1% of the planet's surface as it passed within 6,200 miles. The images were stored onboard a recorder with 330 feet of magnetic tape for later transmission.

As *Mariner 4* passed behind Mars (as seen from Earth), and the radio signal passed through the Martian atmosphere en route to the Earth, its attenuated intensity was measured, and the first hopeful assumption of Mars was dashed—the atmosphere was nowhere near as dense as had been expected; it was less than one percent that of the Earth's. Values ranging from 4 to 7 millibars of pressure were computed as compared to a standard day at sea level on the Earth of 1,013 millibars. There would be no "shirt-sleeve environment" for future explorers here.

When the *Mariner* signal was reacquired after passing behind Mars, the transmission of the taped images to Earth began and continued for almost three weeks; the signals themselves required almost 10 minutes to reach the Earth. The second Mars "myth" was about to be dispelled. The agonizingly slow transmission of each photo, occurring at a data rate of 33 bits per second, heightened the drama as each picture slowly took shape. Each image possessed only 200 scan lines (less than half the average American television set of 485 lines). Each pixel contained six bits of information that allowed up to 64 possible shades of gray (no "red" planet would be revealed). A total of 240,000 bits were necessary to assemble one picture, and all images were transmitted twice to ensure against missing or corrupt data.

The first few pictures were indistinct, and the imaging team (and the press) was held in anticipation that the subsequent ones would be more revealing—and they were. The next dozen images showed a surface scarred with craters. There was no indication of the imagined *canali* that had been the hallmark of Mars since the observations of astronomers Schiaparelli and Percival Lowell. The two had described what they believed could be discerned through the undulating visual images that had filtered through the Earth's atmosphere and into their telescopes decades earlier. The photos scanned several areas where the *canali* had been depicted, and with a resolution 30 times greater than earth-based telescopes, objects as small as 2 miles across were discernable. While the imaging team was quick to point out that the cameras had captured only about 1% of the Martian surface, what people perceived looked little different from the Moon. Later missions would show more interesting geological features as *Mariner 4* had indeed stumbled upon a most barren area.

The spacecraft returned useful data from launch until October 1965, when at a distance from Earth of 200 million miles, the antenna orientation temporarily halted signal acquisition. Two years later, in 1967, *Mariner 4's* elliptical orbit returned it to the vicinity of Earth, and scientists performed a series of tests on the aging spacecraft. In December 1967 the gas supply in the attitude control system was exhausted, and communications with *Mariner 4* were terminated after a flight of more than three years.

Belated Success for the Soviets

The Soviets continued to experience problems with their deep space probes when *Venera 2*, launched in November 1965, ceased operation before the planet was reached, and no data was returned. *Venera 3*, launched two weeks later, actually impacted Venus on March 1, 1966, the first spacecraft to do so. However, the communications systems had failed before any data could be returned. Two more attempts during the November 1965 launch window failed before leaving the parking orbit.

The 1967 launch window started with the June 12th launch of the 2,400 lb. *Venera 4* which experienced the most success. (The June 17th launch failed to leave its parking orbit and was identified as *Cosmos 167*.) On October 18, 1967, the Venera 4 descent vehicle entered the Venusian atmosphere. Electronic signals were returned by the spacecraft, which deployed a parachute after braking to subsonic velocity in the dense atmosphere. Data from two thermometers, a barometer, a radio altimeter, an atmospheric density gauge, and 11 gas analyzers was transmitted until the vehicle reached an altitude of 15 miles when the radio signals ceased.

The 1969 Venus mission window began with the January launch of Venera 5 and Venera 6. The spacecraft were similar to Venera 4, although of a stronger design. When the atmosphere of Venus was approached, capsules weighing 840 lb. and containing scientific instruments were jettisoned from the main spacecraft. During their descent towards the surface of Venus under a parachute, data from the Venusian atmosphere was returned for more than 50 minutes. The spacecraft also carried a medallion with the coat of arms of the USSR and the figure of V.I. Lenin.

The 1970 launch window saw a successful Venera 7 thunder aloft in August and enter the atmosphere of Venus on December 15, 1970. The capsule antenna was extended, and signals were returned for 35 minutes during the parachute descent and another 23 min from the surface on Venus—the first man-made object to return data after landing on another planet.

Solar Space Probes

Pioneer 6 was the first in a series of four solar-orbiting, spin-stabilized, solar-cell-powered satellites that were designed to obtain measurements on a continuing basis of interplanetary data between the orbits of Earth and Venus with slightly varying inclinations to the Sun. The 320-pound probe was launched by a Thor Delta rocket in December 1965. It is currently the oldest NASA spacecraft still communicating, and a few of the science instruments aboard are functioning.

Pioneer 7 was launched in August 1966 into an orbit just outside that of the Earth's. It was similar in size and function to Pioneer 6. When last tracked in March 1995, one of the science instruments was still functioning. In December 1967, Pioneer 8 achieved a solar orbit that crossed the path of the Earth, and, almost 40 years later, one of its science instruments is still functioning. Pioneer 9, the last in the series, went aloft in November 1968. The spacecraft finally failed in 1983.

Mariner Program Concluded

The basic Mariner spacecraft design, coupled with the *Atlas Agena,* continued to provide a successful series of missions in the last half of the 1960s. *Mariner 5* probed the Venusian atmosphere with radio waves, scanned its brightness in ultraviolet light. It sampled the solar particles and magnetic field fluctuations as it flew by on October 19, 1967, at an altitude of 2,500 miles before its operation terminated in November 1967.

Mariner 6 and 7 comprised a dual-spacecraft flyby mission to Mars to establish a baseline for the search for extraterrestrial life, and to demonstrate technologies for future long-duration missions at greater than 2 AUs from the Sun. The two spacecraft were identical, with a total span almost 18 feet for the deployed solar panels that contained 17,472 photovoltaic cells providing 800 Watts of electrical power near the Earth and 449 Watts at Mars. The spacecraft were stabilized in all three axes using the sun and a prominent star (Canopus) as references, three attitude gyros, and two sets of six nitrogen gas jets mounted on the ends of the solar panels. Mid-course adjustments were accomplished with a 50 lb. thrust rocket motor, which used monopropellant hydrazine. Thermal control was achieved with adjustable louvers on the sides of the main compartment.

Three telemetry channels were available for communications: Channel A for engineering data at 8 1/3 or 33 1/3 bps, channel B for scientific data at 66 2/3 or 270 bps, and channel C also for scientific data at 16,200 bps. An analog tape recorder stored the television images for subsequent transmission. Other science data was stored on a digital recorder. The central computer and sequencer provided actuation of specific events at precise times. It was programmed with a standard mission profile but could be (and was) reprogrammed in flight and could perform 62 different commands.

The mission almost came to a disaster ten days before the scheduled launch of *Mariner 6*. A faulty switch opened the main valves of the *Atlas* fuel tank pressurization, and the booster began to deflate and crumple. Two members of the launch team immediately started pressurizing pumps, saving the structure from further collapse and avoiding damage to the spacecraft itself. *Mariner 6* was quickly moved to another *Atlas Centaur*, and launched on schedule. The two individuals, who had acted at personal risk of the 12-story rocket collapsing on them, were awarded Exceptional Bravery Medals from NASA.

Mariner 6 launched from Cape Kennedy on February 24, 1969, and *Mariner 7* thirty-one days later—the first interplanetary mission to use the *Atlas/Centaur*. The Atlas burned for 4 min. 38 sec. followed by a seven and one-half minute *Centaur* burn to inject the spacecraft into Mars direct-ascent trajectory (no parking orbit). A midcourse correction for *Mariner 6* involved a five-second burn of the hydrazine rocket on March 1, 1969. The *Mariner 6* flyby preceded Mariner 7 by five days (different trajectories allowed *Mariner 7* to almost catch-up with its twin), and each carried wide- and narrow-angle television cameras in addition to scientific instruments which were oriented entirely to planetary data acquisition (no cruise mode data were obtained during the trip to or beyond Mars).

En route to Mars, contact was temporarily lost with *Mariner 7* on July 30. When communication was restored, diagnostics revealed that the orientation system for television cameras was malfunctioning. The mission was in danger of being unable to complete one of its most important objectives. The ground controllers reviewed their options and within 24 hours had a manual orientation procedure that allowed *Mariner 7's* cameras to be properly aimed. The successful reprogramming of the imaging system was a tribute to the value of having a programmable computer on the spacecraft.

On July 29, 50 hours before closest approach, the scan platform of *Mariner 6* was pointed towards Mars, and the scientific instruments turned on. Video imaging began two hours later. For a period of 41 hours, 49 scans of Mars were taken through the narrow-angle camera. On July 31st, the near-encounter phase began that included 26 close-up images. The closest approach occurred at a distance of 2,100 miles above the Martian surface. Eleven minutes later *Mariner 6* passed behind Mars (from the perspective of the Earth) and reappeared about one-half hour later. Science and imaging data were then transmitted over the next few days.

Mariner 7 returned 93 far- and 33 near-encounter images with close-ups covering 20% of the surface. The two spacecraft viewed different areas of the planet. With a video system that provided for 704 lines (instead of the 200 lines of *Mariner 4*) that consisted of 945 pixels each (instead of 64), the photos had resolutions ranging as good as 1,000 feet per pixel. Advances in technology allowed transmission rates of almost 2,000 times that of *Mariner 4,* and, unlike its predecessor, these images showed the overall surface of Mars to be similar to the Moon but with indications of frost and the possibility of water erosion-like features. The south polar cap was determined to be composed of frozen carbon dioxide, not water ice—another disappointment in the search for extraterrestrial life. Atmospheric surface pressure was confirmed at between 6 and 7 millibars, and more accurate estimates were made of the mass, radius and shape of Mars.

Mariner-H was actually two spacecraft referred to as *Mariner 8* and *Mariner 9* which were a part of the "Mariner Mars 71" project. They were intended to obit Mars and return images and data. Mariner 8 was launched aboard an *Atlas-Centaur,* but the *Centaur* stage experienced oscillations in pitch and tumbled out of control. The total weight of the spacecraft was 2,200 lb.

Launched on May 30, 1971, *Mariner 9* (also referred to as Mariner 71) combined the mission objectives of Mariner 8 and mapped 70% of the Martian surface while performing a study of seasonal changes in the Martian atmosphere and on the Martian surface. It became the first spacecraft to orbit another planet. A gimbaled engine that used monomethyl hydrazine and nitrogen tetroxide to produce 300 lb. of thrust, and up to five restarts provided midcourse correction. Two sets of six attitude-control nitrogen jets were positioned on the ends of the solar panels. A Sun sensor, a Canopus star tracker, gyroscopes, an inertial reference unit, and an accelerometer provided orientation.

The spacecraft used a central computer and sequencer that had an onboard memory of 512 words.

The command system was programmed with 86 direct commands, four quantitative commands, and five control commands. Data was stored on a digital reel-to-reel tape recorder. The 550-foot 8-track tape could store 180 million bits recorded at 132 kbits/s. Playback could be done at 16, 8, 4, 2, and 1 kbit/s using two tracks at a time.

The 1,116-pound spacecraft circled Mars twice each day for a full year, photographing the surface and analyzing the atmosphere with infrared and ultraviolet instruments. The spacecraft gathered data on the atmospheric composition, density, pressure, and temperature as well as the surface composition, temperature, and topography of Mars.

When *Mariner 9* first arrived, Mars was almost totally obscured by dust storms which persisted for a month. After the dust cleared, *Mariner 9* proceeded to reveal a very different planet than Mariner 4—one that boasted gigantic volcanoes and a grand canyon stretching 3,000 miles across its surface. More surprisingly, the relics of ancient riverbeds were carved in the landscape of this seemingly dry and dusty planet. Mariner 9 exceeded all primary photographic requirements by photo mapping 100 percent of the planet's surface and taking the first close-up photographs of the tiny Martian Moons, Deimos and Phobos.

Mariner 10 was the seventh successful launch in the series of ten, and the first to use the gravitational pull of one planet (Venus) to reach another (Mercury). On November 3, 1973, *Mariner 10* launched into orbit around the Sun by way of an encounter with Venus. The 1,108 lb. spacecraft was insulated with multi-layer thermal blankets and a sunshade (deployed after launch) to protect it from the intense heat of the sun at its relatively close proximity. As it flew past Venus on February 5, 1974, at a distance of 2,610 miles, more than 4,000 photos revealed the planet enveloped in smooth cloud layers. Its slow rotational period was 243 days, and it had only 0.05 percent of Earth's magnetic field. The planet's atmosphere was composed mostly of carbon dioxide. After the Venus flyby, the trajectory was altered by the Venusian gravity and accelerated onward to Mercury. *Mariner 10* flew by Mercury at 438 miles on March 29, 1974, transmitting the first close-up images of the crater-scarred planet. A second pass (after orbiting the Sun) occurred on September 21, 1974, at an altitude of about 29,200 miles. A third and last Mercury encounter, on March 16, 1975, provided 300 additional photographs at an altitude of 203 miles. The Moon-like surface possessed only a faint atmosphere of mostly helium. On March 24, 1975, the supply of attitude-control gas was depleted, and the communication terminated.

During its two-year mission, the spacecraft transmitted more than 12,000 images of Mercury and Venus and employed more course corrections than any previous mission. It was also the first to use the solar wind as a means of locomotion; when the probe's thruster fuel ran low, scientists used the solar panels as sails to make small course corrections. It was also the last in a series of Mariner missions designed to survey other planets in the solar system.

Over the first decade of interplanetary flight, the Soviets had succeeded only three times in 13 attempts in their quest for Venus and had no success in five launches to Mars. The United States had fared much better with three of four launches to Venus (with the Mercury bonus mission) and three of five attempts to Mars. The volume and significance of the data returned forever changed mankind's perception of these three planets. Nevertheless, manned flight would continue to be the dominant factor in the race for the Moon.

Epilogue

The conquest of space resulted from a logical progression of ideas that coalesced into visions. The popularization of these visions inspired theoretical and, ultimately, experimental work that led to new and higher level visions, more theory and further experiments. By the mid-1950s, "rocket science"—a phrase first coined by the noted rocket engineer, Alfred Zaehringer, in 1947—had matured into a well-defined discipline.

The summer of 1957, with the initial tests of the Soviet and American ICBMs, marked that point where rocket science had progressed to where its application brought mankind to the dawn of the space age. That this application of rocket science was a fearful new "ultimate weapon" was regrettable. But, as had been the case from steel to steam, weapons of war are often the initial funding priorities. It is interesting to note that despite the potential for a nuclear holocaust, neither the American nor Soviet ICBMs were ever employed in their design roles, but instead served both nations for half a century as primary vehicles for space exploration and commercialization—essentially "making swords into plowshares."

The impetus of the Cold War between America and the Soviet Union had escalated the arms race into space. The frantic pace of both nations between 1957 and 1961 had moved mankind's reach into space forward by decades. The hostile political climate accelerated the tempo as no breakthrough in technology could have. In the short space of 42 months, the expertise had progressed from simple unmanned artificial satellites to large manned spacecraft, lunar probes, and planetary exploration.

Fifty years later many present-day critics view the space race with contempt, citing the frenzied pace as an indication of the cold war mentality. Those who lived through those years, however, are inclined to reply, *"You had to be there to understand."*

As the space race continued into the early 1960s, the pace and complexity of the missions continued to escalate as both major powers prepared for the conquest of the Moon with their next generation spacecraft.

The following is a list of references used to create and validate Astronautics.

Aldrin, Edwin E., McConnell, Malcolm, Men From Earth, Bantam Books, New York, 1989

Aldrin, Edwin E., Warga, Wayne, Return To Earth, Random House, New York, 1973

Alway, Peter, Rockets of the World, Saturn Press, 1999

Arnold, H.J.P., Man in Space, Smithsonian Publications, New York, 1993

Ashford, David, Collins, Patrick, Your Spaceflight Manual, Crescent Books, London, 1990

Baker, David, The History of Manned Spaceflight, Crown Publishers, 1981

BardwellSteven J., Et al, Beam Defense, Aero Publishers Inc., 1983

Bergaust, Erik, Wernher von Braun, National Space Institute, 1976

Bilstein, Roger E., Stages to Saturn, Univ. Press of Florida, 1996

Bizony, Piers, The Man Who Ran the Moon, Thunder Mouth Press, New York, 2006

Bloomberg, Linlow P., Outer Space-Prospects for Man and Society, Prentice Hall, Inc., New York, 1962

Borman, Frank, Serling, Robert J., Countdown, Silver Arrow Books, New York, 1988

Bower, Tom, The Paperclip Conspiracy, Little Brown & Co, 1987

Boyd, R.L.F., Space Research by Rocket and Satellite, The MacMillan Company, 1960

Brodie, Bernard, Strategy in the Missile Age, Princeton University Press, Princeton, 1959

Bucheim, Robert W., Et al, Space Handbook, Random House, New York, 1959

Burrows, William E., This New Ocean: The Story of the First Space Age, 2001

Burrows, William E., Exploring Space, Random House, New York, 1990

Caiden, Martin, The Astronauts, E.P. Dutton & Company, New York, 1959

Caiden, Martin, War For The Moon, E.P. Dutton & Company, New York, 1959

Chapman, John L., Atlas The Story of a Missile, Harper & Brothers, 1960

Clarke, Arthur C., The Exploration of the Moon, Harper & Brothers, 1954

Clarke, Arthur C., The Making of a Moon, Harper & Brothers, New York, 1958

Clary, David A., Rocket Man - Robert H. Goddard, Hyperion Books, New York, 2003

Collins, Michael, Carrying the Fire, Farrar, Strauss, and Giroux, New York, 1974

Day, Dwayne A., Spaceflight Vol. 36, 1994

Dickson, Paul, Sputnik - Shock of the Century, Walker & Company, New York, 2001

Dornberger, Walter, V-2, Ballentine, New York, 1954

Durant, F.C., Robert H. Goddard - The Roswell Years, National Air & Space Museum, 1973

Dyson, Gerald, Project Orion, Henry Holt & Company, New York, 2002

Editorial Staff, The Viking Mission to Mars, Martin Marietta Company, Denver, 1975

Gatland, Kenneth, Robot Explorers, The MacMillan Company, 1972

Gatland, Kenneth, Manned Space Flight, The MacMillan Company, 1967

Gavaghan, Helen, Something New Under the Sun: Satellites and the Beginning of the Space Age, 1998

Gibson, James M., The Navaho Missile Project, Schiffer Publishing Ltd, Atlanta, 1996

Glenn, John H., Taylor, Nick, John Glenn - A Memoir, Bantam Books, New York, 1999

Godwin, Robert, Whitfield, Steven, Deep Space - The NASA Mission Reports, Apogee Publications, Toronto, 2005

Hall, Al, Et al, Man In Space, Peterson Publishing Co., Los Angeles, 1974

Harford, James, Korolev, John Wiley & Sons, 1997

Hart, Douglas, The Pictorial History of World Spacecraft, Bison Books, New York, 1988

Hart, Douglas, The Pictorial History of NASA, Gallery Books, New York, 1989

Harwood, William D., Raise Heaven and Earth, Simon & Shuster, New York, 1993

Heppenheimer, T.R., The Space Shuttle Decision 1965-1972, Washington DC, 2002

Jenkins, Dennis R., The Space Shuttle The History of Developing the National Space Transportation System, Motorbooks Intl. 1996

Jenkins, Dennis R., Launius, Roger D., To Reach the High Frontier, University Press of Kentucky, 2002

Joels, Kerry Marc, The Space Shuttle Operators Manual, Ballantine Books, New York, 1982

Johnson, David, V1 V2 Hitler's Vengeance Weapons, Stein & Day, New York, 1981

Killian, James R., Sputnik Scientists and Eisenhower, MIT Press, 1977

Kraft, Chris, Flight - My Life in Mission Control, Penguin Putnam, New York, 2001

Krantz, Gene, Failure Is Not An Option, Berkeley Books, New York, 2000

Levine, Alan J., The Missile and Space Race, 1994

Ley, Willy, Rockets Missiles and Space Travel, Vail-Ballou Press, 1959

McDougall, Walter A., The Heavens and the Earth, Basic Books Inc. 1985

McNamara, Bernard, Into the Final Frontier, Harcourt College, 2001

Medaris, John B., Countdown For Decision, G.P. Putnam, 1960

Miller, Walter James, The Annotated Jules Verne - From Earth to Moon, Thomas Y. Crowell Associates, New York, 1970

Muirhead, Brian, Et al, Going to Mars, Simon & Schuster, 2004

Newkirk, Dennis, Almanac of Soviet Manned Space Flight, Gulf Publishing, 1990

Oberg, James E., Red Star in Orbit, Random House, 1981

Ordway, Frederick I., Sharpe, Mitchell R., The Rocket Team, Crowell, 1979

Parkin, Charles M. Jr, The Rocket Handbook for Amateurs, The John Day Company, New York, 1959

Pisano, Dominick A., van der Linden, F. Robert, Winter Frank H., Chuck Yeager and the Bell X-1, Smithsonian Air and Space Museum, 2006

Riabchikov, Evgeny, Russians in Space, Novosti Press, 1971

Russell, John L. Jr., Science Year, Popular Science, 1959

Shepard, Alan, Slayton, Deke, Moonshot, Turner Publishing, Atlanta, 1994

Siddiqi, Asif A., Sputnik and the Soviet Challenge, University Press of Florida, 2003

Siddiqi, Asif A., The Soviet Space Race with Apollo, University Press of Florida, 2003

Stafford, Thomas, We Have Capture, Smithsonian Press, Washington, 2002

Thompson, Neal, Light This Candle, Crown Publishing, New York, 2004

Varfolomeyez, Timothy, Soviet Rocketry that Conquered Space, Spaceflight Vol. 37, 1995

Wagener, Leon, One Giant Leap, Tom Doherty Associates, New York, 2004

Wendt, Guenter, The Unbroken Chain, Apogee Publications, Toronto, 2005

Wilson, Andrew, The Eagle Has Wings, Unwin Brothers Ltd, London, 1982

Yates, Raymond F., Russell, M.E., Space Rockets and Missiles, Harper & Brothers Publishers, New York, 1960

Yeager, Chuck, Janos, Leo, Yeager, Bantam Books, New York, 1988

Yenne, Bill, Secret Weapons of the Cold War, Berkley Books, New York, 2005

Zaehringer, Alfred J., Rocket Science, Apogee Books, 2004

Zaehringer, Alfred J., Solid Propellant Rockets, American Rocket Company, 1955

Zimmerman, Robert, Leaving Earth, Joseph Henry Press, Washington DC, 2003

Zimmerman, Robert, Genesis - The Story of Apollo 8, Dell Books, New York, 1998

In addition, the following URLs have been referenced:

http://www.unitedstart.com/
http://www.worldspaceflight.com/
http://www.astronautix.com/
http://www.hq.nasa.gov/
http://www.aerospace.ru/
http://www.spacedaily.com/
http://science.nasa.gov/
http://www.unitedspacealliance.com/
http://www.colonyfund.com/Viewpoints/Archives/
http://www.xprizefoundation.com/
http://science.ksc.nasa.gov/shuttle
http://www.centennialofflight.gov/essay/SPACEFLIGHT/
http://web.wt.net/~markgoll/
http://www.spacetoday.org/History/
http://liftoff.msfc.nasa.gov/
http://www.shuttlepresskit.com/
http://www.nasm.si.edu/
http://www.designation-systems.net/dusrm/
http://www.biography.ms/
http://www.fas.org/spp/military
http://www.spaceandtech.com/

Index

SAMOS 4 152
SAMOS 5 152
SAMOS 7, 150-153, 167
Samsun, Turkey 98
Sander, Friedrich 79
Sandys, Duncan 44
Sänger, Eugen 32, 51-52, 205
Saturn I 165, 167
Saturn V 25, 30, 43, 46, 51, 126, 165
Sawatski, Albin 56
Schiaparelli, Giovanni 16, 209, 214
Schirra, Wally 7, 173, 190, 200-201
Schneikert, Frederick P. 52
Schriever, Bernard 97, 165
SCORE, Project 6, 135, 138-139, 141, 155, 192
Scout 7, 157-158, 160-161, 192
Scout G 161
SEALAB II 199
SECO 33, 192, 195, 201
Sedov, Leonid 111, 114, 124
Segment-51 196-197
Sergeant missile 66, 69, 99, 112, 158
Seringapatam 15
Sheldon, Charles 103
Shepard, Alan B. (astronaut) 7, 174, 184-186, 189-190, 194-195, 199, 201
Shershevesky, Alexander 24
Shesta, John 33-34
Shklovsky, Iosef 142
Shonin, Georgiy 176
Siberia 35, 177, 182
Sigma 7 201
Silverbird 51, 169, 205
Silverstein, Abe 164-167, 170
Singer, S. Fred 106, 112
Sinkhronnoye Oporozhneiye Bakov 97
Slayton, Donald 174, 198, 201
SM-65 (Atlas Missile) 97
Smithsonian Air and Space Museum 91
Smithsonian Institute 19
Smithsonian Miscellaneous Publication No. 2540 19
SNAP-3 160
Snark missile 98
solar cells 109, 134, 155-156, 159-161, 214
Solovyena 203
sound barrier 42, 49, 82, 84-85, 95
sounding rockets 69-70, 105, 117
Soyuz R 7, 206
space race 116, 126-128, 139, 141, 143, 166-169, 183, 189, 204, 219
Space Shuttle 9, 30, 50-51, 91, 167, 206-207
Space Sickness 7, 190-191, 197-200, 203
Space-Based Infrared System 152

Sparks, Brian 167
spatial disorientation 191
Special Technical Commission 59
specific impulse 18, 25, 74, 89, 96, 98, 112, 118, 131, 157, 162-164, 166
speed of light 14
speed of sound 20, 49, 80, 82, 85-87
Speer, Albert 41, 46, 50
Sputnik I 102-103, 122, 125-127, 129, 131, 134-136, 138, 143, 153, 157, 169-170, 205
Sputnik II 6, 124-127, 134, 136, 163, 169
Sputnik III 6, 103, 134-136, 138-139, 141
SR-71 89, 114, 164
St. Petersburg Society for Physics and Chemistry 18
Stalin, Joseph 34-35, 52, 55, 58-61, 72-73, 95-96, 110, 116, 147-148
Stamer, Fritz 79
Starry Messenger, The 12
Staten Island, NY 33
static test 28, 70, 103, 166
Staver, Robert 56-57, 63, 67, 71
Steinhoff, Ernst 59
Steuding, Herman 39
Stevens, Leslie 105
Stewart, Charles L. 53
Stockton, N.J. 33
storable-propellant 103
Strategic Air Command 148, 152
Strategic Arms Limitation Treaty (SALT) 151
stratonauts 71
Stuhlinger, Ernst 40, 116, 132-133
Summerfield, Martin 65
Sun Tan project 164
Surveyor 167
Sustainer Engine Cut-Off (SECO) 192, 195
Syncom 156
T
T-1 pressure suit 87
Taifun 65
TASS 102, 121, 123, 141, 176-177, 182, 203
Teller, Edward 126
Telstar I 156
Telstar II 156
Tereshkova, Valentina 7, 202-204
terminal velocity 106
TG-180 85
thermonuclear 95, 97, 99, 111-112, 207
Thiel, Walter 39, 41-42, 44, 165
Thing, The 110
Thiokol 66, 158
Thor 98, 112, 129, 136-139, 141, 143, 149-150, 153, 155-160, 162, 217
Thor Able 158-160

Publications & Motion Picture References

Look for Book 2 of

Astronautics
by Ted Spitzmiller

coming soon to all good book stores.

Astronautics
Book 2 - To the Moon and Towards the Future

Book 2 in Ted Spitzmiller's definitive history of space exploration examines the commitment by the American President John Kennedy to land a man on the Moon within the decade of the 1960s.

It details the Gemini, Voskhod, Soyuz and Apollo programs .

It reviews the development of the most complex machine devised by man—the Space Shuttle—and follows the evolution of the space station.

It highlights the effort to find extraterrestrial life and the exploration of the outer planets.

The second book also examines advanced propulsion technologies and speculates on what might lie ahead in space exploration.

Coming soon from Apogee Books!

ISBN 978-1894959-66-7